Semiconductor Nanocrystals

Nanostructure Science and Technology

Series Editor: David J. Lockwood, FRSC
National Research Council of Canada
Ottawa, Ontario, Canada

Current volumes in this series:

Alternative Lithography: Unleashing the Potentials of Nanotechnology
Edited by Clivia M. Sotomayor Torres

Nanoparticles: Building Blocks for Nanotechnology
Edited by Vincent Rotello

Nanostructured Catalysts
Edited by Susannah L. Scott, Cathleen M. Crudden, and Christopher W. Jones

Nanotechnology in Catalysis, Volumes 1 and 2
Edited by Bing Zhou, Sophie Hermans, and Gabor A. Somorjai

Polyoxometalate Chemistry for Nano-Composite Design
Edited by Toshihiro Yamase and Michael T. Pope

Self-Assembled Nanostructures
Jin Z. Zhang, Zhong-lin Wang, Jun Liu, Shaowei Chen, and Gang-yu Liu

Semiconductor Nanocrystals: From Basic Principles to Applications
Edited by Alexander L. Efros, David J. Lockwood, and Leonid Tsybeskov

---

A Continuation Order Plan is available for this series. A continuation order will bring delivery of each new volume immediately upon publication. Volumes are billed only upon actual shipment. For further information please contact the publisher.

Semiconductor Nanocrystals

From Basic Principles to Applications

Edited by

Alexander L. Efros

Naval Research Laboratory
Washington, D.C.

David J. Lockwood

Natural Research Council of Canada
Ottawa, Ontario, Canada

and

Leonid Tsybeskov

New Jersey Institute of Technology
Newark, New Jersey

Springer Science+Business Media, LLC

ISBN 978-1-4419-3402-4 ISBN 978-1-4757-3677-9 (eBook)
DOI 10.1007/978-1-4757-3677-9

Originally published by Kluwer Academic / Plenum Publishers, New York in 2003
Softcover reprint of the hardcover 1st edition 2003

http://www.wkap.com

10 9 8 7 6 5 4 3 2 1

A C.I.P. record for this book is available from the Library of Congress

Preface

Almost fifty years ago, starting from an understanding of the optical and electronic properties of bulk crystalline semiconductors, there appeared not just a new scientific and technological field of solid-state devices but an entire semiconductor industry including electronics, telecommunications, microprocessors, computers and many other critical components of modern technology. More recently, utilization of a new class of man-made structures such as semiconductor quantum wells and superlattices that use a one-dimensional confinement of the carriers has brought us a second round of progress in solid-state device technology. The modern microelectronics industry requirements of miniaturization, lower power, and fast operation make it absolutely clear that the next generation of solid-state devices will use the physics of low-dimensional semiconductor structures such as quantum wires and quantum dots that correspondingly confine carriers in two and three dimensions. The ultimate size limit of such structures that on the one hand remain crystalline but on the other hand demonstrate strong quantum size effects is in the range of several nanometers. For quantum dots this size range has been achieved from two technological approaches: from the top down (e.g., lithography) and from the bottom up (e.g., self assembly). *Semiconductor nanocrystals* is another name used for quantum dots that reflects mainly their morphological structure and their size.

This book, *Semiconductor Nanocrystals: From Basic Principles to Applications* comprises a collection of articles that reflect the most recent trends in the development of realistic devices based on nanocrystals. Utilization of the nanocrystal optical, electronic, and structural properties may revolutionize the market of solid-state devices in the very near future.

Why are semiconductor nanocrystals interesting scientific objects, and why can they be promising for various applications? The most comprehensive answer is: because they can be considered as a new form of matter, and hence they have been named "artificial atoms". The most significant properties of semiconductor nanocrystals are: linear discrete absorption spectra (similar to atoms) and efficient photoluminescence tunable over a wide range of wavelengths. It was suggested and proved earlier that nanocrystal size and shape are two major parameters governing their optical properties. However, there are no particular limitations and great opportunities can be envisioned in future developments of atomically accurate nanofabrication. In principle, one can manipulate and move nanocrystals with precise control to create quantum dot molecules, three-dimensional ordered arrays of artificial atoms, etc. Also, they can be placed in other materials as dopants or joined to a larger molecule to form a super molecule. Therefore, instead of 109 elements in the periodic table, an unlimited number of man-made "atomic elements" can be used for unlimited applications.

The research on semiconductor nanocrystals embedded in glasses was initiated in the early 1980s by Efros and Ekimov [1-4] and later by Borrelli [5]. Simultaneously and independently, Brus started his research on colloidal nanocrystals [6, 7]. This groundwork triggered in the 1990s a worldwide effort that resulted in spectacular progress in nanocrystal fabrication with precise control over their size and shape, as well as a better understanding of their physical properties [8-12]. The effort was further enhanced by Canham's report of visible photoluminescence with unusually high quantum efficiency in porous Si, pointing out that Si nanocrystals can indeed form an optically active material [13].

Where are we now, more than a decade later? This research area has never been limited to pure academic studies of "scaling laws" and "particle in a box" models. Today, tremendous progress has been achieved in both nanocrystal fabrication techniques and the understanding of their physical properties. Thus, a wide class of semiconductor materials can now be prepared in the form of nanocrystals with precise control over their size, shape and surface, and their application as devices is growing rapidly in many different areas of modern technology.

The first chapter, "Photonic applications of semiconductor-doped glasses" by Dr. N. F. Borelli (Corning, Inc.), discusses properties of semiconductor nanocrystals embedded in glasses. Some of these optical properties have been utilized for many years in coloring of optical filters, a long time before scientists started to understand the nature of the technological process. However, the recent understanding of quantum dot physics has created the groundwork for a new class of photonic devices utilizing their strong non-linear optical properties.

Almost from the beginning of scientific and technological interest in quantum dots and other semiconductor nanostructures, it was anticipated that Auger processes should play a critical role in the linear and nonlinear optical properties of semiconductor nanocrystals. Auger processes are considerably enhanced in nanocrystals due to the abruptness of their interfaces, and they affect all aspects of carrier relaxation and recombination: the photoluminescence quantum efficiency, photoluminescence blinking and reversible photoluminescence degradation. All these phenomena are described in great detail in the chapter written by Dr. A. Efros (US Naval Research Laboratory) on "Auger process in nanosize semiconductor crystals". One of the most important conclusions from his theoretical considerations is that the Auger process in semiconductor nanocrystals requires a significant shortening of the stimulated emission rate for lasing in the order to overcome the fast nonradiative decay via two electron–hole pairs. Most likely, a lack of attention to this fact is responsible for almost a decade delay in the first demonstration of a quantum dot laser.

In general, the chapter "Carrier dynamics, optical nonlinearities, and optical gain in nanocrystal quantum dots" written by Dr. V. Klimov (Los-Alamos National Research Laboratory) continues the theme of the importance of Auger phenomena in the optical properties of semiconductor nanocrystals. It presents the experimental work on carrier dynamics, optical non-linearities and lasing in II-VI colloidal quantum dots. In our opinion, the demonstration of a reliable quantum dot laser [14] is the critical development in this field, paving the way for novel photonic devices that utilize the enhancement of optical non-linearities due to three-dimensional quantum confinement in semiconductor nanostructures.

Another class of semiconductor nanocrystals is presented in the chapter "Novel device applications of Stranski-Krastanov quantum dots" written by Dr. K. Brunner (Technical University of Munich) and Dr. A. Zrenner (University of Paderborn). Stranski-Krastanov growth is performed using molecular beam epitaxy (MBE), a 'conventional' tool of modern nanotechnology. This type of growth has been applied to two major categories of materials, III-V and Group IV semiconductors. In particular, Brunner and Zrenner consider the fabrication and properties of GaAs/InGaAs and Si/Ge islands, and several types of novel devices such as optical memory devices, and advanced intraband and interband photodetectors [15].

Compared to colloidal II-VI semiconductor nanocrystals and self-organized nanostructures grown by the Stranski-Krastanov technique, silicon based nanotechnology is significantly less developed. Silicon is the major technological material for modern microelectronics, and modern microelectronics is moving toward nano-scale electron devices (i.e., nanoelectronics). However, right from the beginning the microelectronics industry has strongly focused on photolithography as the major patterning technique. It is obvious that photolithography does not have many options at the length scale of less than 50 nm. In addition, lithography must be combined with some etching procedure, and the use of plasma-induced etching for nanostructure fabrication results in a noticeable interface roughness and interface defect creation. The word-wide enthusiasm and attention to Canham's report in 1990 [13] on visible light emission from Si-based nanostructures created by anodical etching (i.e., porous Si) was, in part, due to a strong hope that the problem can be solved using a very simple and technically inexpensive solution. Light-emission from porous Si and Si nanocrystals has been the subject for numerous reviews and book chapters, and it will not be discussed in this book. On the other hand, the proposed use of porous Si for passive optical devices (dielectric mirrors, filters, etc.) did not receive significant attention until the late 1990s, when it was demonstrated that precise control over the porosity and refractive index of porous Si [16, 17] can be combined with conventional microfabrication for the realization of a 'photonics-on-a-chip' concept. Therefore, the chapter "Porous silicon as an open dielectric nanostructure", written by Dr. J. Diener and Dr. D. Kovalev (Technical University of Munich) is focused on an analysis of the optical and structural properties of porous Si, including control over polarization memory and porous Si layer dielectric anisotropy, together with various applications of these phenomena.

The most recent development in Si nanocrystal fabrication is reflected in the chapter "Nanocrystalline silicon-silicon dioxide superlattices: structural and optical properties", (Dr. D. J. Lockwood, National Research Council of Canada and Dr. L. Tsybeskov, New Jersey Institute of Technology). This work has been initiated by a simple idea that solid phase crystallization of Si can be performed in a layered structure where layers of initially amorphous, nanometer-thin Si are separated by layers of amorphous SiO2. Combining the state-of the art deposition techniques with precisely controlled post-treatment steps (rapid thermal annealing and low-rate oxidation), the authors show that the quality (i.e., periodicity, interface abruptness, etc.) of their nanostructures are approaching that of conventional superlattices. Focusing on detailed structural characterization (TEM, XRD, inelastic light scattering, etc.), they demonstrate that control over the Si nanocrystal size, shape and crystallographic orientation is achievable [18]. In addition, the exceptional quality of the interface between Si nanocrystals and an a-SiO2 separating layer has triggered exploratory applications of these superlattices in electron devices utilizing sequential resonant tunneling and other features of quantum transport.

The last chapter, "Quantum dot photonic crystals" by Dr. D. J. Norris (University of Minnesota) and Dr. Y. A. Vlasov (IBM T. J. Watson Research Center), discusses the possibility of simultaneously combining both electronic confinement and photonic confinement in a single material. Namely, Norris and Vlasov outline possible approaches for incorporating colloidal semiconductor nanocrystals into photonic crystals. Photonic crystals are structures that have a periodic variation in the index of refraction. When this periodicity is on a length scale comparable to the optical wavelength, the density of photonic states can be modified significantly. In fact, if a photonic crystal has a property known as the photonic band gap, it allows control over the photonic density of states. By combining this property with the tunable optical properties of semiconductor nanocrystals, these quantum-dot photonic crystals provide an interesting new material to explore enhanced linear and non-linear optical phenomena and their potential for yielding novel photonic devices [19].

Despite the strongly focused scope of this book, we wish to attract a broad community of readers including researchers, engineers and students. Each chapter has an introductory part, a clear explanation of the motivation and potential application of the specific research subject and an extended list of references to original publications. During our collaborative work on this book we have realized that the field is growing very rapidly and that many other interesting subjects (e.g, biological applications of nanocrystals) could have been included. Therefore, we would like to define this book not as a complete review but rather as a 'milestone' that, we believe, is going to play an important role in the promotion of research and development of functional semiconductor nanostructures.

Alexander Efros
David John Lockwood
Leonid Tsybeskov

November 2002

REFERENCES

1. A. I. Ekimov and A. A. Onushchenko, *JETP Letters* **34**, 345 (1981).
2. AL. L. Efros and A. L. Efros, *Sov. Phys. Semicond.* **16**, 772 (1982).
3. A. I. Ekimov, A. A. Onushchenko, A. G. Plyukhin, and Al. L. Éfros, *JETP* **61**, 891(1985).
4. A. I. Ekimov, A. A. Onushchenko, M. É. Rakh, and Al. L. Éfros, *JETP* **63**, 1054 (1986).
5. N. Borrelli, D. Hall, H. Holland, and D. Smith, *J. Appl. Phys.* **61**, 5399 (1987).
6. L. E. Brus, *J. Chem. Phys.* **79**, 5566 (1983).
7. L. E. Brus, *J. Chem. Phys.* **80**, 4403 (1984).
8. R. Rossetti, J. L. Ellison, J. M. Gibson, and L. E. Brus, *J. Chem. Phys.* **80**, 4464 (1984).
9. M. G. Bawendi, W. L. Wilson, L. Rothberg, P. J. Carroll, T. M. Jedju, M. L. Steigerwald, and L. E. Brus, *Phys. Rev. Lett.* **65**, 1623 (1990).
10. R. W. Schoenlein, D. M. Mittleman, J. J. Shiang, A. P. Alivisatos, and C. V. Shank, *Phys. Rev. Lett.* **70**, 1014 (1993).
11. M. G. Bawendi, P. J. Carroll, William L. Wilson, and L. E. Brus, *J. Chem. Phys.* **96**, 946 (1992).

12. D. J. Norris, A. Sacra, C. B. Murray, and M. G. Bawendi, *Phys. Rev. Lett.* **72**, 2612 (1994).
13. L. T. Canham, *Appl. Phys. Lett.* **57**, 1046 (1990).
14. V. I. Klimov, A. A. Mikhailovsky, D. W. McBranch, C. A. Leatherdale, and M. G. Bawendi, *Science* **287**, 1011 (2000).
15. A. Zrenner, E. Beham, S. Stufler, F. Findeis, M. Bichler, and G. Abstreiter, *Nature* **418**, 612 (2002).
16. D. Kovalev, M. B. Chorin, J. Diener, F. Koch, A. Kux, Al. L. Efros, M. Rosen, N. A. Gippius and S. G. Tikhodeev, *Thin Solid Films* **276**, 120 (1996).
17. D. Kovalev, H. Heckler, G. Polisski, J. Diener and F. Koch, *Optical Materials* **17**, 35 (2001).
18. G. F. Grom, D. J. Lockwood, J. P. McCaffrey, H. J. Labbe, L. Tsybeskov, P. M. Fauchet, B. White, J. Diener, H. Heckler, D. Kovalev, and F. Koch, *Nature* **407**, 358 (2000).
19. Y. A. Vlasov, X. Z. Bo, J. C. Sturm, and D. J. Norris, *Nature* **414**, 289 (2001).

Contents

1

Photonic Applications of Semiconductor-Doped Glasses

N. F. Borrelli

Science and Technology Division, Corning Incorporated

1.1. INTRODUCTION

The technical capability to produce structures in solids that are small and precise enough in dimension to be comparable to the Bohr radius of an electron is a relatively recent technological development. This effort began with the invention of a deposition technique called MBE (molecular beam epitaxy). From this technique came the first layered structures in GaAs where the layer thickness could be controlled to a dimension less than the Bohr radius, $\hbar^2\kappa/m_{e,h}e^2$, where, κ is the dielectric constant of the material, and $m_{e,h}$ is the mass of the electron or hole. These precisely controlled layered structures gave birth to the "quantum well" devices that have played such an important role in the opto-electronics industry. The key property of the structure is the dependence of the energy levels on the dimension of the layers.. The term that applies to the effect where the physical size of the structure is smaller that the normal electron orbit is called "quantum confinement."

The layered structures, or the quantum well devices, involve confinement in only one direction, the layer direction. One refers to this a 1-D confinement. On can imagine then, quantum "wires" with 2-D confinement and quantum "dots" that have full 3-D confinement. The early pioneering work with respect the recognition and description of this phenomenon of three dimensional confinement is attributed to Efros[1], Brus[2] and Weller et al[3]. What has peaked and sustained the scientific interest are the many new methods that have brought to bear to produce such structures. Clearly, the combination of high-resolution electron-lithography and MBE has played a major role.

Not surprisingly, nature had provided us with quantum dots even before we knew what to call them, let alone recognize their possible technical application. They appeared as tiny nanocrystals embedded in a glassy matrix. The use of Au, Ag, Cd, S, Se, among others were well known among glassmakers of antiquity as additives to produce color. The most recent vivid demonstration of this was a set of sharp cut-off optical filter glasses made by Corning Incorporated[4] (formally Corning Glass Works). The spectrum of the set is shown Fig. 1.1. The additives were Cd, S and Se. The glass as melted was clear and a subsequent thermal treatment produced the coloration. The thermal treatment brought out of solution a CdS_x Se_y nano-phase, which imparts the coloration. We will deal specifically with this phenomenon later on in a subsequent section.

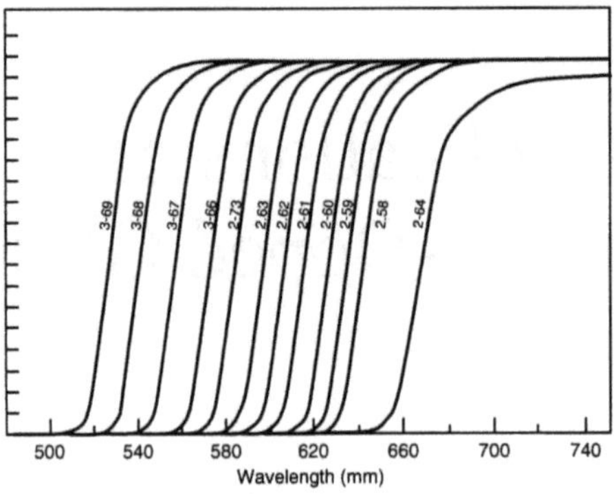

FIGURE 1.1. Transmission spectra of a series of Corning filters based on CdS-CdSe.

What was not appreciated until lately was that the subtle change in transmission cut-off through the series of glasses was not entirely due to the amount, or composition of the crystalline phase, but also due to the particle size distribution. As a matter of fact, the glasses with the deepest red cut-off were all pure CdSe-containing. The difference between them was the thermal schedule that was used to develop the color. What was actually happening as a consequence of the different thermal schedule was the changing of the particle size distribution to those with smaller crystallites, and consequently, the edge shift was coming from the quantum confinement effect.

There is another glass technology that pre-dated the quantum dot and that is the one that dealt with "photochromic" glass[5]. This was a glass system that contained a nano-phase made up of silver halide, and or copper halide crystallites. Although, the inherent photochromic effect (glass turned absorbing under uv-light) was not a quantum confinement effect, nonetheless it was later found that one could observe excitons whose behavior was size-dependent in these glass systems.

As we will see in the next section, it has turned out that using glass as the host for the growth of quantum dots of CdS, CdSe, CuCl, and PbS is quite convenient and can be controlled sufficiently to produce a narrow particle size distribution. Of course, there are other methods of quantum dot preparation, each with their individual advantages and disadvantages. They include chemical preparation techniques[6,7], e.g., sol-gel. The unique advantage of these techniques is that the particle is accessible to surface treatment, and chemical manipulation before the encapsulation into the matrix occurs[8]. This allows one to pacify the surface states so that there influence is minimized. This is in contrast to the process where the semiconductor phase is developed within the glass. Here the nature of the interface is largely uncontrollable.

In this chapter, we will be primarily concerned with the host being made of glass and the processes essentially derived from the thermal development method. This particular bias comes from the practical point of view that commercial utilization of quantum dot phenomenon will require a robust host that is impervious to the environment. That is not to say that the glass host approach is without other disadvantages, not the least of which is the possible concentration of dots that can be achieved. On the other hand much that is said applies irrespective of the fabrication method.

There is much material to cover in the chapter. The applications are based on phenomena, which in turn are based on the physical understanding. We will only include as much of the underlying physics as required for the reader to follow the content without confusion. We will very briefly describe the origin and basis of the quantum dot effect primarily to establish the basic ideas and the terminology that inevitably follows. After this we will cover the quantum dot systems that have been studied and review the basic spectroscopy for each system. We will then move to the glass fabrication and review the basic thermodynamic/kinetic process, which allows one to control the growth. The final sections will attempt to deal with the various and sundry optical effects that one can produce from these materials. This is in fact the main purpose of the chapter. We have divided the optical phenomena into sub-headings, like saturable absorption, and then further into the applications like Q-switching and mode locking. The same for the phenomena related to lasing and amplification. Finally we will deal with the nonlinear effects. In each section there will be a bit of theory so as to make the presentation coherent and understandable.

It should be pointed out that this chapter is not a comprehensive review, but rather a brief introduction to the field. Because of this approach we have neglected many worthy contributors, and we apologize beforehand.

1.2. BRIEF REVIEW OF PHYSICS OF QUANTUM DOTS

Because of the following discussion of the various optical phenomena involves quantum dots embedded in glass, it is necessary to very briefly review the physical solid-state physics basis. This is for the reader who is primarily concerned either with the applications, or the material itself, and not explicitly the physics of quantum dots. I am sure that a deeper and more sophisticated treatment of the physics of quantum dots can be found in other chapters of this book. The intention here is to make the contents of the chapter as self-contained as possible. The inclusion of a simple view of the physics and the associated terminology is given below so that when the specific experiments and applications are discussed, the basic ideas will be in context.

1.2.1. Electrons in Solids

1.2.1a. Free Electrons

The starting for the treatment of electrons in solids is to consider the behavior of free electrons. The Hamiltonian is simply written as the following where for simplicity we restrict the motion to one dimension. (For the 3-D treatment, one uses the vector **r**, rather than x and replaces the second derivative with ∇^2)

$$\frac{\hbar^2}{2m}\frac{d^2\varphi}{dx^2}+[E-V(x)]\varphi=0 \tag{1.1}$$

The free electron condition means the potential, V(x) can be set to zero and thus one can immediately solve Eq. (1) for the wavefunction which turn out to be simple plane waves of the form,

$$\varphi = A\exp(ikx) \tag{1.2}$$

Here, k is the wavenumber (vector) and defines the energy of the allowable states through the expression,

$$k^2 = \frac{2m}{\hbar^2}E \tag{1.3}$$

This solution shows that the energy of the electron is continuous and follows the simple parabolic form.

1.2.1b. Electrons in a Periodic Potential

In a crystal the electron is not free. It experiences a periodic potential provided by the positively charged nuclei as sketched in Fig. 1.2a. Because of the periodicity of this potential, "d" in the diagram, one has the condition that

$$V(x) = V(x+d) \tag{1.4}$$

The solution of Eq. (1} with this potential is accomplished with the so-called "Bloch function," of the form

$$\varphi(x) = u_k(x)\exp(ikx) \tag{1.5}$$

Here, one has the condition $u_k=u_k(k+d)$. Physically, one can think of this as a plane wave (free electron form) modulated by a function that reflects the periodicity of the potential function.. As a consequence of this modulation of the plane wave, the allowable energy of the electron is now restricted to "bands" as shown in Fig. 1.2b. This representation is called the extended Brillioun zone. Because of the periodicity, the shape of the energy band repeats in each interval of $2\pi/d$ for the simple one-dimensional structure. The influence of the potential on the wavefunction is the greatest when the spatial frequency is the same. In the Brillouin diagram this is at the zone boundaries. The gap in energy arises at these points because there are two possible solutions for the wavefunction which are degenerate in the absence of a periodic potential. One solution is where the electron probability wave is in phase with the positive potential. This lowers the energy relative to the case in the absence of the potential. The other solution must be orthogonal to it (all solutions of Eq. (1) are mutually orthogonal) and thus must be out of phase with the positive potential. Consequently its energy will be higher. This is indicated in Fig. 1.2b by the downturn of the lower band and the upturn of the upper band at the zone boundary.

For real crystals, the Brillouin zone will reflect the full symmetry of the crystal structure. The Brillouin is the reciprocal space version of the unit cell. Because of the

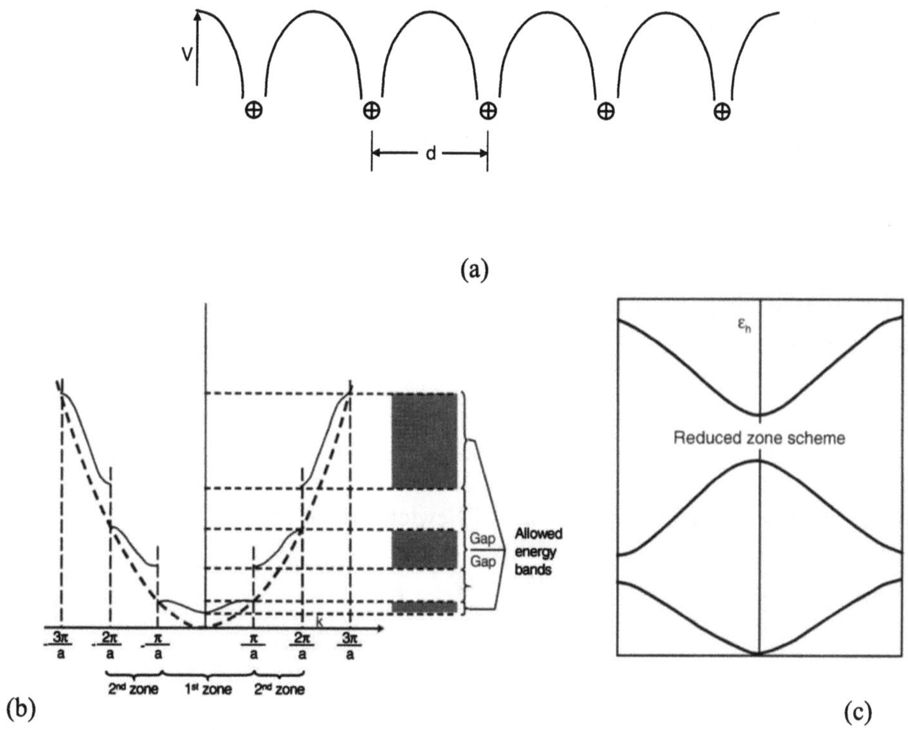

FIGURE 1.2. (a) Representation of an electron in a periodic potential produced by positive nuclei; (b) extended zone representation of allowed bands as a consequence of the periodic potential; (c) reduced zone representation of allowed bands.

periodicity, it is easier to portray the Brillouin within an interval of the reciprocal lattice constant, $2\pi/a$. One merely takes the bands shown in Fig. 1.2b and reflects them back into the first interval as shown in Fig. 1.2c. This is called the "reduced" zone and the one commonly used to show the allowable energies for different crystal structures.

1.2.3. Motion of Electrons in Bands

To determine how best to treat the motion of electrons/holes in their respective bands, one has to consider a number of factors. First, Eq. (3) is a simple first order approximation of how the energy of the bands depends on the crystal momentum. In general, as we will see below, it will be dependent on the over all crystal symmetry as well as the shape of the bands.

A second very important factor is how to deal with the electron as represented by Eg 5. It is useful to consider the relationship expressed by Eq. (5) and the wave motion in classical systems. The wave-packet solution to Eq. (1.1) corresponds to the equations of motion of classical particles with the appropriate substitution of the operator equivalents. For example, the classical expression for the velocity and momentum of a particle are the following

$$v = \frac{dr}{dt} = \frac{\partial H}{\partial p}$$
$$\frac{dp}{dt} = -\frac{\partial H}{dr} \tag{1.6}$$

The first equation yields the expression recognized as the group velocity of a wave packet where we have used the expression for H=E(p) + V(r), and $\hbar k = p$

$$v = \frac{1}{\hbar}\frac{\partial E(k)}{\partial k} \tag{1.7}$$

From the second equation, one gets a form of Newton's equation

$$\hbar\frac{dk}{dt} = \frac{\partial V}{\partial r} = F \tag{1.8}$$

If one takes the time derivative of Eq. (6), the relationship relating the force to the acceleration and thus defines what is termed the "effective mass" is obtained.

$$m^* = \frac{\hbar^2}{\partial^2 E(k)/\partial k^2} \tag{1.9}$$

One simple way to look at this formulation is that the effect of the periodic potential on the electron is taken into account by the effective mass in such a way as to be able to treat it as quasi-free electron. For example Eq. (3) is now written as

$$\mathrm{E} = \frac{\hbar^2}{2m^*}k^2 \tag{1.10}$$

For a free electron, and the corresponding parabolic dependence of E(k), one gets the expected result m=m*. In three dimensions, Eq. (8) defines a 3x3 tensor whose components reflect the complexity of the actual band structure.

1.2.4. Wave-Packets

The concept of a wave-packet follows directly from the way one treats waves that contain a band of frequencies centered about some central frequency. They form a

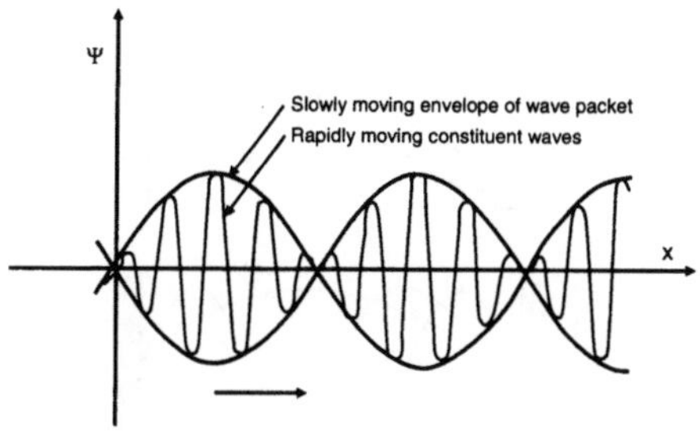

FIGURE 1.3. Representation of motion of an electron as a wave packet.

beat-frequency as shown in Fig. 1.3. The beat, or wave packet, moves with a velocity, called the group velocity, defined as the following.

$$v_g = \frac{d\omega}{dk} \tag{1.11}$$

With the equality, $E = \hbar\omega$, one sees this is the same expression as Eq. (6). According to the Heisenberg uncertainty principle in order for the electron to be spatially localized, there must be a corresponding spread in its momentum. This means that the wavefunction representation of our electron is made of a collection of terms like Eq. (5). In the limit of infinitesimally different values of k one can write the wave-packet wavefunction as the following where the a(k) defines the width of the packet, usually taken as a gaussian function.

$$\varphi(r) = \int_{-\infty}^{\infty} a(k) u_k(x) \exp(ikx) dk \tag{1.12}$$

This completes our simple description of the way one deals with an electron in a crystal and will serve as the lead in for the following discussion of confinement.

1.2.5. Confinement

With the above description of the electron as a quasi-free electron, one can immediately proceed to the simple case representing quantum confinement, namely the electron in a "box." This corresponds to writing the Schrodinger equation with a potential function that has the following properties

$$\frac{\hbar^2}{2m^*}\nabla^2\varphi(r) + [E - V(r)]\varphi(r) = 0 \tag{1.13}$$

$$\begin{aligned} V(x,y,z) &= 0, x < a, y < b, z < c \\ V(x,y,z) &= \infty, x > a, y > b, z > c \end{aligned} \tag{1.14}$$

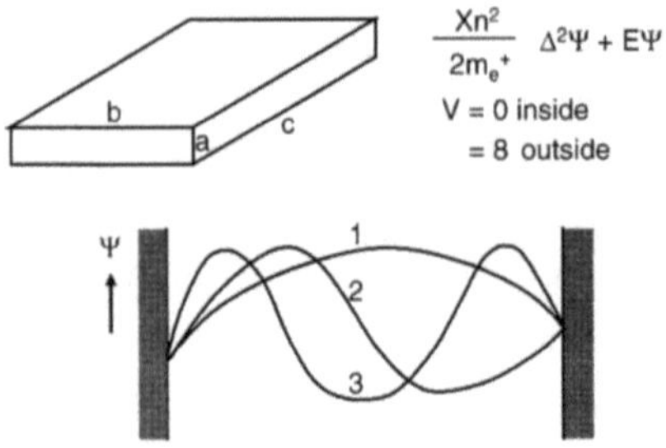

FIGURE 1.4. Representation of an electron in a box. Potential is infinite everywhere outside the box. Wavefunctions solutions must have nodes at the boundary of the box as shown.

One immediately finds that the solution yields all the wavefunctions whose amplitudes are zero at the boundary (Fig. 1.4) and the discrete energies associated with these solutions are given by the following,

$$E = \frac{\hbar^2 \pi^2}{2m}\left(\frac{n_1^{\ 2}}{a^2} + \frac{n_2^{\ 2}}{b^2} + \frac{n_3^{\ 2}}{c^2}\right) \tag{1.15}$$

Here, n_j take on integer values. One thus obtains discrete energy levels and the spacing between the levels will be inversely proportional to the dimensions of the box. If a=b=c, then the separation between the energy levels would be inversely proportional to the square of the cube dimension.

1.2.6. Quantum Dots

One can produce the physical confinement situation described above in real structures by exercising sufficient control of the dimensions of the particle by the appropriate processes. An idea of the required dimension of a crystal that must be achieved to permit the discrete evolution of the energy levels can be obtained by comparison to the Bohr radius,

$$a_{e,h} = \frac{\hbar^2 \kappa}{m_{e,h} e^2} \tag{1.16}$$

Here, we have used the effective mass of the electron, or hole and κ is the dielectric constant. As mentioned above, if one dimension satisfies this size requirement then one has a quantum well structure, two dimensions, a quantum wire, and if all three are sufficiently small, one has a quantum dot. It is the last, the quantum dot that we will be exclusively concerned with.

The formulation of the actual physical situation is more complicated than the electron in the box example because we must consider the electron and the hole separately, as well as their interaction. As a consequence, the Hamiltonian must also include separate energy terms for the electron and the hole, as well as a coulombic term to account for the mutual attraction of the pair.

This leads to a number of different limiting situations relative to the degree of confinement that occurs.

$$H = -\left(\frac{\hbar^2}{2}\right)\left(\frac{\nabla_e^{\ 2}}{m_e{}^*} + \frac{\nabla_h^{\ 2}}{m_h{}^*}\right) + V(r_e) + V(r_h) - \frac{e^2}{\kappa\,|\,r_e - r_h\,|} \qquad (1.17)$$

From Eq. (1.17) one can associate the first two terms with the confinement energy of the electron, $\hbar^2 / m_e{}^* a^2$, of the hole, $\hbar^2 / m_h{}^* a^2$, and the last term with the coulombic energy, $e^2 / a\kappa$. This allows one to classify the nature of the confinement according to the dominance of one, or more of the energies. This is shown in TABLE 1.1. The value of this classification is that of the quantum dots that can be derived from glass, they fall into three distinct categories and this will largely determine the way they will interact with light[1]. As seen from the expression for the confinement energy, the effective mass plays the dominant role, and this is a function of the particular semiconductor crystal.
In the strong confinement limit, one can solve Eq. (1.17) directly for the case of an infinite spherical potential. The envelope portion of the wavefunction is expressible in the following form, where the j's are the spherical Bessel functions, and $Y(\theta,\phi)$ is the spherical harmonic.

$$\phi_{n,l}(r) = \left(\frac{2}{a^3}\right)^{1/2} Y_{l,m}(\vartheta,\varphi)\frac{j_l(k_{n,l}r/a)}{j_{l+1}(k_{n,l})} \qquad (1.18)$$

Here, $k_{n,l}$ is the nth root of the lth order Bessel-function. The total wavefunction is a Bloch function expressed as the product of the form expressed in Eq. (1.18) multiplied by the function containing the periodic part of the bulk wavefunction, $U_v(r)$. The eigenvalues, $k_{n,l}$ are the zeros of $j_l(k_{n,l}r/a)$ (which is equivalent to $J_{l+1/2}$) so that the energy levels for the electron(hole) can be expressed as the following.

$$E_{n,l}{}^{(e,h)} = \frac{\hbar^2 k_{n,l}{}^2}{2m_{e,h}a^2} \qquad (1.19)$$

The initial state, n=1, l=0, is labeled 1S, and similarly the next state 1P corresponds to, n=1, l=2. (Actually one can estimate the coulombic contribution to the energy by perturbation theory[12] and obtain a correction term to Eq. (1.18) of the following form $\delta E = 1.765e^2 / \kappa a$. In the next section we will discuss the transitions from the hole states to the electron states.

TABLE 1.1

Regimes of Confinement Classification	*Size**	*Effect*
Weak confinement		
- $a > a_e, a_h$, $E_{coul} = \dfrac{e^2}{\kappa a} >> E_{conf}\ \dfrac{\hbar^2}{m_{e,h} a^2}$		
- exciton confined with effective $m^* = m_e + m_h$		
Intermediate		
- $a_{e,(h)} > a > a_{h,(e)}$, confinement energy of the electron(hole) $> E_{coul}$		
- electron (hole) confined with $m^* = m_{e,(h)}$, only conduction (valence) states are discrete		
Strong		
$-a_e, a_h > a$, confinement energy for both electron and hole $> E_{coul}$		
- both electron and hole are confined, $(1/m^*) = 1/m_e + 1/m_h$ both valence and conduction levels are discrete.		

$$a_{e(h)} = \hbar^2 \kappa \Big/ m_{e(h)} * e^2$$

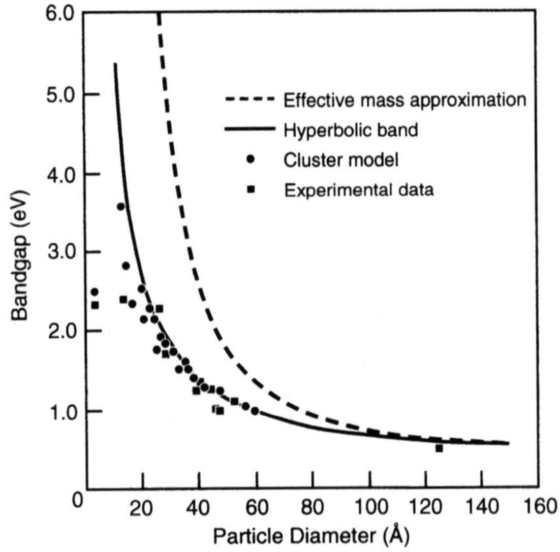

FIGURE 1.5. Band gap of PbS as a function of particle size. Dashed line is simple effective mass approximation. Solid line is that obtained using hyperbolic band model. Squares are experimental data. For details see Wang et al, reference 9.

The deviation from pure parabolic band shapes affects the expression for how the energy of the levels changes with dot size. In the simple effective mass derivation of an electron in a infinite central potential, one predicts the splitting of the levels to be proportional to $1/a^2$ as indicated by Eq. (1.19). In fact the real relationship is more complicated, and the extent of the deviation depends on the exact band structure of the particular semiconductor in question. As a good example is PbS where the energy gap is relatively small, 0.3eV, the effective masses are small, and the gap is not at the center of the Brillouin zone. All of this contributes to a more complicated expression for the band splitting vesus dot size. This has been addressed by Wang et al[9] and the results are shown in Fig. 1.5 for PbS. In the case where the coulombic term is significant,, the deviation from the ideal classification appears to be a poor estimate of at what size relative to the Bohr radius do confinement effects begin to occur. Estimates are that even at sizes equal to 10x the Bohr radius, one can still see size dependence of the energy of the confined exciton.[10]

1.2.7. Spectroscopy of Quantum Dots in Glass

Unfortunately, in the real world things are not so ideal as presented above in the previous section. As a consequence, the measured optical properties deviate from the idealized model predictions. The extent of deviation from the simple model predictions depends on the particular semiconductor, the strength of confinement, and the host itself. An example is the effects introduced by interactions in a real band structure. In the wurtzite and zincblende structures, the valence band is six-fold degenerate and is split into two bands due to spin-orbit interaction.

In the simplest approximation of no coulombic interaction, one can solve the Hamiltonian for the electron and the hole separately. They are non-interacting species. What we are interested in is the nature of transitions between the electron and hole state. One obtains this information based on the evaluation of the matrix element $<\phi_{n,l}(r)U_v(r)\,|\,r\,|\,\phi_{n',l'}(r)U_c(r)>$ where use is made of the wavefunctions of the form of Eq. (1. 18). One can show that the evaluation of the matrix element reduces to the overlap between the electron and hole envelope functions and thus leads to the result that l=l'. In other words, in the zero-th approximation of no coulombic interaction between the electron and the hole, only transitions between hole and electron states with the same a quantum number are allowed. For example, $1S_h$-$1S_e$, $1P_h$-$1P_e$ would be allowed transitions. The inclusion of the coulombic interaction allows mixing of the states so now transitions such as 1S-1P are allowed. This means more transitions allowed among more states

One can write the expression for the absorption coefficient for the transition from a valence band state to a conduction state as the following.

$$\alpha(\hbar\omega)=\frac{4\pi^2\omega}{n_0 c}N_0<\mu_{cv}>^2\sum_{n,l}2(2l+1)[\hbar\omega-E_g-E_{n,l}] \qquad (1.20)$$

N_0 is the number density of dots and μ is the dipole moment matrix element connecting the two states. The best example of this case is in the PbS system which exhibits strong confinement. A typical set of spectrum corresponding to different dot size are shown in Fig. 1.6.

The situation in the intermediate confinement systems like CdS and CdSe, is more complicated in that the electron is confined but the hole is not. The Coulombic interaction is sufficiently strong to allow the mixing of the states. In Fig. 1.7, we show the

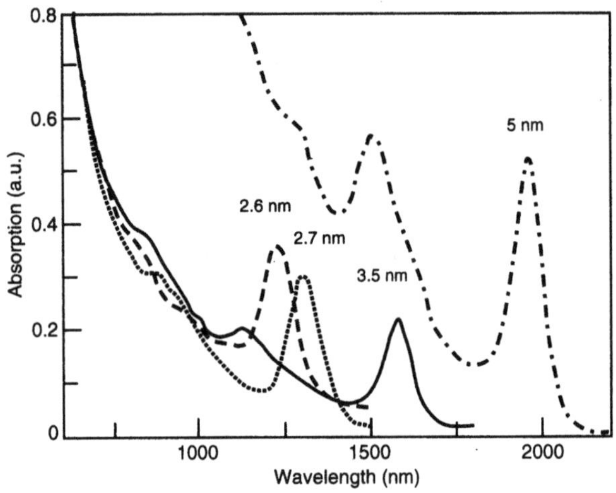

FIGURE 1.6. Representative optical absorption spectra of PbS-quantum dots in glass.

assignments made to the various peaks in the case of CdSe.[13] Here, one has further complications from the simple theory in that the uppermost band is split into what is called a "light" and "heavy" hole band.[11] This means that the shape of the bands yields different effective masses through Eq. (1. 8). Clearly this complicates the interpretation of the absorption spectrum, not only because there are more levels, as well as the rigid selection rules being bridged.

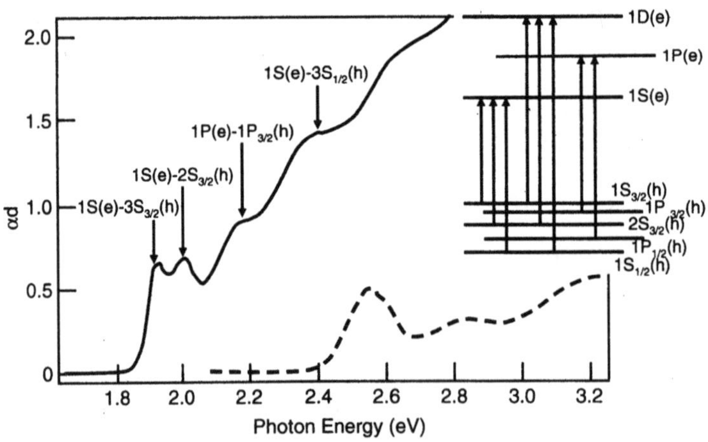

FIGURE 1.7. Allowed optical transitions (arrows) between the several lowest electron and hole levels in model assuming valence band mixing. (After Ekimov et. al., Ref. 11).

The situation for weak confinement as occurs is CuCl and CuBr is quite different. Here, the particle size relative to the Bohr radius is not small enough to produce distinct valence and conduction states. However, the particle is small enough to allow the exciton to feel the presence of the particle boundary. As a consequence one does see the spectral position of the exciton shift with the particle size. This is the situation where one can think of the exciton itself as confined, as mentioned above. A representative spectrum showing this behavior measured at 4 K is shown in Fig. 1.8. [13]

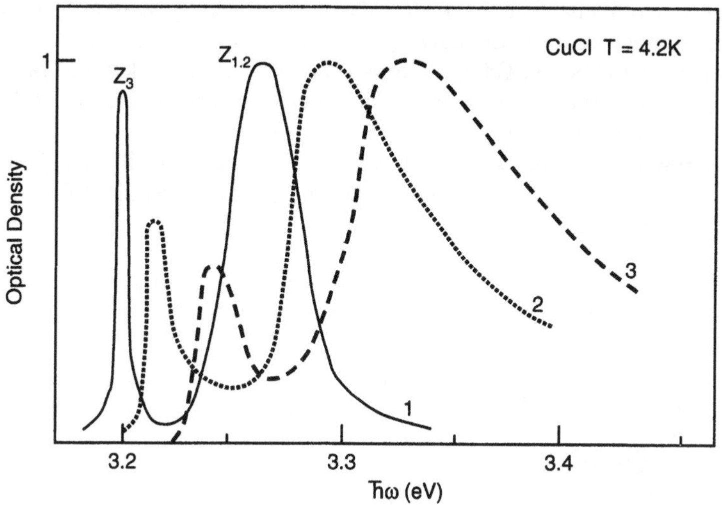

FIGURE 1. 8. Spectra of CuCl nanocrystals for three sizes, (1), 31nm, (2), 2.9nm, and (3), 2nm. (Reference 13).

1.3. PRECIPITATION OF QUANTUM DOTS IN GLASS

1.3.1 Quantum Dots in Glass Matrix

The fact that quantum dots can be formed in an inorganic host at high temperature presents a number of unique advantages and disadvantages. Certainly, the robust inorganic host provides the ability for a variety of sample preparations, such as ion-exchange and even optical fiber.

Another advantage is that the dots prepared in glass are impervious to ambient changes up to relatively high temperatures. The down side to this condition is that the dots are not accessible for surface modification. Since surface states and the associated defects contribute in a major way to the optical properties, one is left to whatever the process provides. In the fabrication process where the dots are prepared chemically and then embedded in a host, there is the opportunity to chemically modify the surface for specific purposes [14].

It is difficult to say how serious a problem the surface state issue is. Clearly for the dots with a smaller Bohr radius, the number of states with a lattice constant of the surface becomes that much greater relative to the states within the volume. Wise[15] estimates that for CdSe with a radius equal to the Bohr radius that 30% of the atoms can be considered as surface atoms for CdSe. On the other hand, for PbSe with a larger Bohr radius, this number reduces to only 5%. Photoluminescence often indicates the presence and extent of these surface states. These states provide a multitude of radiative and non-radiative paths to the ground state and yield a non-exponential radiative lifetime. As we will see the kinetics of the excitation for the various optical effects discussed later, like saturation absorption and laser/amplifier, will be largely governed by these states.

The electronic bandgap of oxide glasses is sufficiently large to make the infinite potential approximation a reasonably good one. However, for the larger bandgap semiconductors like CdS and CdSe, there is observed a photodarkening phenomenon[16] associated with high exposure intensity where an electron can escape into the conduction band of the glass through a Auger process. This process is shown schematically in Fig. 1.9. The absorption associated with this trapped electron accounts for the photodarkening effect. Photodarkening can be a problem when the excitation required for saturation, or laser pumping is high.

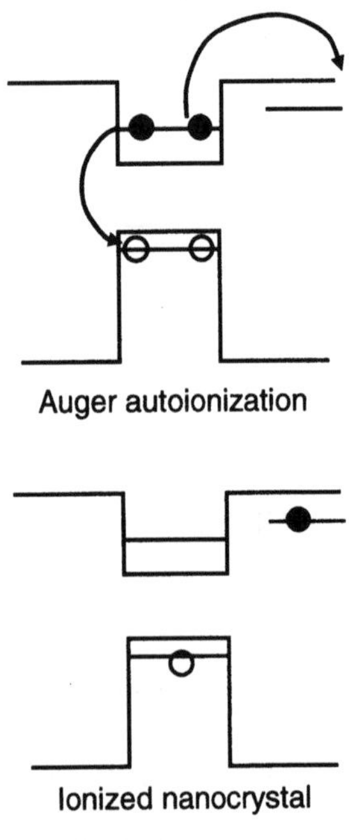

FIGURE 1.9. Schematic representation of photo-darkening mechanism. Auger process produces electron with sufficient energy to escape into conduction band of host and ultimately get trapped into color-center site

The following brief review of the physical basis for the formation of quantum dots is given below so as to give some context to the reader who may not be familiar with glass, or the process by which one can produce a crystalline phase, or phases therefrom. Some understanding of the underlying phenomenon is important to the extent that it explains why glass can be the ideal host for the formation of quantum dots.

1.3.2. Nucleation and Growth

Because of the large temperature dependence of the viscosity of glasses, especially silicate glasses, the structure can be easily frozen in a homogeneous state by means of a rapid quench from the melt, even though formation of a multiphase system may be more stable thermodynamically in the low temperature regime. In some cases, precipitation (e. g. of the semiconductor phase) can then be induced by subsequently heating the glass to some intermediate temperature at which the glass viscosity is very high but at which the rate of precipitation is sufficiently rapid. Precipitation in a very viscous medium also affords a measure of control of the rates of nucleation and growth of the colloids. Therefore, judicious choice of matrix glass composition and careful control of the thermal history of the system allows one to precipitate colloidal particles and to control the morphology of the minor phases formed.

The rate at which the minor phase precipitates from a glass depends on two factors. One factor is the thermodynamic driving force . The driving force is the decrease in the free energy of the system that results from the formation of a precipitate. For driving forces that are not too large, the rate of a process is proportional to the decrease in free energy resulting from the process. The second factor is a kinetic factor (mobility) whose magnitude increases approximately exponentially with the temperature. The kinetic factor is the constant of proportionality between the driving force and the rate of particle formation. The dependence on temperature of these two factors is shown schematically in Fig. 1.10. In the temperature regime in which the glass is very viscous, the magnitudes of all kinetic factors are very small. Therefore, very strong thermodynamic driving forces are required for a satisfactorily fast precipitation of colloids. In a large number of glasses this requirement is not met.

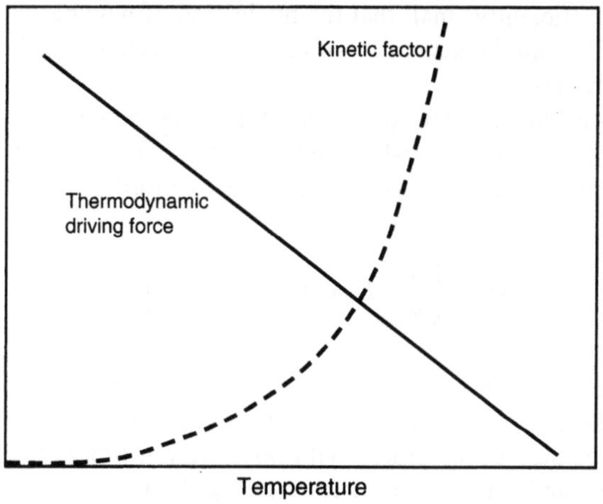

FIGURE 1.10. Schematic diagram indicating the kinetic rate and the thermodynamic rate as a function of temperature

One wants a glass composition in which the solubility of semi-conductor constituents is strongly temperature dependent. Unfortunately, little is known about the free energy of solution of these components in most glasses. Therefore, theory provides little guidance in the search for a stronger temperature dependence of solubility, and extensive empirical work is often required to identify a suitable glass.

An example of a reasonably well-studied host is the family of alkali borosilicate glasses. In many of these glasses, the manner in which boron is bonded into the glass structure varies strongly with temperature. This imparts a much stronger temperature dependence of the solubility of silver halides, cuprous halides, or cuprous oxide in these glasses than that observed in many other glasses. Unfortunately, it is not known if structural changes that occur in other glasses exert a strong influence on the temperature dependence of solubility of specific semi-conductors.

In general, in order to grow particles to the same size, one should nucleate the homogeneous glass at a temperature as low as possible (determined by the amount of time available). By doing this, one hopes to produce the desired number of nuclei (particles larger than the critical size at some higher temperature) which grow larger only extremely slowly so that all the nuclei are, to a satisfactory degree of approximation, the same size. One then hopes to find a higher temperature at which the nuclei grow at a sufficiently rapid rate but at which no new nuclei are formed. In order most nearly to achieve these ideal goals, one must optimize both the chemical composition of the glassy solvent and the thermal treatment.

1.3.3. Ripening

There is yet another phenomenon which can play a critical role in the ultimate control of the particle size and distribution of the precipitated phase which is termed "ripening", or "coarsening." This process is generally observed in the later stages of a precipitation process after the supersaturation of the solvent with respect to the solute atoms has become sufficiently small that the nucleation of new particles is unlikely. It is at this stage one deals with the instability of the smaller particles, and thus the shift in the size to the larger particles.

The theory of this diffusive phase decomposition process in supersaturated solid solution has been developed by Lifshits and Slyozov [17] The fundamental result of this analysis is the prediction of the average particle size as a function of growth time.

$$< a >= \left(\frac{4}{9} \alpha t D_0 \exp(-E_D / kT) \right)^{1/3} \qquad (1.21)$$

D is the diffusion coefficient with activation energy, E_D, and α is a parameter having to do with the interfacial surface energy. Ekimov et al have shown that this relation describes the average size development with post heating time quite well in their glass system for both CuCl and CdS which is shown in Fig. 1.11.

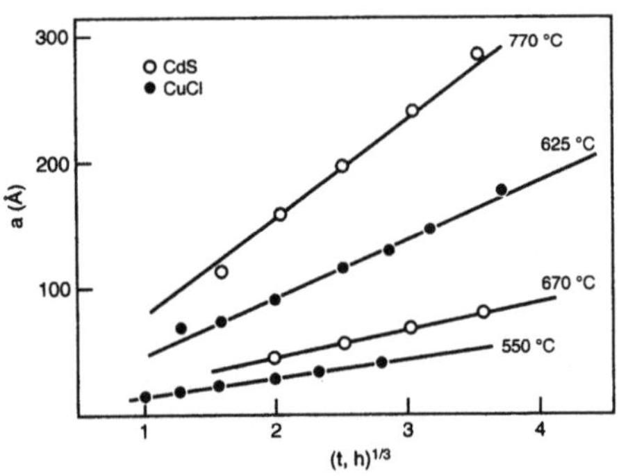

FIGURE 1.11. Dependence of the average radius, of CdS and CuCl quantum dots as a function of the time at the noted temperature of thermal treatment (Ekimov et. al., Ref. 12).

Potter et al [10] give an excellent critical review of the ripening approach and find that it indeed does apply to CdS in an alkai-boro-silicate glass composition. On the other hand they found that in the mixed system CdSSe, it does not hold indicating that there is also a stoichiometric component to the growth process (see Fig. 1.12).

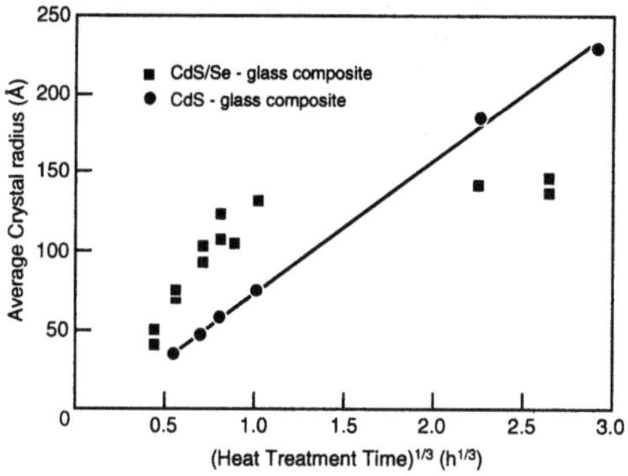

FIGURE 1.12. Dot size vs cube root of heat treatment temperature for CdS, circles, and CdSSe, diamonds (Ref.10).

TABLE 1.2
Energy shift versus thermal treatment of PbS QDs[a]

Treatment (degrees/h)	Particle Size[b] (nm)	ΔE (eV)
500/16+550/0.5	7.6	0.29
500/16+550/1.5	8.0	0.24
500/16+600/0.5	12.6	0.17
550/3	(9.6)[c]	0.16
550/5	13.3	0.12
450/16+535/1	-	0.43
450/87+525/1	-	0.65

[a] Based on bulk gap of 0.4eV
[b] Estimated from line broadening of x-ray pattern
[c] Very weak intensity

For the PbS system the particle size as a function of the thermal treatment also appears to be of a more complex nature. Typical results[27] are listed in TABLE 1.2.

1.3.4. Quantum Dot Compositions

In the Following TABLE III-2 we list the semiconductor compounds that have been obtained in glass by various investigators. We have given the bulk bandgap, the effective masses, and the approximate Bohr radius where known.

As mentioned above the specific glass composition for the formation of quantum dots depends on the particular semiconductor through the solubility of the constituents and its temperature dependence. For sulfides, the problem is compounded by

TABLE 1.3
Semiconductor Compounds Made in Glass

Composition	Energy Gap eV	Effective Mass electron	hole	Bohr radius nm
CuCl	3.35	0.5	2	<1
CuBr,		0.3	1.4	1
Cu_2O	3.0			(2)
Cu_xS	2.0			4
AgCl	3.6			
CdS	2.5	0.24	1.35	1.5
CdSe	1.84	0.12	-	3
CdTe	1.44			
ZnO	3.0	0.34	-	1
ZnS	3.6			
In_2O_3	3.7			
PbS	0.4	0.1	0.1	>10
PbSe	0.2	-	-	>20
PbTe	0.29			

TABLE 1.4

Typical glass compositions

Oxide	Wt%				
SiO_2		59	56	56	48
	-				
P_2O_5		.	.	.	
	45				
AlF_3		.	.	.	
	2.5				
Al_2O_3		4	6	6-	
	-				
Ga_2O_3		-	-	-	
	4.5				
B_2O_3		.	18	17	17
Na_2O		15	4	10	
	15				
K_2O		-	6		30
CaO		-		9	
Li_2O	-	2			
ZnO		15			5
	22				
Semicond Constituents					
PbO		4	Cu 0.2		
S		1.5	Cl 0.2		

the sulfur retention in air melting conditions. We list in the TABLE 1.4 above some representative glass compositions that have been used to develop the semiconductor phases. It is by no means complete, but only a sampling of the base glass compositions. To these would be added the constituents that would ultimately make up the semiconductor phase after the appropriate thermal treatment. The first two entries give examples of the constituents that are added that would lead to the PbS phase, in the first case and CuCl in the second.

1.3.5. Summary

The physical and chemical processes leading to the precipitation of a discrete second crystalline phase from a glass matrix are qualitatively known, but are difficult to quantify in terms of properties that determine the specific glass composition and post heat treatment schedule. In one case where the process is dominated by diffusion, there is a quantitative agreement with heat treatment time. In the simplest terms, one wants a glass that has a high solubility of the desired crystalline constituents at high temperature (melting), and a low solubility at low temperature (post heat treatment). This solubility factor is a strongly dependent on what crystalline phase is desired.

FIGURE 1.13. Electron photo-micrograph of PbS dots in glass at the magnification of 200,000 X. The average dot radius is 5 nm.

As the constituents come out of solution at the lower temperature, the largely unknown (obtained by trial and error) nucleation and growth kinetics parameters come into play to produce the particle size distribution. In spite of the complexity of the processes involved, glass provides a stable host for a wide variety of semiconducting quantum dots. Glass processing lends itself to the production of a stable and controllable fabrication to narrow particle size distribution for a wide variety of semiconductor/insulator particle phases. A typical photo-electron micrograph of the PbS dots as they develop in the glass matrix is shown in Fig. 1.13.

1.4. OPTICAL EFFECTS

1.4.1. Introduction

In the following sections we will present a brief review of the nonlinear optical effects observed in semiconductor-doped glasses that contain particles with dimension sufficiently small to manifest quantum confinement effects. Further, we will briefly cover the possible optical devices derived from such materials. The nonlinear effects will basically originate from the saturation of the excitonic absorption through a state-filling mechanism. In other words because of the discrete confinement-produced states, the promotion of one electron per dot produces a non absorbing state and further excitation an inverted state. The former corresponds to saturable absorption-based devices and the latter to laser and amplifier applications.

The unique and interesting aspects of quantum dot-glasses, (we will signify these as simply QD-glasses) as well as the larger class of quantum dots fabricated in different ways, is the "tunabilty" of the energy band, and the ability to dynamically alter the optical properties with the application of a control beam near resonance with the dot size-controlled bandgap. One can think of this as a optical nonlinear effect in that the optical absorption, for example will be a function of the intensity.

$$\alpha(I) = \alpha_0 + \beta I \tag{1.22}$$

Here, α is the intensity dependent refractive index with β characterizing the nonlinear effect. In the nonlinear optical formalism, one can see that this is nothing more than the imaginary part of the third order susceptibility, χ_3,

$$P_i(\omega) = \chi_{ijkl} E_j(\omega) E_k{}^*(-\omega) E_l(\omega) \tag{1.22}$$

The magnitude of these resonant nonlinearities, deriving from transitions between real states, is three orders of magnitude larger than that obtainable from non-resonant phenomenon which are driven by virtual transitions. Correspondingly, the time domain impulse response function that measures the response of the medium to the external field is essentially a δ-function for the non-resonant mechanism. This is to be contrasted to the resonant mechanism in which real states have finite the lifetimes. In the case of the QDs this is either the exciton, or carrier recombination time which is the order of 1-ns, or the lifetime of some trap state which is longer.

As briefly mentioned above, the optical nonlinearity of strongly confined QD-glasses primarily stems from what is called "state-filling."[18] A related effect is observed in bulk semiconductors, where one supplies sufficient intensity to empty the top of the valence band and occupy the bottom of the conduction band. In bulk systems it is hard to produce significant inversion because of the large number of electron (hole) states that are available, and the natural thermalization between the closely spaced levels. In the strongly confined QD-glass regime, the size effect produces widely separated energy states, which makes the bleaching effect very likely. Here, the lowest electronic level is occupied by two electrons with opposite spins. The promotion of a single electron effectively bleaches the lowest optical transition. This bleaching effect on the absorption produces a refractive index change at longer wavelength (real part of χ_3) through dispersion. Optical devices are based on this induced refractive index change. We will specifically deal in the next section with the derivation of the expression describing bleaching, or "absorption saturation," as it is often called.

In the weak confinement limit, the origin of the optical nonlinearity stems from the alteration of the dominant coulombic interaction between the electron and the hole.

Here, at high intensity, the exciton formation is inhibited due to the screening caused by the electron-hole plasma produced by the ionization of the excitons. The excitation into the exciton absorption can destabilize exciton formation and thus produce the nonlinear optical effect.

Aside from the interesting physics presented by semiconductor quantum dots in glass and their interaction with light, there is an implied if not explicit interest in a possible practical application. The new exploding area of "photonics" presents an intriguing venue for ideas for devices based on the saturable absorption phenomenon. There are essentially two approaches to possible device applications. The first is to use the optically-induced transmittance change directly, as in a Q-switch, or mode locking operation, while the second is to use the refractive index change associated with the induced absorption change at some longer wavelength. The latter is a consequence of dispersion whose details will be derived and discussed in the next section. A whole host of nonlinear effects are thus made available through the intensity dependence of the refractive index.

There is a third area of interest as well that is based on the "tunability" of the band gap through the dot size, and correspondingly the wavelength of the photoluminescence. Much as quantum well structures are designed to produce narrow band laser sources, so too it possible to use the QD-glass as a laser/amplifier medium. The wavelength region can be as wide as the control of the particle size would allow. As we will see, this application requires an understanding of the nature of the radiative decay processes as they are influenced by quantum confinement.

In the following section, a brief review of the progress made toward application will be given for each of the topics listed. Again, this is not to be taken as an exhaustive review but merely a representative sampling of the field.

1.4.2. Saturable Absorption

It is worthwhile to initially set a theoretical framework for the saturation phenomenon before we start discussing specific experimental results. The phenomenological expression for the absorption saturation of a simple two level system is quite a familiar one, and easily derived from a simple kinetic model.

$$\alpha(I) = \frac{\alpha_0}{1 + I / I_{sat}} \tag{1.22}$$

In the simple kinetic model the quantity $I_{sat} = \hbar\omega / \sigma\tau$, with σ being the absorption cross-section, and τ being the upper state lifetime.

Although this is the correct form we are looking for to represent the simple quantum dot system, it really is not sufficient to allow one to estimate how much refractive index change at a given wavelength would result from what amount of absorption change. For this one needs a more fundamental (quantum mechanical) expression for the complex susceptibility.

The most straightforward approach is that using the density matrix formalism to calculate the ensemble average of the dipole moment (ultimately the susceptibility) for a simple two level system. The expectation value of the dipole moment is written as the trace of the product of the density matrix and the dipole moment matrix.

$$<\mu>=Tr(\rho\mu) \tag{1.23}$$

The diagonal elements of the density matrix measure the probability of finding an atom in state "n", and the off-diagonal elements measure the interaction through the perturbation term of the Hamiltonian, H' where the total Hamiltonian is, $H=H_0+H'$. The equation of motion (equivalent to the time-dependent Schrodinger equation) takes the following form.

$$\frac{d\rho}{dt}=\frac{i}{\hbar}[\rho H-H\rho] \tag{1.24}$$

For the simple case of two levels ρ is a 2x2 matrix. Since $\mu_{12}=\mu_{21}=\mu$ one can easily show from Eq. (1. 24) that

$$<\mu>=\mu(\rho_{12}+\rho_{21}) \tag{1.25}$$

For the interaction with light, one takes the time-dependent perturbation of the form $H'=\mu E(t)$.. One can expand Eq. (1.25) into the four linear differential equations using the expression for the Hamiltonian as the following.

$$H_{m,n}=\begin{pmatrix}\hbar\omega_1 & \mu E\\ \mu E & \hbar\omega_2\end{pmatrix} \tag{1.27}$$

Further one can show that $\rho_{12}=\rho_{21}{}^*$, thus an equation for ρ_{21}, and the difference ($\rho_{22}-\rho_{11}$) is sufficient

$$\begin{aligned}\frac{d\rho_{21}}{dt}&=-i\omega_0\rho_{21}+\frac{i\mu}{\hbar}E(t)(\rho_{11}-\rho_{22})-\frac{\rho_{21}}{T_2}\\ \frac{d(\rho_{11}-\rho_{22})}{dt}&=2i\frac{\mu}{\hbar}E(t)(\rho_{21}-\rho_{21}{}^*)-\frac{(\rho_{11}-\rho_{22})-(\rho_{11}-\rho_{22})_0}{T_1}\end{aligned} \tag{1.28}$$

To these equations we have added a term to each to reflect the decay process. If one turned the perturbing field off, these terms would be the route by which the system would return to equilibrium. In the first equation a term was added to measure the dephasing of the dipole moment as say would occur from collisions if one were talking about a gas. It is also valid in a more general sense in representing all the ways that fluctuations in the environment can cause the radiating dipoles to lose phase coherence. This term is called T_2 because of the early development in the treatment of electron spin resonance. To the second equation which measures the population difference we added the lifetime.T_1.

If the perturbing field is light at frequency ω, then we can write the density matrix elements in a "slowly varying envelope " (SVE) approximation form.

$$E(t) = E_0[\exp(i\omega t) + \exp(-i\omega t)] \tag{1.29}$$

$$\rho_{mn}(t) = \sigma_{mn}\exp(i\omega t) \tag{1.30}$$

If we substitute Eqs. (1.29) and (1.30) into Eq. (1.28) and then take the steady state solution, the desired expression for the density matrix elements is obtained. Further, according to Eq. (1.26), one now has the expression for the dipole moment. Multiplying the dipole moment by the number of oscillators yields the polarization. We can that extract the real and imaginary parts of the susceptibility in the following form.

$$\chi'(\omega) = \frac{\mu^2 T_2 \Delta N_0}{\hbar\varepsilon_0}\frac{(\omega_0-\omega)T_2}{1+(\omega_0-\omega)^2 T_2^{\,2} + 4\Omega^2 T_2 T_1} \tag{1.31a}$$

$$\chi'' = \frac{\mu^2 T_2 \Delta N_0}{\hbar\varepsilon_0}\frac{1}{1+(\omega_0-\omega)^2 T_2^{\,2} + 4\Omega^2 T_2 T_1} \tag{1.31b}$$

In these expressions we have used the parameters, $\Delta N_0=(\rho_{11}-\rho_{22})_0$ and, $\Omega=\mu E_0/2h$. The real part of the susceptibility is related to the refractive index, and the imaginary part is related to the absorption coefficient

$$(n^2-1)=4\pi\chi', \text{ and } \alpha=4\pi k_0\chi'' \tag{1.32}$$

In the above equation $k_0 = 2\pi/\lambda_0$. We can simplify the expressions somewhat by realizing that in the limit of E and $(\omega_0-\omega)$ being very small, one can define the value of χ_0'', and hence α_0. In addition we can convert the last term in the denominator that contains the field squared to an intensity in the form

$$\alpha(\omega) = \frac{\alpha_0}{1+(\omega_0-\omega)^2 + I/I_{sat}} \tag{1.33}$$

Here, we have defined $I_{sat}=(c/4\pi)(\hbar^2/\mu^2 T_1 T_2)$, and $\alpha_0 = (4\pi k_0)(\mu^2 T_2 \Delta N_0/\ \hbar)$. One can simplify the expression for the refractive index somewhat as well in that it is really the difference in index that is produced by the light that we are interested in. In this case we can subtract the refractive index at small I from that of Eq. (1.31a) to obtain the expression for the induced index change.

$$n^2 - n_0^{\,2} = -\frac{\alpha_0}{k_0}\frac{\delta\,(I/I_{sat})}{(1+\delta^2)(1+\delta^2+I/I_{sat})} \tag{1.34}$$

Here we have defined $\delta = (\omega_0-\omega)T_2$ as the detuning factor. The experimental procedure would be to measure the absorption change and estimate from the data a value of I_{sat}. From this one can estimate how the refractive index would change as a function wavelength using Eq. (1. 34).

It is interesting to inquire as to whether the expressions 31a and b are consistent with the Kramers-Kronig relation which relates the real and imaginary part of a complex function that is analytic in the lower (or upper) portion of the complex plane through a Hilbert transformation. The complex function in this case is

$$\chi(z) = \chi' - i\chi'' \tag{1.35}$$

One can show that in the limit of low exposure $I \longrightarrow 0$, the condition is met, however for the general case it is not.

1.4.2a. Experimental Results

There is considerable data in the literature on the saturable absorption measured in QD-glasses. The bulk of the data is for CdS and CdSe,[19] with the more recent data on PbS.[20] As outlined above, the basic mechanism is thought to be state-filling which is considered to be enhanced in strongly confined dots, all other things being equal.

The results are usually presented, as we will see, in one of two ways. The first is to plot the change in transmittance at a given wavelength within the exciton absorption feature as a function of the exposure intensity. In the other method, the entire spectrum is captured as a function of a given exposure intensity. Here a white light source is used as the probe. The latter gives a more complete picture of the bleaching phenomenon.

This section is intended to very briefly review some of the representative experimental data to give the reader an idea of the saturation phenomenon as it has been observed in QD systems in glass. It is not intended in any way to be a survey of the existing literature There are a number of excellent review articles that cover the experimental literature in more depth, (Klimov,[19,21] and Hennenberger[22]), as well as a large number of articles dealing with specific details. Some of these will be cited as we mention the effect. The proposed idea here is to introduce the reader to the general concept of the saturation phenomenon in QDs and show how this exhibits itself in a number of optical effects. In the sections below, we will discuss a representative sampling of the experimental results dealing with saturation absorption for three major QD systems.

II-VI (CdS, CdSe)

The earliest data are for these glass systems because of their ready availability as filter glasses. Subsequently, more size-controlled samples became available for more specific and fundamental studies. In particular, a major question to be answered was the role of the dot size relative to the confinement limit as it affects the extent of nonlinear behavior.

One of the more complete studies on CdSe was done by Park et. al.[23] which essentially covers all the points that have been mentioned above with respect to saturation absorption. They studied the spectral bleaching as a function of ns-pulse pump intensity for three different dot sizes. Data were taken both a room temperature and 10K. The linear absorption of the three samples is shown in Fig. 1.14. The dot sizes corresponding

to the three thermal treatments were 3.0, 4.4, and 7.9 nm from lower to higher temperature. The room temperature differential transmission spectrum for the three samples labeled A, B, and C in Fig. 1.14 are reproduced here as Fig. 1.15a-d. These data are obtained for a number of exposure intensities ranging from 1-7MW/cm^2. The pump wavelength is listed on the inset. The same results are shown with the temperature at 10K in Fig. 1.16. They were able to fit the low temperature data for all three samples to the 2-level expression of Eq. (1.23). The I_{sat} values were 1.9, 0.7, and 0.1MW/cm^2 respectively for the 600C, 650C and 700C samples. This fits to Eq. (1.23) for the 10K data are shown in Fig. 1.17. It should be noted that photo-darkening as previously defined, occurred at room temperature sufficient to limit the amount of data that could be taken at the high intensity. This is a real practical problem for CdSe.

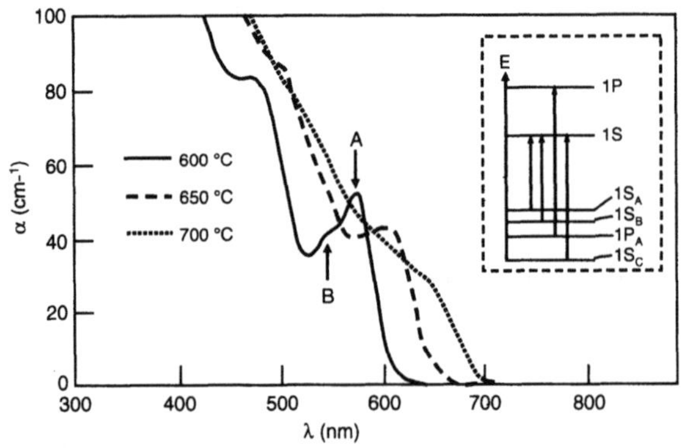

FIGURE 1.14. Linear absorption coefficient vs wavelength for CdSe-doped glass developed at three different temperatures. The higher temperature corresponds to the development of a larger average dot size. The inset shows the absorption assignment taking into account the split valence band. (After Park et al, Ref. 23).

Hall[24] et al looked at the extent of saturation absorption as a function of dot size using ps-pulses up to energies of 5mJ/cm^2 (GW/cm^2). In Fig. 1.18a is shown the initial absorption spectrum of two commercial CdSSe-based filter glasses and an experimental glass with a large measured dot size. (See TABLE. 1.4.) The transmittance at the wavelength indicated by the dotted lines in (a) is shown as a function of fluence in Fig.18b. In addition they show the saturation results for three experimental glasses with much smaller measured dot sizes. The initial absorption curves are shown in Fig. 1.19a. The transmittance vs fluence are shown for the three samples at a number of wavelengths in Fig. 1.19b-d.

In their treatment they assumed that electron-hole relaxation time, T_1, was much longer than the pulse duration, τ. so that the saturation was essentially a fluence, rather than an intensity dominated mechanism. This simply means that they assumed no significant relaxation during the pulse so that once the excitation energy roughly corresponds to one photon per dot, bleaching should be complete. Using a modified 2-level scheme, the overall result indicated that the saturation fluence increased with decreasing dot size. This essentially agreed with the Park et al[23] result, in that it is harder to saturate the smaller dots.

What is interesting in comparing the results from these two studies is that they both obtained about the same degree of bleaching with roughly the same pulse energy.

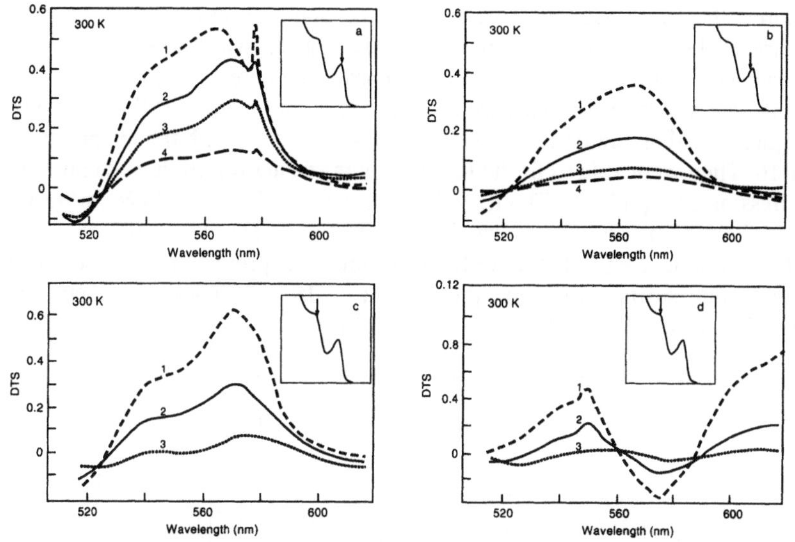

FIGURE 1.15. Room temperature DTS (differential transmittance) vs wavelength for a CdSe quantum dot glass (dot size of 3nm) as a function of pump intensity ranging from 1-7MW/cm^2. The sequence a-d are for different pump wavelengths as shown in the respective insets. (Ref. 23)

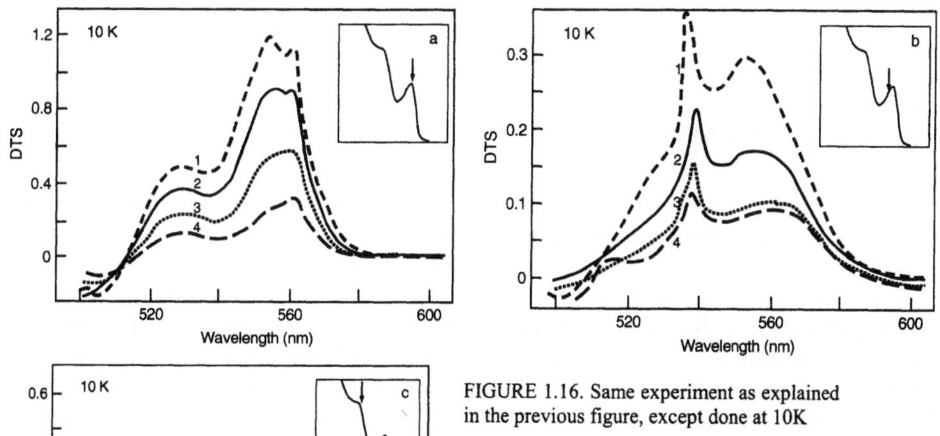

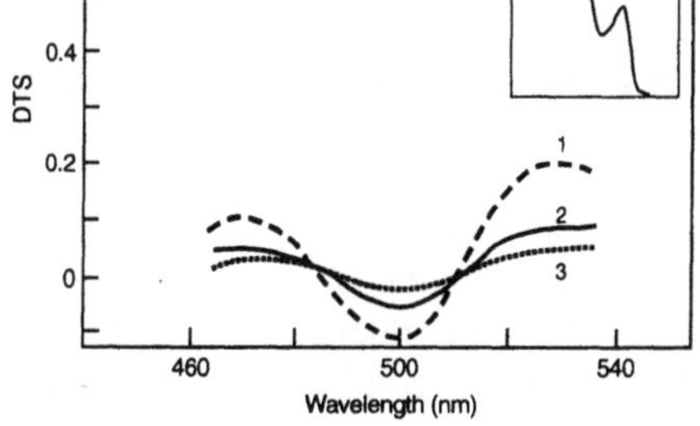

FIGURE 1.16. Same experiment as explained in the previous figure, except done at 10K

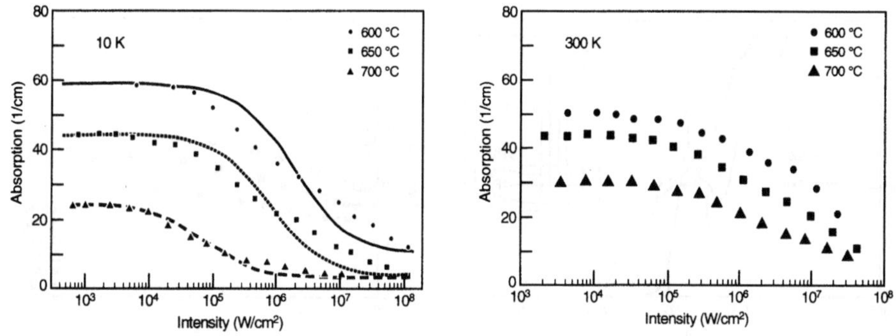

(a) (b)

FIGURE 1.17. Absorption vs Intensity for the three CdSe sample, (a) for the low temperature data of Fig. 1.16 along with the two-level fit according to Eq. (1.16); (b) the room temperature

However ns-pulses were used in the former and ps-pulses were used in the latter study. This means that the saturation intensity is of the order of MW/cm^2 in one case and GW/cm^2 in the other. One possible explanation of this is that there a finite number of relatively long-lived traps which control the recombination kinetics, traps whose lifetime are longer than 1-ns. Once these traps are filled, then the much faster recombination time of the electron-hole recombination come into play. We will see a more definite example of this in the section dealing with PbS.

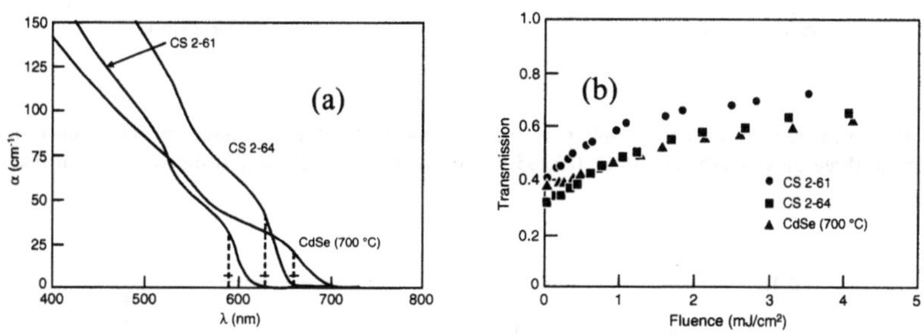

FIGURE 1.18. Linear absorption of commercial filter glasses and experimental CdSe glass (a), and transmittance vs exposure fluence (b) taken at the wavelength shown by the dotted lines in (a). Hall et al reference 24.

Klimov[19] shows differential spectral transmission data for a 2.3nm CdSe dot in glass using an unspecified pulse exposure. This is shown here as Fig. 1.20. The interesting aspect of these data are the appearance of two distinct bleached transitions which they attributed as originating from the split hole states. They point out a hint of bleaching of a third hole state as well which appears as a shoulder at about 2.15eV. They were able to fit this data with a modified two level expression incorporating all of the transitions which are shown in Fig. 1.21. The reader is referred to the original reference for the details. What they have shown fairly conclusively is that all of the bleaching phenomenon is attributable to a state filling mechanism

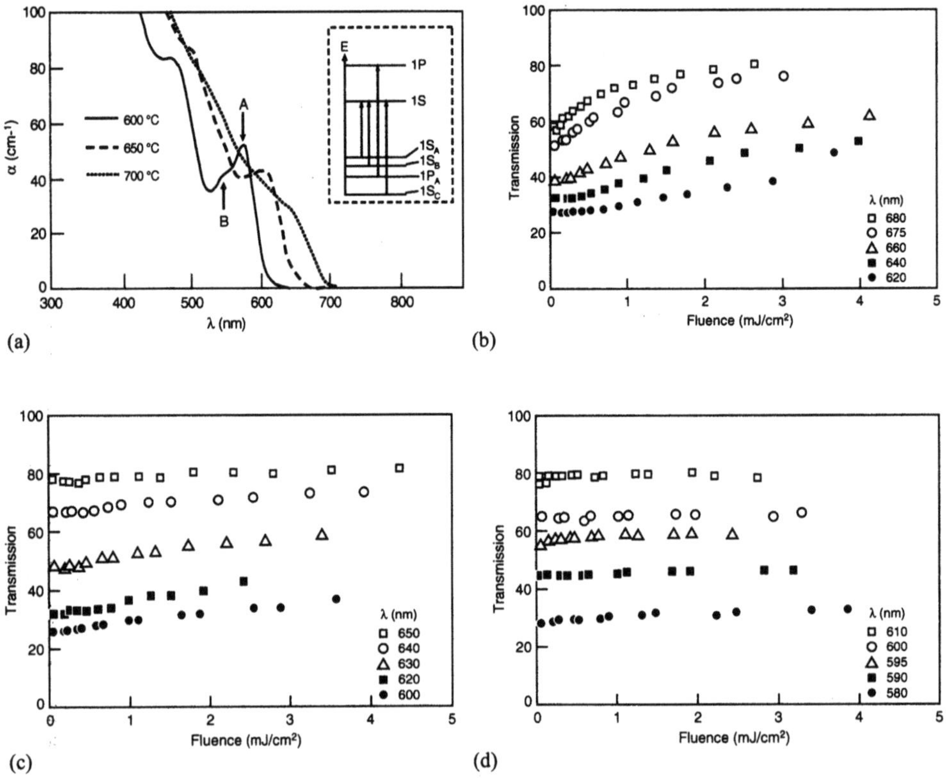

FIGURE 1.19. Linear absorption of a number of experimental CdSe-doped glasses (a), and transmittance vs exposure for the three glasses labeled 700C, (b), 650C. (c), and 600C, (d) each at the wavelength listed. (Ref. 24)

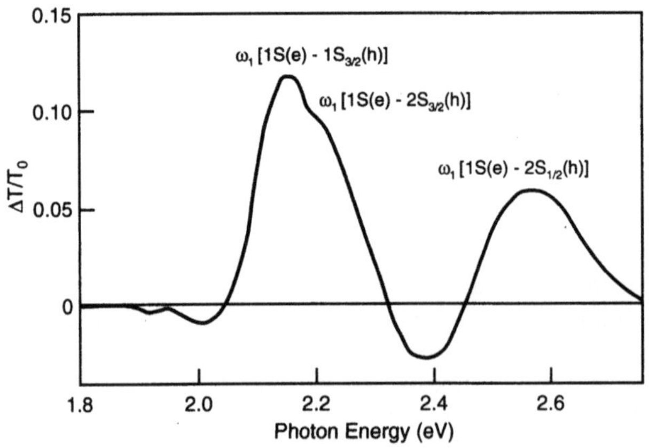

FIGURE 1.20. Differential transmittance of CdSe-doped glass showing two distinct bleached transitions as labeled. (Klimov et. al., Ref. 19)

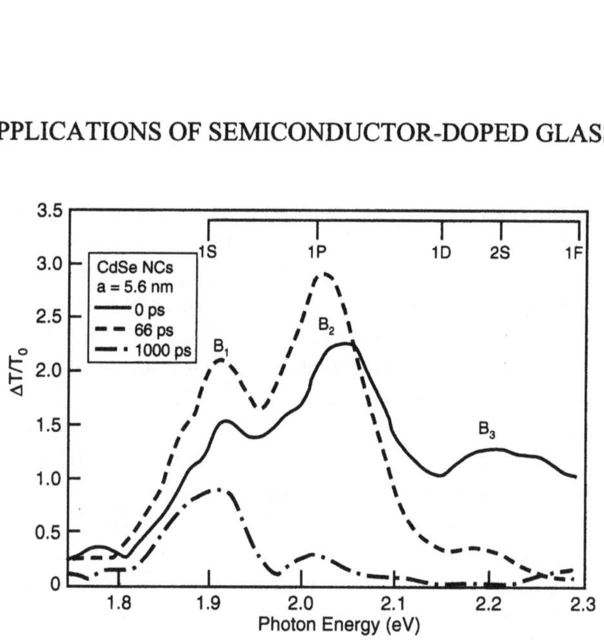

FIGURE 1.21. Differential transmittance vs wavelength of CdSe-doped glass at various delay times after pulse showing the number of transitions involved in the bleaching. The upper designations refer to the assigned transitions. (Ref. 19).

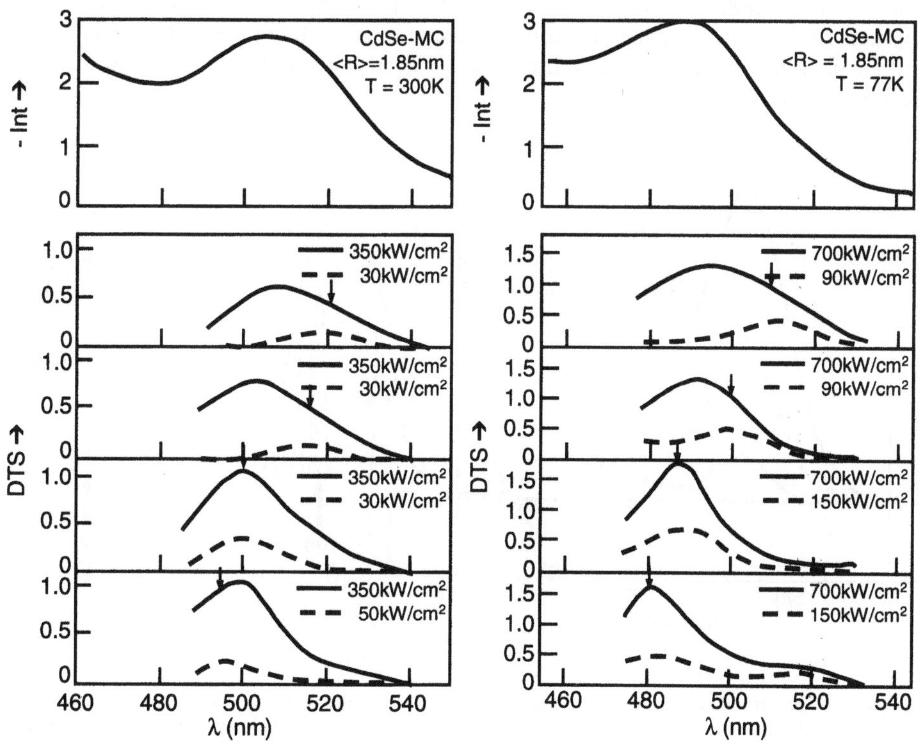

FIGURE 1.22. Differential transmittance vs wavelength for CdSe-doped glass as a function of two values of the incident intensity. The arrows point to the wavelength of the excitation. This was done at both room and nitrogen temperature. The top traces are the respective initial linear absorption curves. The data shows that at the lower intensity, the bleaching peaks near the excitation, while at the higher intensity is broadens. (Hennenberger et al, Ref. 22.)

The role of the distribution of particle size as naturally occurs in the fabrication certainly has to be taken into account in any quantitative modeling of the bleaching mechanism. It has been shown that spectral "hole-burning" can occur. This was shown by Henneberger[22] et al where they show bleaching results for excitation wavelengths through the absorption band as indicated by the arrows in Fig. 1.22. It is seen that at low intensity the bleaching is more restricted to the wavelength region around the pump wavelength indicating selective excitation. Peyghambarian et al[25] also indicate the difference in the nature of the spectral bleaching as a function of pump intensity from the data shown in Fig. 1.23a. However, they conclude that the homogeneous linewidth is comparable to the inhomogeneous width as schematically shown in Fig. 1.23b.

DeRougemont et al[26] show strong evidence for the role of Auger recombination in CdSe-QD-glass although unfortunately in a glass that has relatively large dots (commercial Schott CdSSe-based filter glass,RG715). This fact notwithstanding they showed that through the ad hoc inclusion of a term representing the Auger process into the density matrix formulation-derived equations of the excitation process, then a good fit is obtained for the intensity dependence of the index change produced by the absorption saturation process. The significance of this result is that at high intensity one can obtain a faster recombination time beyond that provided by the limit of normal electron-hole recombination times. This is important since the upper limit to the data rate is just the reciprocal to this time. Normal electron-hole recombination times are of the order of few hundred picoseconds, thus the maximum data rate that could be achieved in a device would be of ~ 10 GHz.

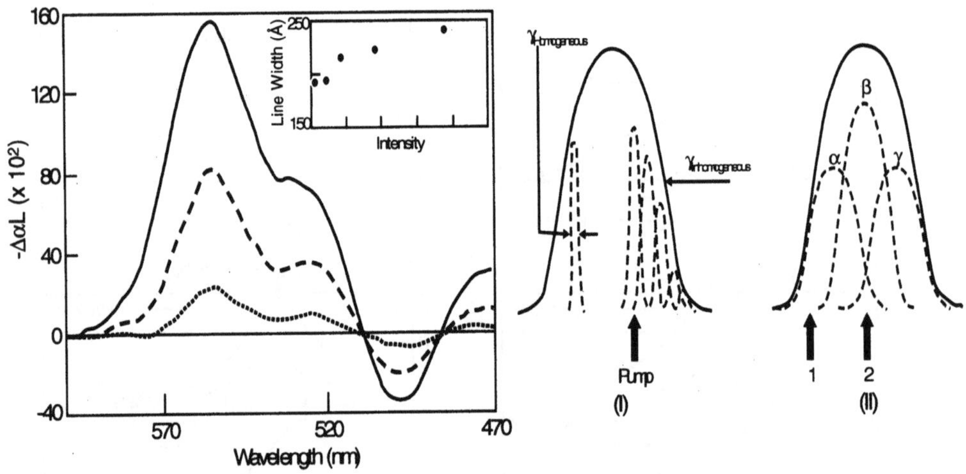

(a) (b)

FIGURE 1.23. (a) Differential absorbance vs wavelength for three values of the incident intensity. Inset is a plot of linewidth of bleaching as a function of intensity. (b) From the data in (a) the diagrams on the right suggest that the homogeneous linewidth is more like the right-most diagram, namely comparable to the overall inhomogeneous linewidth. (Peyghambarian et al, Ref. 25).

TABLE 1.5
Physical and Optical Data of the Commercial and Experimental Glasses CdSSe-doped Glasses

Parameter		CS2-61	CS 2-64	700C	650C	600C	
Se fraction	0.48		0.74		1.0	1.0	1.0
Avg. Diam. (nm)	10		11		7.9	4.4	3.0
Distribution Width (nm)	2.5		2		2.2	1.2	-
E_g Bulk (eV)		2.06		1.92		1.72	1.72
1.72							
E_{ass} (eV)	2.02		1.89		1.80	1.97	2.03
Sat (mJ/cm^2)		0.75		0.70		1.5	2
(10)							

IV-VI (PbS, PbSe)

The PbS QD/glass system is a more recent development and has several significant advantages over the II-VI-based semiconductors. The first is a practical one in that the bulk band gap is much smaller at 0.4eV, and consequently, the quantum confined states can overlap the telecommunication bands at 0.8eV, so that the device utilization might be more likely. The second advantage is that because of the much smaller effective mass for both the electron and hole the system is truly strongly confined for both. This makes well defined and separated hole and electron states. This is in contrast to the II-VI system where the levels are relatively congested due to closely spaced hole levels and are complicated by valence band mixing. Further, as a consequence of the smaller effective mass in the II-VI system, the Bohr radius is large, thus the dot size can still be relatively large, but with significant confinement. The argument can be made that one can minimize surface states relative to volume states because of the larger dot size.

The PbS quantum dots developed in glass show sharp absorption features throughout the spectral region of 1000-2000nm. A set of spectra representing different average dot sizes are shown in Fig. 1.24. Wise[15] calculated the position of the first three transitions (1S-1S, 1P-1P, and 1D-1D) as indicated in Fig. 1.25a as a function of the dot size.. We reproduce his result in Fig. 1.25b. This result represents a better agreement to the data than that provided by the simple effective mass approximation.

The saturation behavior of PbS/QD glass was first reported by Guerreiro[20] et al where they did a single beam measurement at 1300nm which corresponded to the peak of the 1S-1S transition of the sample. They used 30-ps pulses from a Nd:YAG pumped OPA. Their result is shown in Fig. 1.26. A reasonable fit was obtained in the form of Eq. (1.23) yielding a value of I_{sat} = 0.18MW/cm^2 This is comparable to the ns-pulse result in CdSe mentioned above, although considerably lower than the ps-pulse result.

Toward the understanding of the kinetics of the bleaching mechanism, Wundke[28] et al used a pump probe technique to measure the change in transmittance as a function of delay time after the 120-fs pulse. This is shown in Fig. 1.27 for three different pulse energies ranging from 0.5mJ/cm^2 to 6mJ/cm^2. The decay time decreases with increasing pulse energy and is bi-modal with a fast initial decrease over first 20ps followed by a 10x slower component. The intensity in this case is in the TW/cm^2 regime. So again as in the CdSe case, comparable changes in absorbance occur with similar energy per pulse, but with much different intensity (pulse duration.)

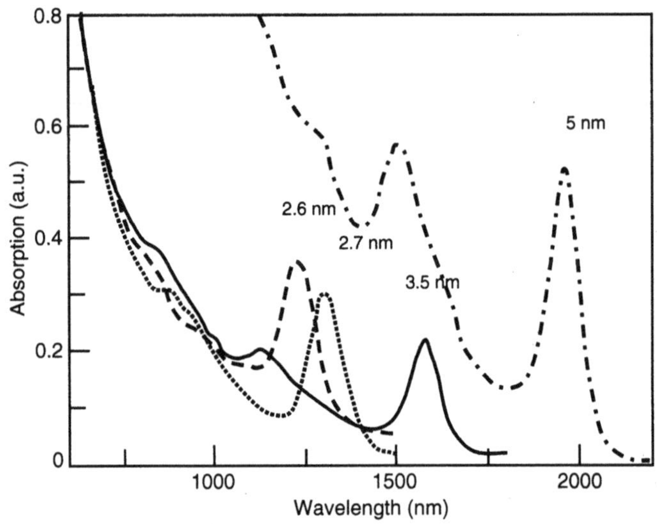

FIGURE 1.24. Absorption spectrum of PbS-doped glasses for various dot sized.

Tamulaitis[29] et al have measured similar behavior as indicated in Fig. 1.28 and 29. The linear absorption is shown in Fig. 1.28 along with the induced change at three delay times. Their z-scan results of the transient response shown in Fig. 1.29 are for a number of excitations that are labeled in the previous figure. They essentially cover the 1S-1S transition. They also show the results at a given excitation as a function of fluence up to70mJ/cm^2. They report in addition to bleaching, a broad induced absorption, although the origin is not specified. They attribute the bimodal decay in terms of a model where the electrons can be excited to higher states (process not specified, although could mean Auger). They argue this relaxation is the faster of the two and causes a bottleneck for the bleaching.

Another possible explanation for this behavior is that the initial decay represents the fast trapping of the carriers at shallow levels. This situation is essentially a three level scheme where the trapping rate is sufficiently fast the upper electron level is essentially empty. The absorption saturation is with respect to the trap level. When the traps are filled then the recombination is dominated by the electron-hole recombination time which is of the order of hundreds of picoseconds. If one further assumes as Wise has suggested that the trap state decay is slow (spin forbidden), then saturation kinetics are dominated by the rather long-lived trap decay. We shall see later on that the photoluminescence has a lifetime of a few microseconds.

I-VII (CuCl, CuBr)

Justus et al[30] have shown a saturable absorption effect in the weakly confined CuCl QD/glass system. At 77K they show well resolved exciton peaks as shown in Fig. 1.30 (a) for three different dot sizes. As discussed above, in the weak confinement limit, the effect of the particle dimension is on the exciton itself, thus the spectral position of

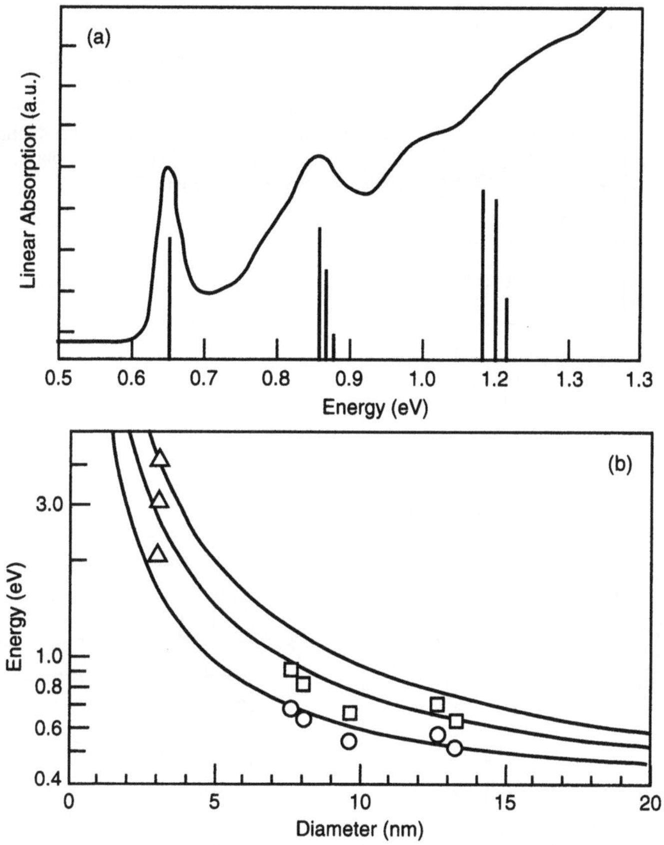

FIGURE 1.25. (a) Absorption spectrum of PbS quantum dot glass indicating position of the first three transitions, and (b) results of the calculation of the position of these transitions as a function of dot size. Wise, reference 15.

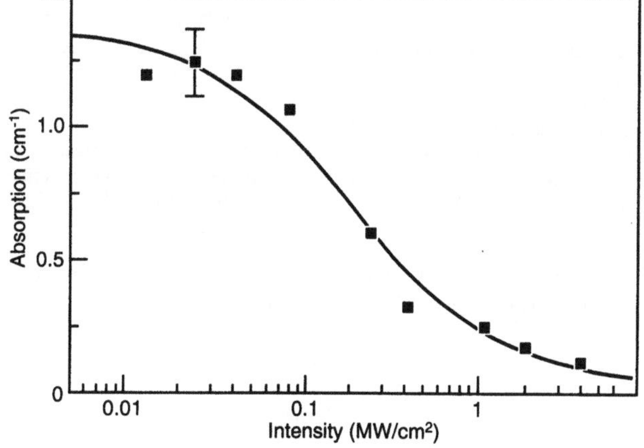

FIGURE 1.26. Absorption saturation behavior of PbS quantum dot glass. Behavior fit to two-level saturation of Eq. (1.23). (Guerrerio et al, reference 20).

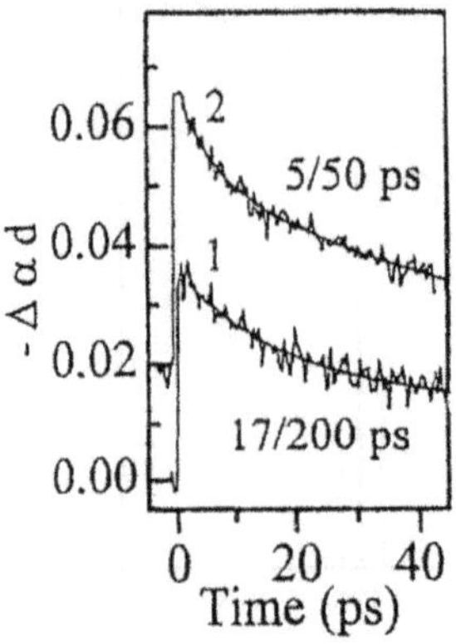

FIGURE 1.27. Time response of excitation of PbS quantum dot glass at two different values of the excitation fluence; (1) 0.5mJ/cm2, and (2) 4mJ/cm2. The exponential constants for the initial and long-term decay are noted on the curves. Pulse duration was 120fs. Wundki et al, reference 28.

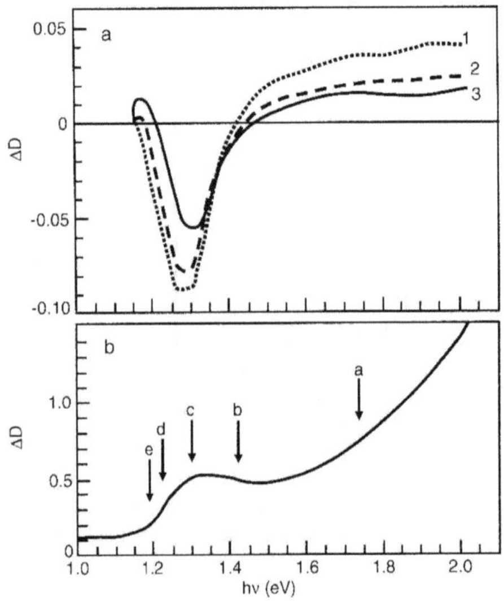

FIGURE 1.28. Linear, (b), and difference spectrum, (a) of PbS-doped sample exposed at 12mJ/cm^2. The number designation in (a) refers to the time delay of the measurement of 0, 7 and 100ps. The letters on the curve (b) refer to the spectral position of the kinetic data shown in next figure. (Tamulaitis et al, reference 29).

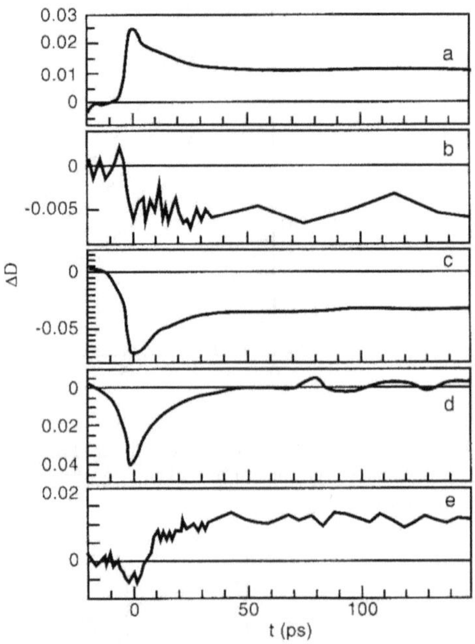

FIGURE 1.29. Z-scan response time measurement of sample described above at the noted energy. (Ref. 29)

the exciton peak depends on the CuCl particle size. The weak confinement expression given by Ekimov[12] yields the energy shift of the exciton.

$$\Delta E = \frac{\hbar^2}{2Ma^2} \tag{1.36}$$

Here, M is the translational mass of the exciton and a is the particle diameter. Ignoring the coulombic term in the Hamiltonian makes the quantitative aspect of this suspect, but the significant part of the study is the bleaching of the exciton, not the exact particle size relationship. They exposed at the nominal wavelength of the exciton at 77K with 25-ns pulses from a excimer pumped dye laser. The bleaching result is shown in Fig. 1.31. The fit the bleaching data to 2-level expression as expressed by Eq. (1.23) and obtained the values of I_{sat} of the order of 1-8kW/cm^2. It is not at all clear why this is the appropriate form since clearly the mechanism for this bleaching is different from that of the strongly confined cases. This fact not withstanding, it does give a quantitative comparison with particle size. They also estimated the refractive change near the edge as a consequence of the bleaching using a Kramers-Kronig technique.

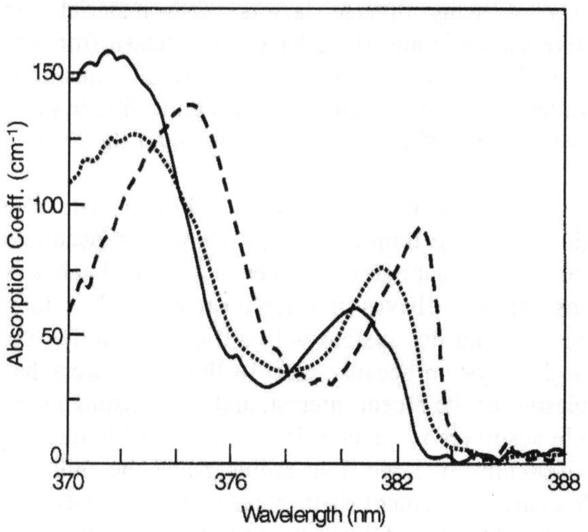

FIGURE 1.30. Absorption spectrum of CuCl-doped quantum dot glass at 77K: solid line corresponds to dot diameter of 2.2nm, dotted to 2.7nm, and dashed to 3.3nm. (Justus et al, reference 30).

As the authors suggest, the mechanism is not simple, although clearly it must involve a screening effect provided by the ionized excitons. Why this should depend on the particle radius is not clear.

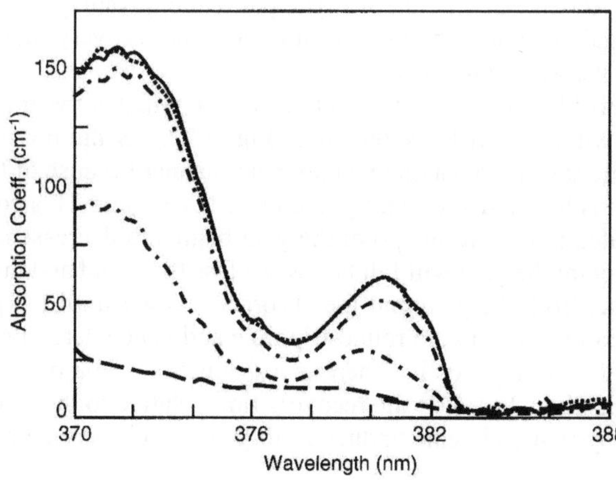

FIGURE 1.31. Saturated absorption spectrum of 2.2nm CuCl dot at 77K. Incident intensity in the order from the top, 0.4, 1.2, 3.2, 14, and 27kW/cm^2. (Reference 30).

There is also a report of two-photon enhanced non-resonant nonlinearity by the same group. Using a Z-scan method at 800-nm which corresponds to roughly half the exciton energy they measured an enhanced nonlinear refractive index in the CuBr QD/glass. The argument made is that the presence of the exciton increases the TPA which in turn enhances the nonlinear refractive index.

1.4.3. Q-Switching and Mode Locking

The kinetic or dynamic behavior of a system experiencing gain is quite complicated because of the interplay of many diverse factors, both material and geometric. Relatively fast and sensitive saturable absorbers have been chosen from the large class of organic chromophores, but few, if any organics have electronic transitions in the near IR. Equally so, readily available QD dot systems like CdS and CdSe in glass fall in the visible wavelength range where they offer no real advantage over the organic dyes in terms of speed or performance.

Our interest here is simply to introduce the role of the QD/glass saturable absorber as a robust medium to produce a regular train of short pulses when placed in a laser cavity. Toward this end, the operation of a saturable absorber as described below is generic, in the sense that that the transmittance behaves in a nonlinear way. Therefore, the typical behavior of one particular quantum dot system will suffice to describe the application. We have picked the PbS/glass system because most of the recent work has been devoted to it . One significant reason for the recent interest, and what distinguishes it from other QD system, and saturable absorbers in general, is that it is useable through the near IR wavelength region 1000-2000 nm. Its large Bohr radius makes the ability to adjust the confined transitions much easier to accomplish in glass than other solvents. This would lead to a narrower dot size distribution and thus narrower absorption features. The idea of using a semiconductor as a saturable absorber to initiate the mode locking behavior is not new.[32] There has been success in using epitaxially grown semiconductors quantum wells as passive intra-cavity saturable absorbers in such lasers as Cr:LiSrF, Ti:sapphire, Cr^{+4}:YAG and Cr^{+4}:Forsterite The particular advantage depends heavily on the lifetime of the gain medium, the rate of saturation, the recovery time of the saturable absorber as well as the wavelength itself.

The general idea of how a saturable absorber initiates the train of shot pulses with cw pumping within a cavity is shown in Fig. 1.32. As the cw pump intensity is raised, the loss in the cavity decreases in a nonlinear manner because of the saturation. At some point the loss is lowered sufficiently that there is net gain. This corresponds to the onset of the generated pulse. At this point the gain begins to decrease due to stimulated emission. At some point the gain will fall below the loss threshold and this corresponds to the end of the pulse. Both the gain and the absorber recover until the pulse makes one round trip which corresponds to the round-trip time and hence determines the frequency of the pulse train $f = 1/T_R$ = c/2L There is a distinction to be made as to "fast" and "slow" saturable absorbers based on the recovery time relative to the round trip time. For the PbS QD/glass system and semiconductors in general, one is invariably in the slow regime.

An easy way to look at the typical behavior of the device is as a cavity with the output mirror having a saturable absorber attached as shown in Fig. 1.32. One can then plot the reflectivity of the mirror/absorber assembly as a function of the intensity seen by

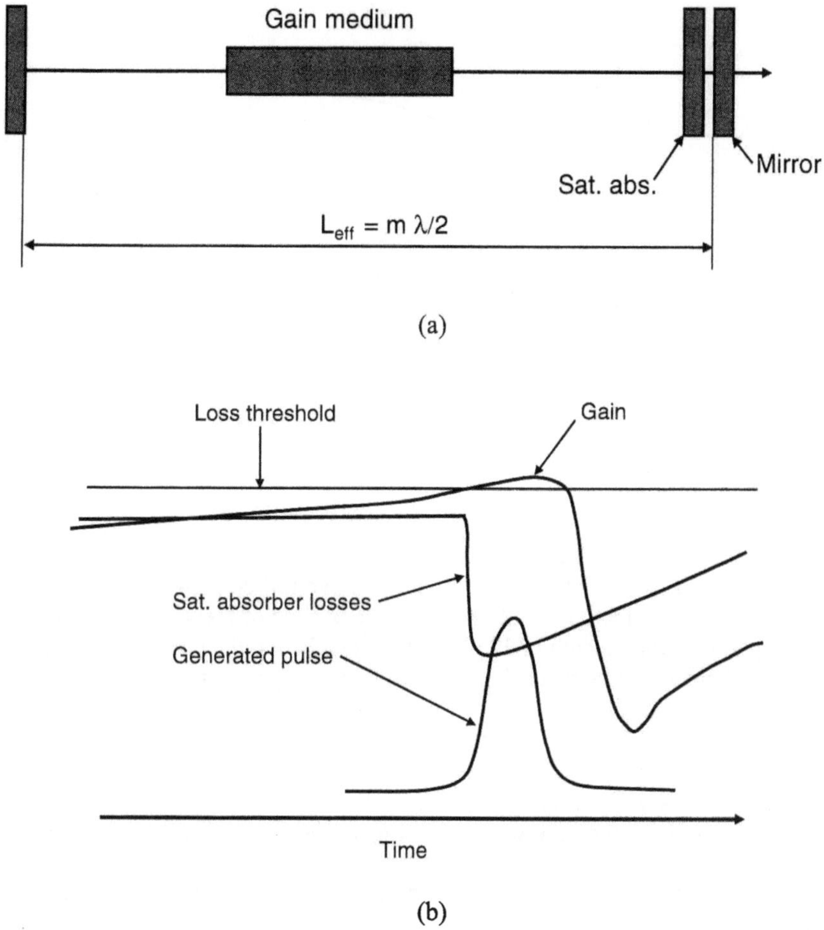

FIGURE 1.32 Schematic representation of (a) the experimental setup and (b) how mode locking through saturable absorption produces short pulses.

the absorber.[33] This is schematically represented in Fig. 1.33. The ideal condition for mode locking is where the value of I_{sat} is balanced with respect to the gain recovery time such that the following inequality holds.

$$\left|\frac{dR}{dI}\right| I < \frac{T_R}{\tau} \tag{1.37}$$

Here, T_R is the round-trip time and τ is the upper state laser lifetime (corresponds to the gain recovery curve in Fig. 1.32). If this condition is not met one will operate in a Q-switch mode. So, what one has as the saturable absorber material parameter is the value of I_{sat} (scales with the reciprocal of dR/dI) with the recovery time always being fast compared to the gain recovery. It is through the enhancement of the saturable absorption property that one hopes the use of QDs would be advantageous.

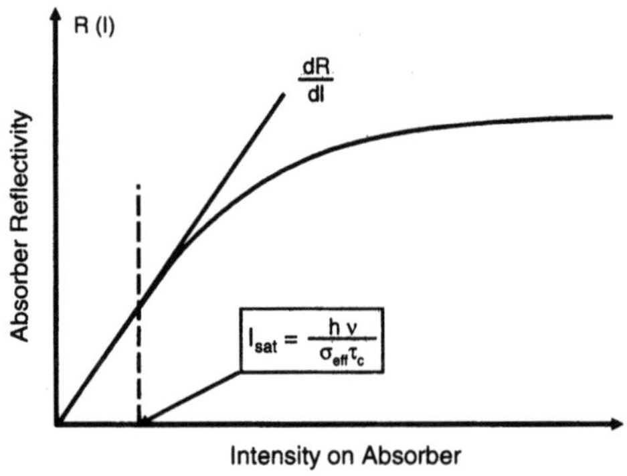

FIGURE 1.33. Schematic representation of the absorber reflectivity as a function of the intensity on the absorber. (Keller et al, reference 33).

1.4.3a. Experimental Results

A PbS QD glass with saturation behavior as shown in Fig. 1.34a, was used as a saturable absorber in the optical configuration as shown in Fig. 1.34b which incorporates Cr:forsterite as the laser medium.[20] The 150μm thick slice of the glass (transmittance of 96% at the operating wavelength of 1250nm) was attached to the output mirror as shown. The saturation behavior was fit to the standard 2-level approximation. As indicated by Eq. (1.23) above, using the parameters α_B=0.03cm^{-1}, α_0=1.34cm^{-1}, and I_{sat}= 0.18MW/cm^2. A 110MHz train of pulses with up to 74mW of average power was obtained for a pump power of 7.2W. A typical temporal pulse in shown in Fig. 1.34c and the spectral shape is shown in Fig. 1.34(d). If one assumes the transform-limited expression fore the pulse then the width was 4.6ps. The spectral width was measured to be 0.4nm.

A more detailed analysis of the mode locking performance of a similarly prepared PbS-QD glass was carried out by Wundke et al..[28] Their analysis showed that transform limited performance was not being achieved because the GVD (group velocity dispersion) was not being fully compensated by the nonlinearity in the gain medium. The comparison of the computed transform-limited pulse shape from their analysis to the measured shape in shown in Fig. 1.35. A higher intensity in the cavity should have the consequence of increasing the small-signal gain. This would have reduced the pulse width as shown in the inset to Fig. 1.35, where the normalized pulse width is plotted against the small signal gain. Unfortunately, with the higher cavity intensity glass nonlinearity began to play an interfering role. It would appear that with sufficient care and design that the PbS QD saturable absorber could produce sub-picosecond pulses.

There have been a few other studies done as well. Malyarevich et al[34] have reported mode-locking and Q-switching of Nd-doped lasers with PbS-QD glasses. Pulse duration in Nd:YAG was reported as 70ps with 18μJ energy. The saturation intensity was reported as 2.3MW/cm^2, and a bleaching relaxation time of 23ps.

Philipps et al[35] reported using a PbS quantum dot sample to Q-switch a Er-Yb-doped phosphate and fluorophosphate. A typical Q-switch result using the PbS QD glass as the saturable absorber (initial absorption spectrum of the QD-glass with 3nm average dot size is shown in Fig. 1.36a) is shown in Fig. 1.36b for two different thickness of the

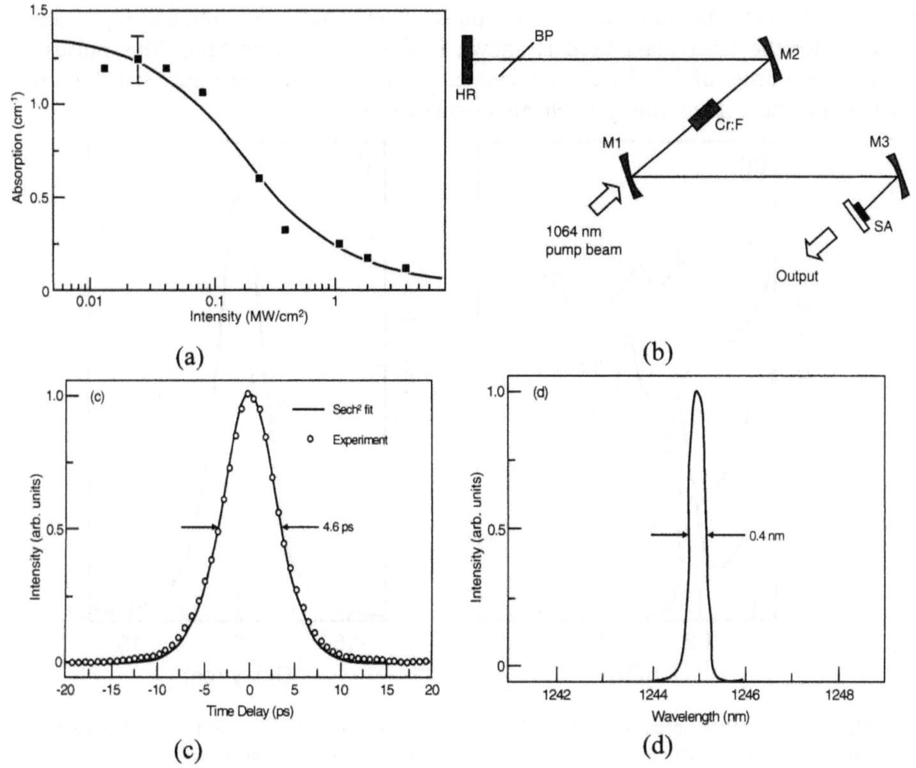

FIGURE 1.34. Experimental results on the use of PbS QD glass as mode locker for Cr:forsterite at 1300nm. (a). the measured saturation behavior; (b) the experimental configuration. The measured pulse is shown in (c) and the spectral shape is shown in (d). (Reference 20).

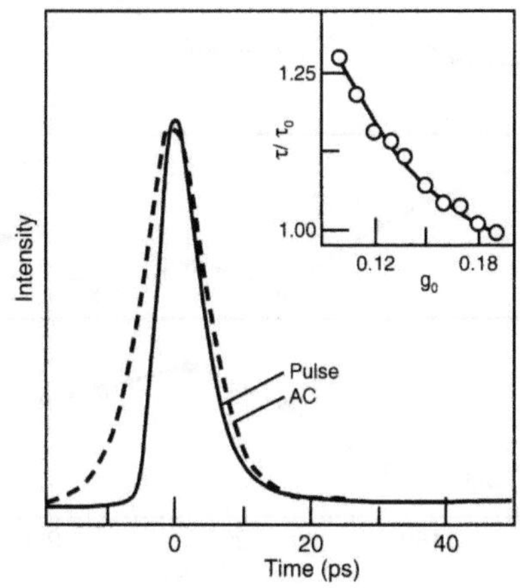

FIGURE 1.35. Comparison of calculated pulse width compared to the measured width. For a mode locking application of a PbS-QD glass (Reference 28).

absorber. Explicit measured data for the pulse energy, pulse duration and repetition rate obtained for the laser glass used is shown in Fig. 1.37. In general, 30ns pulses with energy of the order of 1.5μJ were obtainable. The inability to mode lock at 1500-nm was attributed to the unfavorable gain characteristics of Er^{+3}.

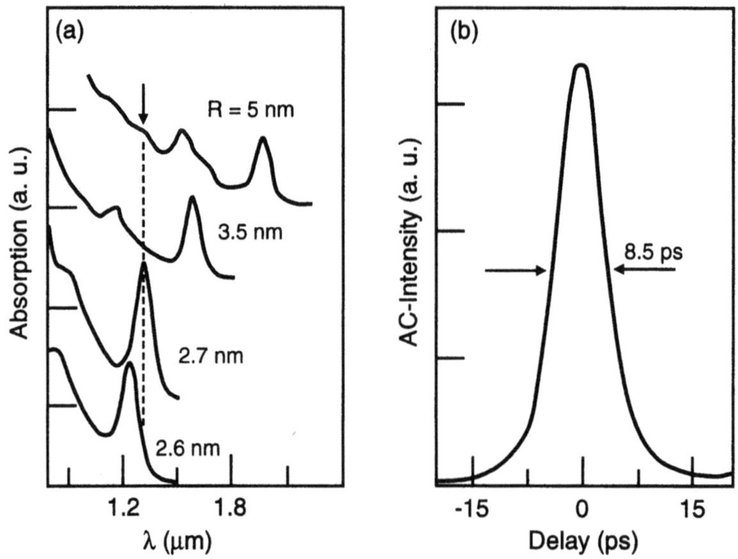

FIGURE 1.36. Absorption spectrum of PbS-QD glass used as Q-switching element for Er-Yb glass laser (a), pulse duration obtained for two different absorber thickness (b). Philipps et al, reference 35.

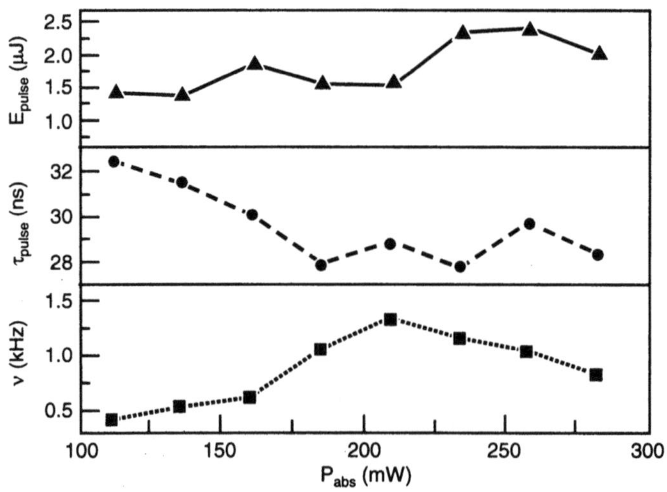

FIGURE 1.37. Pulse energy, pulse width, and repetition rate using PbS-QD glass with Yb-Er-phosphate glass (Reference 35).

1.4.4. Laser/Amplifier

This is a potentially important application area because of the broad emission wavelength range due to the ability to control the energy gap through the dot size. Unfortunately, the electronic processes are more complicated than the simple idea that one is dealing with the dipole-allowed transitions between two quantum confined levels with well defined quantum numbers. To make matters worse, the degree of deviation from the idealized situation, and hence the specific gain mechanism, depends on the particular semiconductor system. The QD systems that have been specifically looked at as possible gain media span the range from the weakly confined CuCl-based system, to the intermediately confined CdSe, to finally the strongly confined PbS-based glasses.

1.4.4a. Experimental Results

CuCl

Masumoto et al[37] report the occurrence of stimulated emission at 77K in a CuCl-doped NaCl crystal. The doping level was 0.3 % (mole). The photoluminescence spectrum is shown in Fig. 1.38a. The significant feature is that the band labeled as M grows in with increased excitation intensity. It is attributed to the biexciton recombination. With nitrogen laser pumping at 337-nm at 77K, lasing at 391-nm was observed at a threshold of $2MW/cm^2$. The input-output results are shown in Fig. 1.38b and the emission line in 38c. The explanation of the emission is given in terms of an inversion of biexciton level to an exciton transition. The argument is that the confinement effect produces sufficient separation of the biexciton states to allow them to be the dominant transitions. The question remains what properties of the biexciton transition relative to the exciton decay make them more likely to produce inversion.

CdSe

There have been a number of papers dealing with gain in the CdS/CdSe quantum dot systems. Dneprovskii, et al[38] reported stimulated emission from a CdSe-doped glass at 80K using 25-ps pulses from a doubled Nd/YAG laser. The energy threshold was somewhat below $0.5mJ/cm^2$. The result is shown in Fig. 1.39 where the stimulated emission is compared to the spontaneous emission. The inset shows the energy threshold. They also measured time-resolved differential transmittance as a function of wavelength. It shows three distinct bleaching features in spite of the fact that the initial absorption spectrum is relatively featureless. They indicate gain over a broad region indicating the involvement of both the 1S and 1P transitions.

On the other hand, Giessen et al[39] show results for CdSe in glass (dot size of the order of 25-nm) suggesting that the gain arises from biexciton transitions, as shown in Fig. 1.40a. The experiment corresponds to measuring the absorption 5ps after the 115fs pump pulse. It was done at 10K. They argue the biexciton origin of the gain from the fact that the gain occurs predominantly on the long wavelength side of the initial absorption and thus arises from biexciton to exciton transitions as indicated in Fig. 1.40b, here is shown all the possible emission lines from the !S 1P and 1D single exciton states and all the possible transitions from the upper biexciton states to the single exciton states. They

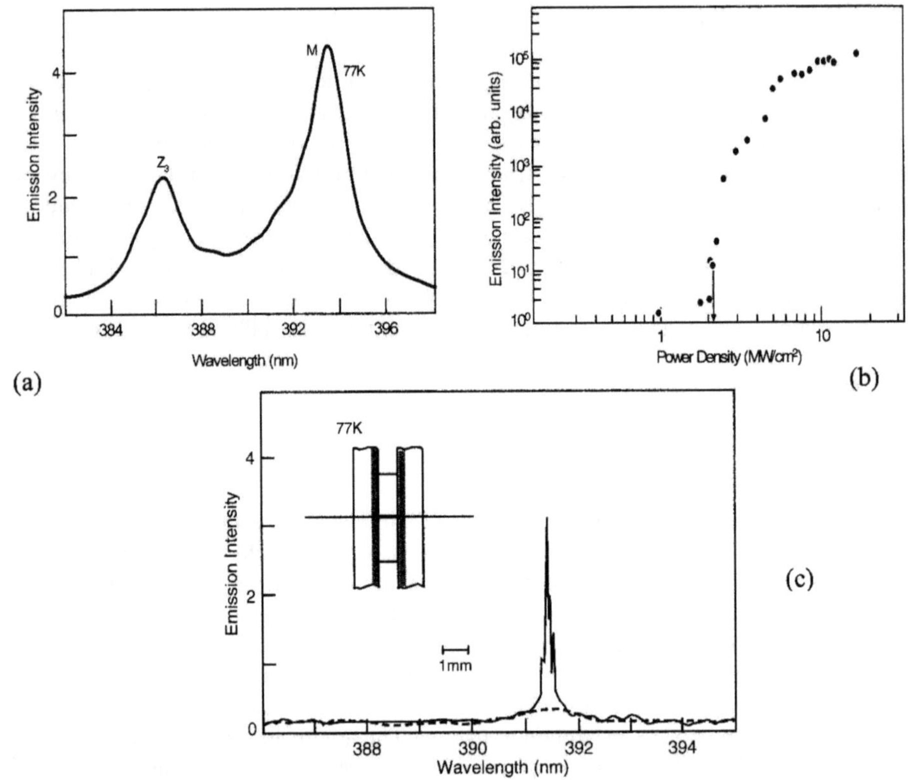

FIGURE 1.38. (a) Luminescence of CuCl embedded in NaCl at 77K. M-band corresponds to biexciton recombination, (b) emission intensity of laser versus pump intensity at 77K, (c) emission spectrum of laser at 77K, the solid line is obtained at $1.08I_{thr}$, and the dashed line at $0.86I_{thr}$ Pump wavelength was 530-nm using 25ps pulses. The inset is the cavity used. Masumoto et al, reference 37.

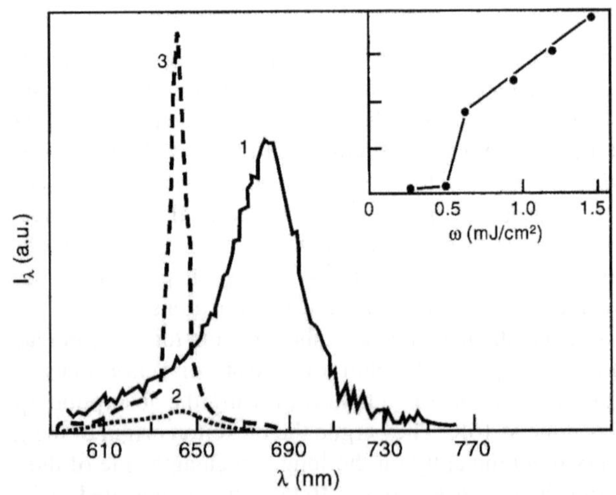

FIGURE 1.39. Spectra of spontaneous emission (1), and stimulated emission (2 pumped ar 0.5mJ/cm^2) and (3, pumped at 0.8mJ/cm^2) for CdSe-doped glass at 80K. The inset shows the threshold behavior Dneprovskii et al, reference 38.

present an extensive model to substantiate their explanation of the gain as stemming from the biexciton transitions. These results and interpretation were essentially substantiated by Woggon et al.[40]

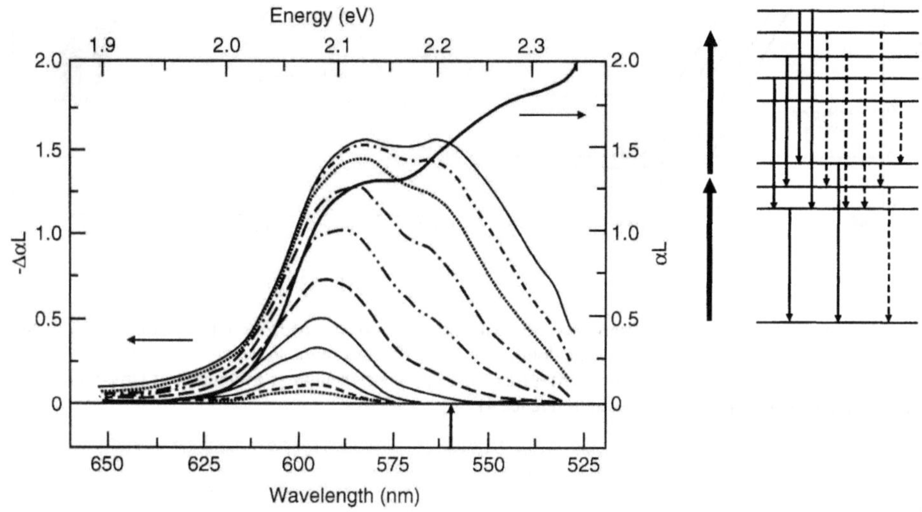

(a) (b)

FIGURE 1.40. (a) Experimental differential absorption as a function of increasing pump intensity at 10K for CdS QD in glass. Also plotted is the linear absoprtion, +αL on the right hand vertical axis. Gain is present when differential absorption (-αL) exceeds the absorption (+αL). Note gain is to the long wavelength side for the low pump intensity. (b) The schematic of the discussed optical transitions (Hu et al, reference 42).

Butty et al[41] have observed gain up to room temperature in a CdS-doped sol-gel glass. The estimated size of the dot was slightly smaller than the Bohr radius of 2.9nm. Their results are shown in Fig. 1.41a where the absorption spectrum is shown as a function of pump intensity. The gain vs wavelength as a function of temperature is shown in Fig. 1.41b, although the reason for the large temperature dependence is not given. The inset shows the measurement arrangement. The photoluminescence data indicated a significant long wavelength band in addition to the so-called intrinsic band corresponding to the absorption edge. The distribution of the particle size was quite broad which certainly complicates the explanation. The lifetime of this emission was reported to be quite short at 50ps. They too, attribute the gain ($33cm^{-1}$) to biexciton to exciton transitions primarily on the basis of its preferential development on the low energy side of the absorption edge.

The results presented above do not paint a very clear picture of what is going on with respect to the gain mechanism. Although Hu, et al[42] present a comprehensive model it is a large number of levels and parameters to which numerical values must be assigned. Another problem with the CdSe, CdS systems studied is that the absorption seems quite diffuse indicative of a broad inhomogeneous dot size distribution. This is not taken into account in the respective analyses. From a practical point of view it would appear that rather high pump intensity is required, and that low temperature is required for efficient gain to be realized.

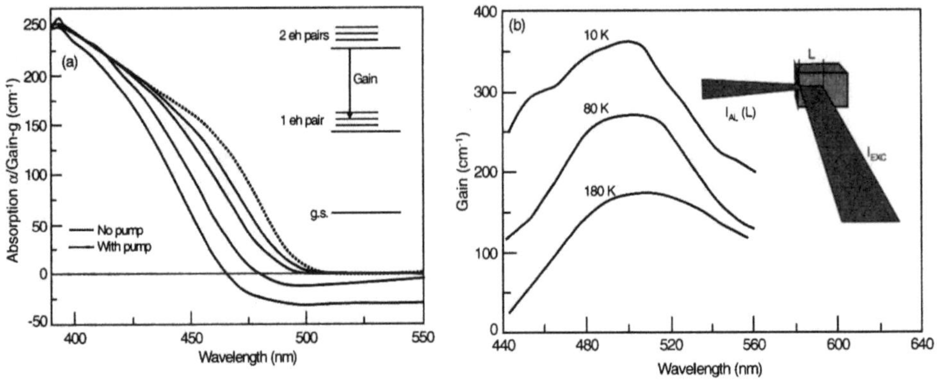

(a) (b)

FIGURE 1.41. (a) CdS QD glass absorption/gain vs wavelength, dotted line is the initial absorption. The pump intensity is from $0.15 MW/cm^2$ shown as the dotted line, to $32 MW/cm^2$, corresponding to lowest solid-line trace. Inset shows the proposed gain transition; (b) measured gain vs wavelength as a function of temperature. Inset shows the pumping geometry. Butty et al, reference 41.

PbS

This compositional system provides a more defined example of a strongly confined quantum dot with well-resolved spectral features indicating a relatively narrow dot size distribution. This is not to say that the salient issues are the same for obtaining gain in the PbS system as opposed to CdSe, or CdS, but at least the basic simplifying assumptions seem to be better met, for example, non-mixed well separated hole states as well as electron states. With the larger Bohr radius, of PbS one can use larger dots and lessen the role of surface states.

Wundke et al[43] have demonstrated room temperature gain at 1300nm in a PbS QD glass. The pumping scheme was to pump into the 1P transition (130fs pulses) and to probe the signal at the wavelength corresponding 1S transition. The gain coefficient calculated on the basis of the dot diameter, $80 cm^{-1}$, is somewhat misleading. More realistically the gain is small considering the actual dot number density.

The time-resolved absorption data is shown in Fig. 1.42 for two probe wavelengths, the one resonant with the absorption peak at 1317, and the other at 1352 nm. Two pump wavelengths were used, 900-nm and 980-nm. It is interesting to note that gain could be achieved only within a given portion of the distribution, that being where the 1S and 1P transitions were in the same particle. This is shown by the fact that the 900nm pump/1317nm signal showed gain, but the combination 980nm/1317nm did not and the reverse situation in the case with the probe at 1352nm. This indicated that even for what appears to be a sharply defined absorption feature, one is still dealing with a inhomogeneous system. In a real way, the strong confinement (wide separation of the energy levels) makes this problem worse. With closer spaced levels one accesses more dots for the same distribution.

Auxier et al[44] looked into the possibility of 980 nm cw pumped gain in the PbS QD glass. It was observed that there was a rather intense photoluminescence slightly

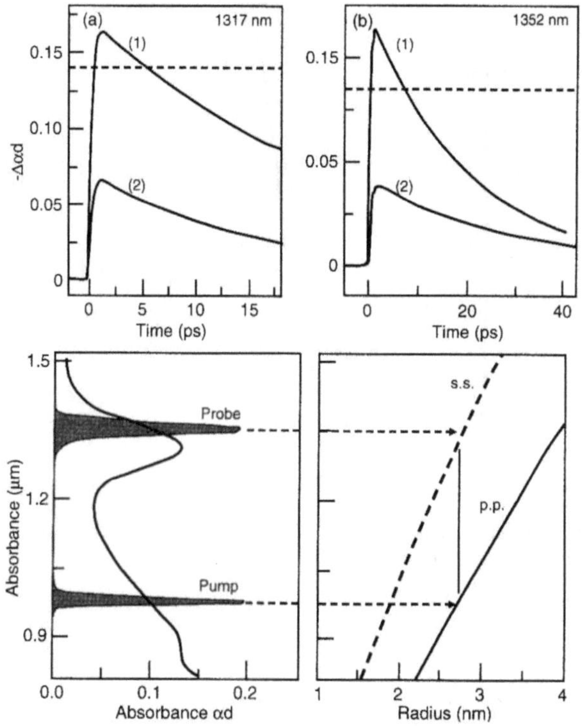

(c)

FIGURE 1.42. Transient absorption of PbS-QD glass at RT, (a) measured at 1317nm with pump at 900nm at 4mJ/cm^2 curve 1, which corresponds to !P transition; and pump at 980nm; (b) measured at 1352nm with 980nm pump, curve 1 at 4mJ/cm^2, and curve 2 at 2mJ/cm^2. Dotted line is the linear absorption at the respective wavelengths. Wundke et al, reference 43. (c) data plotted in a different way to show the correspondance of the pump and probe levels.

red-shifted from the absorption maximum as shown in Fig. 1.43. The normalized integrated intensity as a function of the cw pump power is shown in Fig. 1.44 and shows a distinct saturation behavior. The arrow points to the fluence corresponding to the pulse threshold gain reported by Wundke above. The evidence points to luminescence from two sources, free excitons and shallow trap states. The former have a recombination rate in the 20-ps range while the latter ais of the order of microseconds. The saturation behavior occurs when the traps are filled.

As in the CsSe and CdS, cases discussed above, the likelihood of producing a practical efficient gain medium from the PbS QD-glasses is not high. The difficulty appears to stem from the very fast exciton recombination rate, thus making it difficult to obtain a sufficient inversion. The defect states on the other hand are easier to invert because of their much slower recombination rate, but are insufficient in number to produce sufficient gain to overcome the inherent losses. The quantum yield at low excitation intensity is estimated to be around 10%. It is not clear as to whether the overall situation gets better or worse when confinement is dominant.

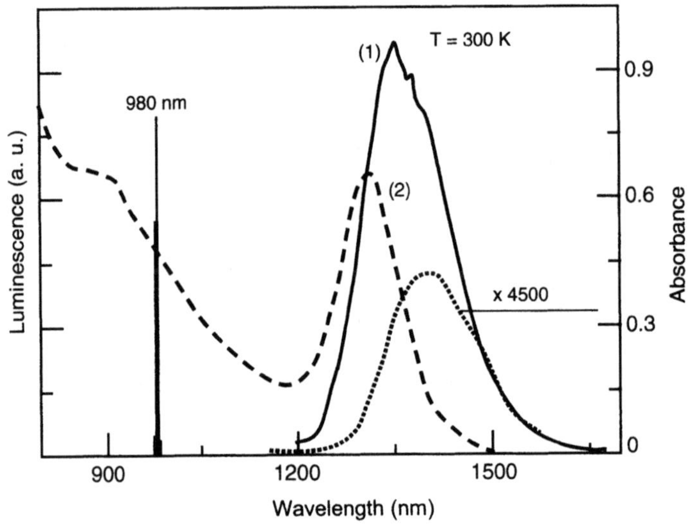

FIGURE 1.43. 980nm cw-pumped luminescence in PbS-QD glass, curve I at 75mW, and curve 2 at 75μW. Dotted line is the linear absorption. Auxier et al, reference 44

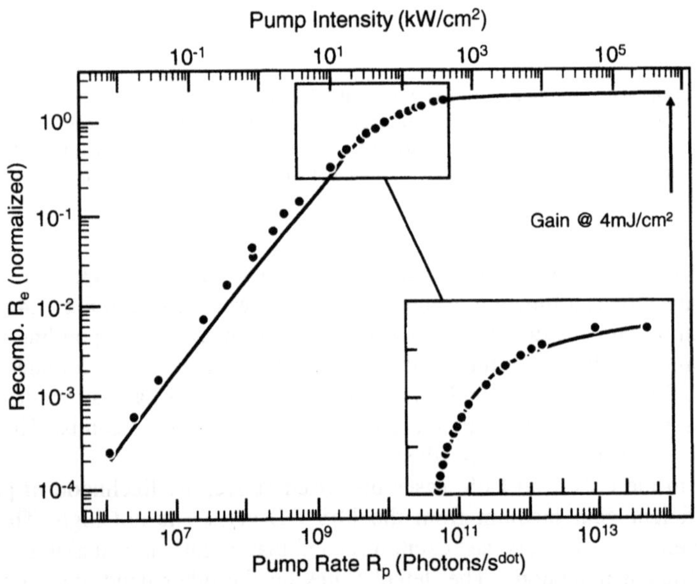

FIGURE 1.44. Normalized intensity vs cw-pump rate in photons/s per dot) Inset is a blow-up of saturating region. Arrow points to where gain threshold was for pulsed exposure. Reference 44.

1.4.5. Other Nonlinear Optical Devices

There is another useful class of optical devices derived from the optical nonlinearity of semiconductor-doped glasses that is based on the utilization of the optically induced phase change that accompanies the absorption saturation.. An excellent example of such an application is the device which is schematically represented in Fig. 1.45, after Mayweather et al[45]. It is a light activated gate based on a Mach-Zehnder configuration. The idea is that the direction of the output beam is controlled by the presence or absence of the pump. The pump induces a refractive index change in the upper arm through the nonlinear behavior of the semiconductor-doped glass. This is turn, produces a relative phase change of the form

$$\delta = \frac{2\pi}{\lambda} \Delta n L \tag{1.38}$$

In this particular configuration, the amount of intensity in the output arm (pump on) is just

$$T_{out} = \sin^2 \delta \tag{1.39}$$

Clearly what is important is the level of intensity that is required to produce a significant phase shift in the active element. Toward this end Horan and Blau[46] measured the values of CdSeS semiconductor-doped glasses by a degenerate four-wave-mixing technique. This technique splits the beam to essentially create a real-time grating. The nonlinear coefficient derives from the measurement of the grating efficiency. They used a tunable dye laser as the source and obtained the results shown in Fig. 1.46 for three Schott filter glasses. The absorption is plotted as well. The behavior is monotonic with the nonlinearity following the absorption. Typical values for χ_3 are 4-8 x 10^{-11} cgs esu., or 3-6 x 10^{-12}cm^2/W.

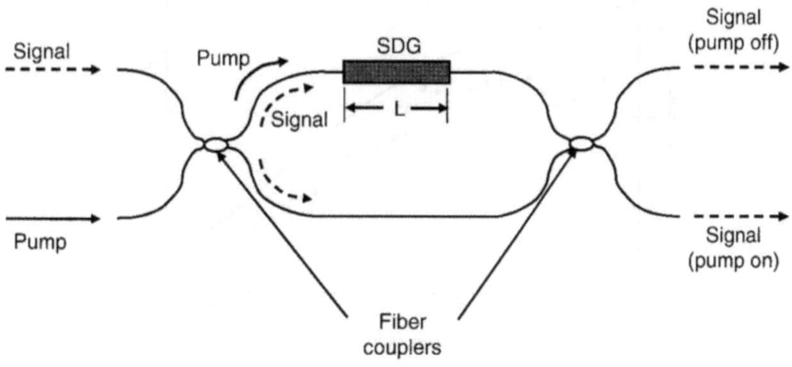

FIGURE 1.45. Schematic representation of a nonlinear switch based on a Mach-Zehnder configuration with the QD-material as the nonlinear element. Mayweather et al, reference 45.

Mayweather performed computational estimates for the nonlinear coefficients for commercial Cd(Se/S) filter glasses. It is not clear how representative they are in light of they are based on commercial filter glasses, not controlled quantum dot glasses. Nonetheless they give an order of magnitude of the extent of effect. Their estimates are reproduced here in Fig. 1.47 The values given are reasonably similar to those listed above by Horan and Blow. This is not surprising since they are measuring essentially the same commercial glasses. For a practical value of the pump intensity at 488nm of $1MW/cm^2$, the magnitude of the induced refractive index change at 565nm would be the order 10^{-5}, (10^{-11} cm^2/W x 10^6 W/cm^2.) This would mean the path length or thickness of the optical element would have to be the order of 6cm for a full-wave phase change.

In both of these studies one does not come to grips with two major issues. The first is a practical issue of loss. For an application as a photonic switch, one could define a figure of merit which would be the nonlinear coefficient divided by the loss per unit length. In this case the use of a silica fiber would be more efficient in spite of its low nonlinear coefficient because its loss is so low. The χ_3 of the QD-glass is 1000x that of silica, but the loss is well in excess of that number.

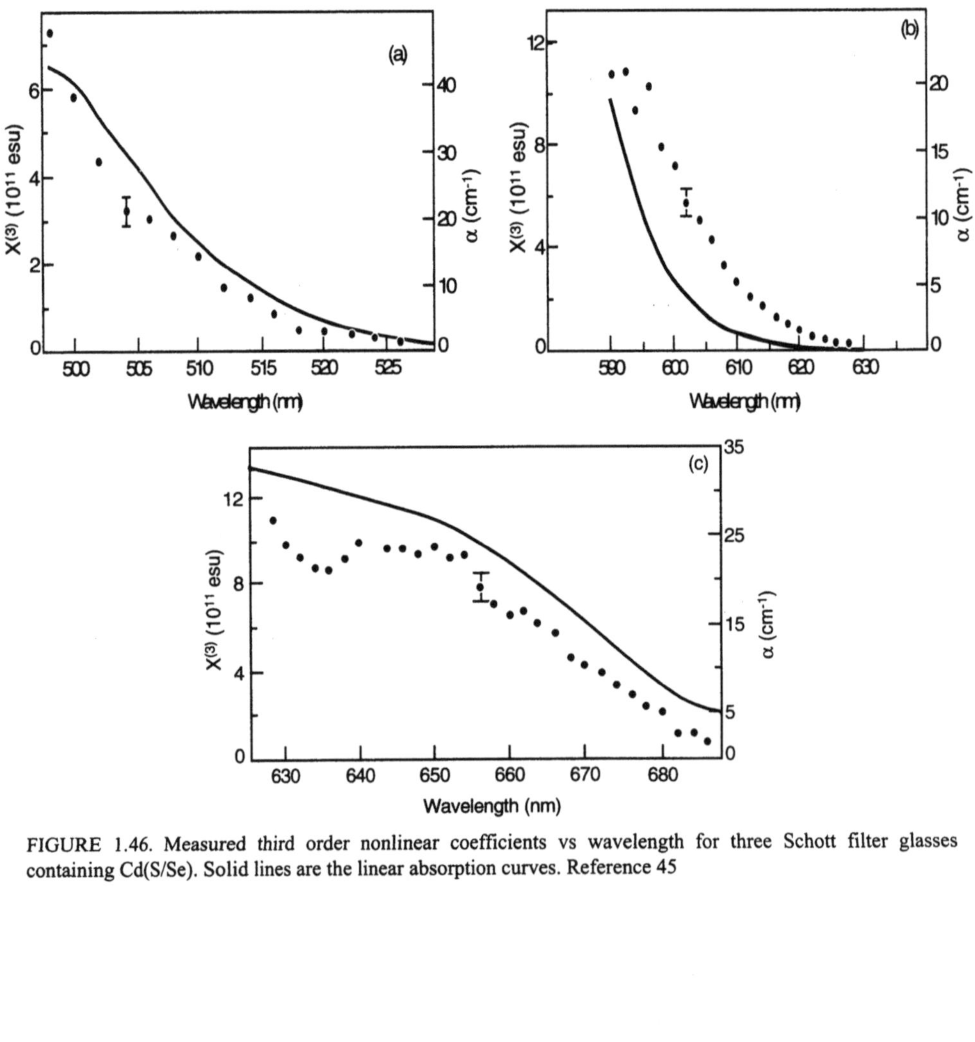

FIGURE 1.46. Measured third order nonlinear coefficients vs wavelength for three Schott filter glasses containing Cd(S/Se). Solid lines are the linear absorption curves. Reference 45

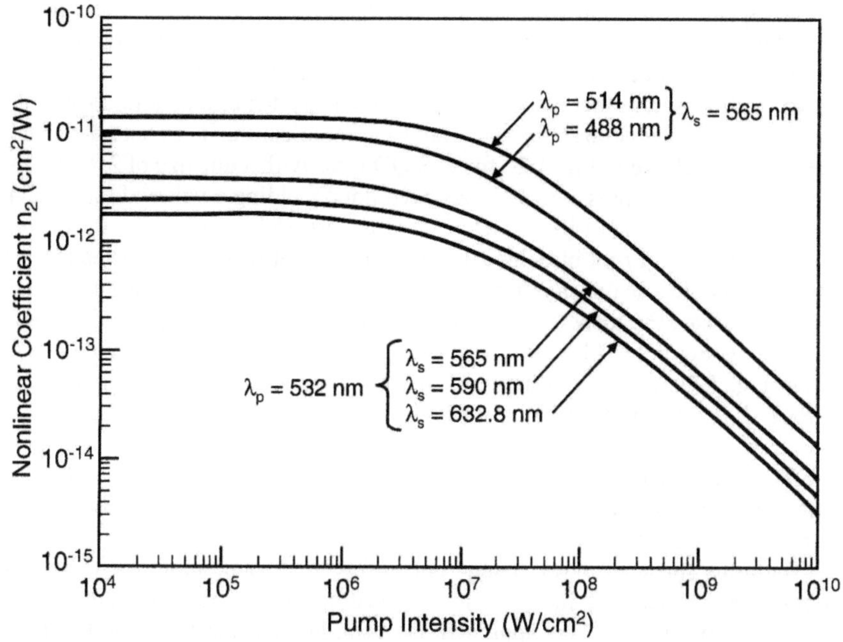

FIGURE 1..47. Nonlinear coefficients versus pump intensity in a Schott filter glass OG550 at different pump and signal wavelengths. Horan et al, reference 46.

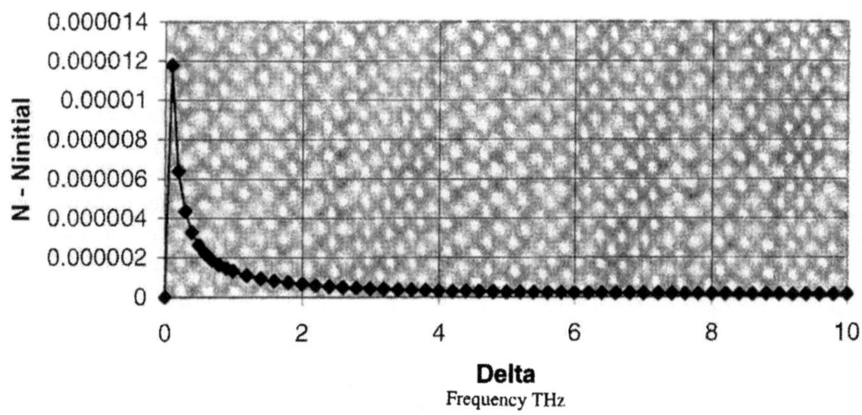

FIGURE 1.48. Estimate of dispersion of the induced refractive index change of PbS-QD glass showing the saturation behavior of Fig. 1.26. Assumed $I = I_{sat.}$ and $T_2 = 0.025$ THz.

The second has to do with the fact the role of quantum confinement has really not been addressed. One could argue that the way to look at this is through the way I_{sat} varies with dot size since it appears that saturable absorption is the dominant nonlinear mechanism. Unfortunately there is little consistent and systematic data to conclude what the overall effect is. One straightforward way to accomplish this is to take the saturation data such as shown above in Fig. 1.26 for PbS QD glass with a dot size of 2.7nm and fit it to the standard 2-level expression. This has been done yielding a value of α_0 of $1.34cm^{-1}$ and a value of I_{sat} of $0.18MW/cm^2$. One has the form of equation for the resonance condition, namely $\delta=0$. From bleaching vs intensity data at other frequencies away from resonance, one could obtain an estimate the $\delta^2=(\omega_0^2-\omega^2)T_2$. With this one could use Eq. (1.33) to estimate the dispersion of induced refractive index change. This is done for a case corresponding to the PbS saturation data given above and the following result was obtained in Fig. 1.48. For this example, if the pump were at 1430nm the induced index change for $I=I_{sat}$ is the order of 10^{-8} at 1500-nm. The value of T_2 that was used was that from the experimental absorption curve thus corresponded in to the inhomogeneous width. Likely, the de-phasing time is somewhat shorter. Of course, with a higher concentration of dots (larger peak absorption coefficient) this number could be increased, but one must be aware of the absorption tail.

With this small an index change one is forced to think about a QD near single mode optical fiber where the I_{sat} is easily achieved because of the confinement and the length is less than a meter. This approach has been pointed out by Pantell et al[47] who conjecture that a few watts of power in a few centimeters of fiber would be sufficient under their criteria for switching Attempts to make optical waveguide fiber using the QD glass have as yet not been successful for reasons having to do with the optical fiber process and the thermal treatment involved in developing the semiconducting phase.

1.5. ACKNOWLEDGEMENT

I gratefully acknowledge the support of Corning Incorporated, and in particular Lina Echeverria, for allowing me the time to write the chapter. I thank R. J. Aruajo for his great help in writing the section on nucleation. I am also grateful to Nasser Peyghamabarian and his group at the Optics Science Center, University of Arizona, and Frank Wise of Cornell University for our collaborations over the years and the helpful discussions that ensued. I am indebted to Charlene Delaney Smith for her critical reading of the manuscript, and to Nancy Foster for the drawings. I also wish to thank my good friend Sasha Efros for the invitation (arm twisting) to participate in this effort.

REFERENCES

1. Al. L. Efros, and A. L. Efros, *Sov. Phys. Semicond.* **16**(7), 772-775 (1982).
2. L. E. Brus, *J. Phs. Chem.*, **90**, 2555, (1986).
3. H. Weller, H. Koch, U. Guterrez, and M. Henglein, *Phys. Chem.* **88**,549, (1984).
4. "Glass Color Filters" CF-3 Corning Glass Works page 10, sharp cut filters, 1965.
5. R. J,. Araujo, and N. F. Borrelli, Photochromic glasses in "Optical Prop. of Glass,", eds. D. R. Uhlmann, and N. J. Kreidl, Am Cer Soc., 1991.
6. H. S. Zhou, I. Honma, J. W. Haus, H. Sasabe, and H. Komiyama, *J. Lum.*, **70**(1-6), 21-34, (1996) and references contained.

7. D. J. Norris, Al. L. Efros, M. Rosen, and M. G. Bawendi, *Phys. Rev.* B,**53**(24), 16 347-16 354 (1996).
8. M. L. Stiegerwald, A. P. Alivsatos, J. M. Gibson, T. D. Harris, P. Kortan, A. J. Muller, A.M. Thayer, T. M. Duncan, D.C Douglass, and L. E. Brus, *J. Am Chem Soc*, **110**, 3046, (1988).
9. Y. Wang, A. Suna, W. Mahler, and R. Kasowski, *J. Chem Phys.* **87**, 7315, (1987).
10. B. G. Potter, and J. H. Simmons, *Phys. Rev.* **37**(18), 10 838, (1988).
11. A. I. Ekimov, F. Hache, M. C. Schanne-Klein, D. Ricard, C. Flytzanis, A. I, Kudyravtsev, T. V. Yazeva, A. V. Rodina, and Al. L. Efros, *JOSA-B*, **10**, 100, (1993).
12. L. E. Brus, *J. Chem. Phys.*, **80**, 4403 (1984).
13. A. I. Ekimov, Al. L Efros, and A. A. Onushchenko, *Sol. State Comm.*, **56**, 921, (1985).
14. C. Murray, D. Norris, and M. Bawendi, *J. Am Chem. Soc.* **115**, 8706, (1993).
15. F. W. Wise, *Accounts of Chem Res.* **33**(11), 773, (2000).
16. P. Roussignol, D. Ricard, J. Lukisak, C. Flytzanis, *JOSA*-B **4**, 5 (1987).
17. I. M. Lifshitz, and V. V. Slyozov, *JETP*, **35**, 331, (1959).
18. S. Schmitt-Rink, D. Miller, and D. Chemla, *Phys. Rev.* B, **35**, 8113, (1987).
19. V. I. Klimov, "*H'book on Nanostructured Materials and Nanotechnology*," ed H. Nalwa, Academic Press 1998.
20. P. T. Guerreiro, S. Ten, N. F. Borrelli, J. Butty, G. E. Jabbour, and N. Peyghambarian, *Appl. Phys. Lett.*, **71**(12) 53, (1997).
21. V. I. Klimov, *Phys. Chem* B **104**, 6112, (2000).
22. F. Hennenberger, J. Pulis, Ch. Spiegelberg, A. Schulzgen, H. Rossman, V. Jungnickel, and A. I. Ekimov, *Semicond. Sc. Tech.* **8**, A41-A50, (1991).
23. S. Park, R. Morgan, Y. Hu, M. Linberg, S. W. Koch, and N. Peyghambarian, JOSA B **7**, 2097, (1990).
24. D. W. Hall, and N. F. Borrelli, *JOSA-B*, **5**, 1650 (1988).
25. N. Peyghambarian, B. Fluegel, D. Hulin, A. Migus, M. Joffre, A. Antonetti, S. Kock, and M. Lindberg, *IEEE JQE* **25**(12), 2516, (1989).
26. F de Rougemont, R. Frey, P. Roussignol, D. Ricard, and C. Flytzanis, *Appl. Phys Lett.* **50**(23), 1619, (1987).
27. N. F. Borrelli, and D. W. Smith, *J. Non-Cryst. Sol.* **180**, 25, (1994).
28. K. Wundke, S. Potting, J. Auxier, A. Schulzgen, N. Peyghambaian, and N. F. Borrelli, *Appl. Phys. Lett* **76**(1), 10, (2000).
29. G. Tamulaitis, V. Gulbinas, G. Kodis, A. Dementjev, and L. Valkunas, *J. Appl. Phys.* **88**(1), 178 (2000).
30. B. L. Justus, M. E. Seaver, J.A. Ruller, and A.J. Campillo, *Appl. Phys. Lett.* **57**(14), 1381, (1990).
31. B. L. Justus, and J. A. Ruller, *Opt Mat.* **2**, 33, (1993).
32. K. J. Blow, and D. Wood, *JOSA* **5**(3), 629, (1988).
33. U. Keller, K. J. Weingarten, F. X. Kartner, D. Kopf, B. Braun, I. D. Jung, R. Fluck, C. Honninger, N. Mansuchek, and J Aus de Au, *IEEE Selected Topics in QE*, **2**(3) 435, (1996).
34. A. M. Malyarevich, V. G. Savitski, P. V. Prokoshin, N. N. Posnov, and K. V. Yurnashev. TBP.
35. J. E. Phil\ipps, T. Topfer, H. Ebendorff-Heidepriem, D. Ehrt, R. Sauerbrey, and N. F. Borrelli, *Appl. Phys.* B, **72**, 175, (2001).
36. V. I. Klimov, A. A. Mihailovsky, S. Xu, A. Malko, J. A. Hollingsworth, C. A. Leatherdale, H. J. Eisler, and M. G. Bawendi, *SCIENCE*, **290**, 314, (2000).
37. Y. Masumoto, and T. Kawamura, *Appl. Phys. Lett.* **62**(3), 225, (1993).
38. V. S. Dneprovskii, V. I. Klimov, D. K. Okorokov, and. Y. V. Vandyshev, *Sol. State Comm.* **81**(3), 227, (1992).
39. H. Giessen, U. Woggon, B. Fluegel, Y. Z. Hu, S. W. Koch, and N. Peyghambarian, *Opt. Lett.*, **21**(14), 1043, (1996).
40. U. Woggon, O. Wind, F. Gindele, E. Tsitishvili, and M. Muller, *J. Lumin.* **70**, 269, (1996).
41. J. Butty, N. Peyghambarian, Y. H. Kao, and J. D. Mackenzie, *Appl. Phys. Lett.*, **69**(21), 3224, (1996).
42. Y. Z. Hu, S. W. Koch, and N. Peyghambarian, *J. Lumin.* **70**, 186, (1996).
43. K. Wundke, J. Auxier, A. Schultzgen, and N. Peyghambarian, *Appl. Phys. Lett.* **71**(20), 3060, (1999).
44. J. Auxier, K. Wundke, A. Schulzgen, N. Peyghambarian, and N. F. Borrelli, *Proc. CLEO'00 OSA Technical Digest Series* 385 (2000).
45. D. Mayweather, M. G. F. Digonnet, and R. H. Pantell, *J. Lightwave* Tech., **14**(14), 60, (1996).
46. P. Horan, and W Blau, *Semicond. Sci Techn.* **2**, 382, (1987).
47. R. H. Pantell and M. J. F. Digonnet, *J. Lightwave, Tech.* **12**(1), 149, (1993).

2

Auger Processes in Nanosize Semiconductor Crystals

Alexander Efros
Nanostructure Optics Section, Naval Research Laboratory

2.1. INTRODUCTION

In this chapter I discuss the role that various types of Auger processes play in the linear and nonlinear optical properties of semiconductor nanocrystals (NCs). The nanocrystal form of semiconductors was discovered by Ekimov [1], Hengeline [2], and Brus [3] independently in semiconductor doped glasses and colloidal solutions almost 20 years ago. However, rapid progress in semiconductor nanocrystal research, both in chemical synthesis and in physical understanding has been made only in the last ten years. A wide class of semiconductor materials can now be prepared in nanocrystal form including: covalent Si and Ge, III-V compounds (GaAs, GaP, InP), II-VI compounds (CdSe, CdS, ZnSe, CdTe, PbS and their alloys), and I-VII compounds (CuCl, CuBr, AgBr) (see for review Ref.[4]). Furthermore, in many cases, technology allows one to control the size (from 2 to 30 nm), shapes and surface of these nano-size semiconductor crystals (for an example, see the excellent reviews by Brus [5] and Alivisatos [6]). Realistic calculations of the nanocrystal properties allow us to make a serious analysis of experimental data (for example, see Ref.7). This together with progress in chemical synthesis has brought us to the stage of making different applications of NCs.

Why are nanocrystal quantum dots especially interesting? Because they are a truly new form of matter that can be considered as "artificial atoms." They have linear discrete absorption spectra (like atoms) and photoluminescence tunable over a wide range, from the far infrared to the deep ultraviolet. The nanocrystal size and shape are practically the only two parameters that govern optical transition energies.

On the other hand we can move the NCs around and so form quantum dot molecules, and 3D ordered arrays of these atoms. We can put them in other materials as dopants or join them to a larger molecule to form a super molecule. As a result, instead of 109 elements, we have at our disposal, in principle, an unlimited number of "atomic elements" to form new materials.

These exciting properties lead to numerous potential and actual applications. I mention only two of them that already have interested the industrial world. The first is observation of tunable gain and stimulated emission in CdSe nanocrystal quantum dot solids at room temperature.[8] The second, the successful labeling of a biological molecule by nanocrystals.[9]

Today when we talk seriously about NC applications, the significant role played by Auger processes is especially clear, because they affect all aspects of carrier relaxation and recombination. I will discuss Auger quenching of the photoluminescence (PL), Auger autoionization of the nanocrystals, and Auger-like thermalization of the carrier in the nanocrystals. The impressive experimental data on the PL of the nanocrystal materials will be discussed and analyzed within theoretical models that consider the various Auger processes.

2.2. QUENCHING OF THE NANOCRYSTAL PHOTOLUMINESCENCE BY AN EXTRA CHARGE

Extremely impressive evidence of Auger processes is observed in the PL of a single quantum dot. They are clearly seen in the PL of single nanosize quantum dots that show a random intermittence in their PL intensity, an effect that is washed out in ensembles by the contributions from many quantum dots.

The effect was first observed in the PL of a single CdSe quantum dot at room temperature Ref.10 and, since that time, by many other groups at various temperatures [11] and in many other types of nanocrystals: CdS [12], CdTe[13], InP [14] and porous Si [15]. Figure 2.1 from Ref. [10] shows the time dependence of the photoluminescence intensity of a single CdSe nanocrystal with a radius of 2.1 nm under CW excitation of intensity 0.7 $kWcm^{-2}$.

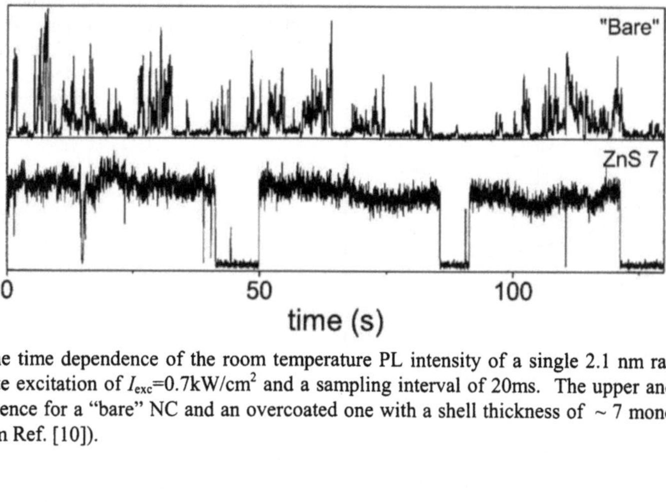

FIGURE 2.1. The time dependence of the room temperature PL intensity of a single 2.1 nm radius CdSe NC under steady state excitation of I_{exc}=0.7kW/cm^2 and a sampling interval of 20ms. The upper and lower panels show the dependence for a "bare" NC and an overcoated one with a shell thickness of ~ 7 monolayers of ZnS respectfully (from Ref. [10]).

The time dependence of the photoluminescence intensity exhibits a sequence of "on" and "off" periods. For particles coated by a 7-monolayers-thick ZnS shell the duration of the "on" periods is much longer than that for "bare" particles, which have a

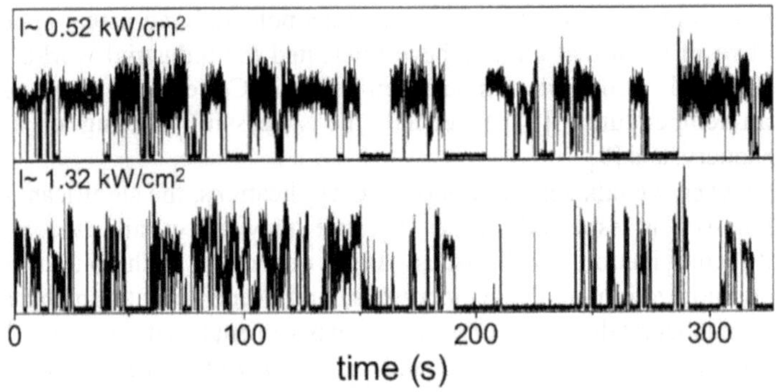

FIGURE 2.2. Comparison of the time dependence of the room temperature PL of a single 2.1 nm radius CdSe NC overcoated by a 4 monolayer thickness ZnS shell. The upper and lower panels show the dependence measured at excitation intensity I_{exc}=0.52kW/cm^2 and I_{exc}=0.1.32kW/cm^2, respectively. The average on/off times at the two intensities are: <τ_{on}> ~0.97s, <τ_{off}> ~0.44s at I_{exc}=0.52kW/cm^2 and <τ_{on}> ~0.32s, <τ_{off}> ~0.43s at I_{exc}=1.32 kW/cm^2, respectively.

very low quantum efficiency they are dark most of the time. Figure 2. 2 from Ref. [10] shows that the duration of the "on" periods decreases with excitation intensity, while that of the "off" periods is almost independent of it.

This unexpected "blinking" behavior of the PL under steady state (CW) excitation condition raises a question about the cause of the "off" periods in the PL intensity. We propose that it is a result of Auger quenching of the photoluminescence in quantum dots that contain an extra charge, i.e., they have become ionized.

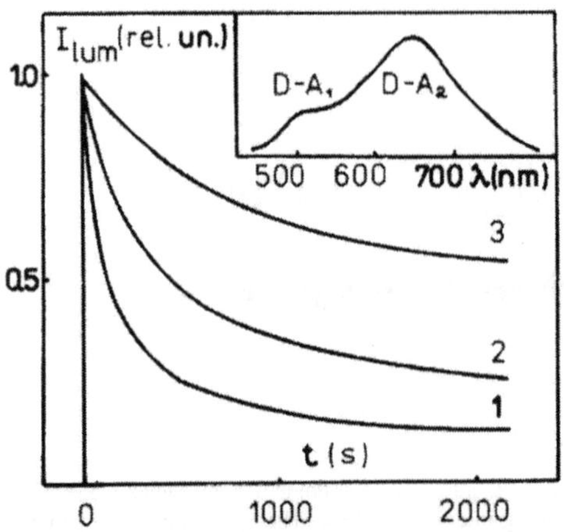

FIGURE 2.3. The time dependence of the PL intensity I_{lum} of 2.2nm radius CdS NCs at different intensity of excitation: curves 1, 2 and 3 corresponds to I_{exc}, 0.45 I_{exc}, and 0.2I_{exc}, respectively. The insert shows the PL spectra of this sample.

Actually, the idea that Auger processes quench PL in a nanocrystal with an extra charge was first proposed to explain the photodarkening effect in semiconductor doped glasses.[16] The decrease in the photoluminescence intensity with time from nanocrystals embedded in glasses under steady state excitation conditions [17] is a general property of all glasses containing nanosize crystals. The effect is reversible: heating the sample above room temperature usually restores the initial PL intensity. We have studied the light-induced photodarkening in glasses doped by CdS nanocrystals. These samples do not show any band edge PL. The PL spectra of these samples are completely dominated by the two transitions from the shallow donor to the two deep acceptors.[18] The inset in Fig. 2.3 shows a typical PL spectrum of CdS nanocrystals with an average radius of 2.2 nm.

Figure 2.3 shows the time dependence of the PL intensity of this sample under a steady state excitation condition. We have studied the decrease of luminescence intensity of CdS nanocrystals at a very low excitation intensity. At the maximum intensity we used, I_{exc}, the fraction of excited crystals was 10^{-5}. Figure 2.3 shows that the rate of the PL degradation as well as the total decrease in the luminescence intensity depends strongly on the exciting intensity. In a first approximation, we characterize this dependence by the time τ_i at which the intensity of luminescence decreases by factor of two.

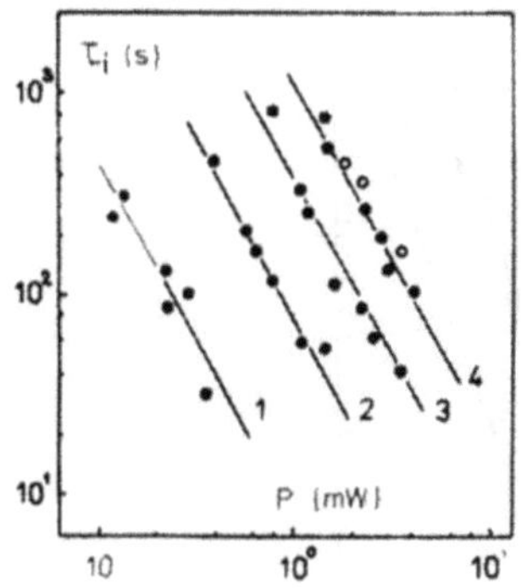

FIGURE 2.4. The dependence of ionization time τ_i on absorbed power for different size CdS NCs. The straight lines: 1, 2 , 3, and 4 approximate these dependencies for NCs with average radius: 1.3, 1.6, 2.2 and 3.9 nm, correspondingly. Filled and open circles show experimental data measured at the excitation by the light with the wavelength of 400 nm and 476 nm respectfully.

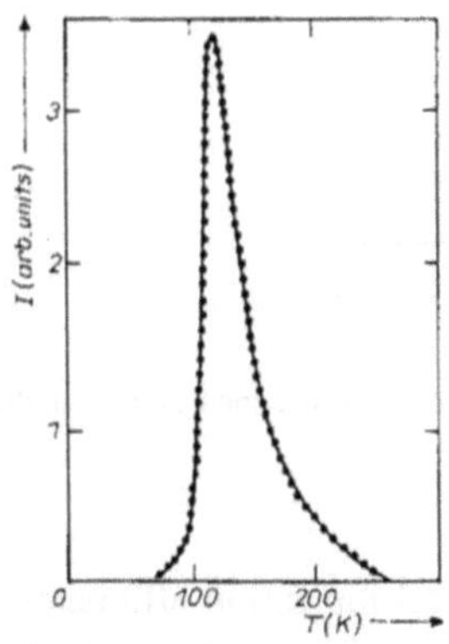

FIGURE 2.5. The thermostimulated luminescence (TSL) curves measured in CdS doped glass (solid lines) and in undoped glass (dotted line). Both samples were irradiated at 80K for 1 h.

The dependence of τ_i on absorbed power for nanocrystals with different average sizes is shown in Fig.2.4. One can see that τ_i decreases rapidly with crystal size and that for all samples the time τ_i is inversely proportional to the square of absorbed power [16], P: $(\tau_I)^{-1} \sim P^2$. This dependence clearly shows that the PL degradation is connected with two photon excitation of the nanocrystals.

Important information about the degradation mechanism was obtained in the thermo-stimulated luminescence (TSL) experiments with these samples.[19] Fig. 2.5

from Ref. [19,20] compare the TSL spectra of an undoped glass and the glass doped with 3.0 nm radius CdS nanocrystals. Both samples were initially irradiated at T=80 K for 1 h. The undoped sample was irradiated by light with energy $\cong\omega_{ex}$=6eV, which corresponds to interband absorption of the glass, and the TSL was measured within the spectral range of intrinsic luminescence of the glass at $\cong\omega_{em}$=3.5eV. The TSL curve of the last sample is shown in Fig.2.5 by the dotted line. It is a typical TSL curve for the well-known E_1 capture center of the glass [21]. The doped sample was subjected to light irradiation with $\cong\omega_{ex}$=3.6eV, which is in the spectral range of interband absorption of nanocrystals, and the TSL spectrum was measured within spectral range of the nanocrystal impurity PL at $\cong\omega_{em}$=1.7eV. The TSL curve of the doped sample shown in Fig. 2.5 by the solid line coincides completely with that of the undoped glass. This shows directly, that electrons photoexcited in the nanocrystal have a finite probability to leave it and to be captured by the E_1 traps in the glass.

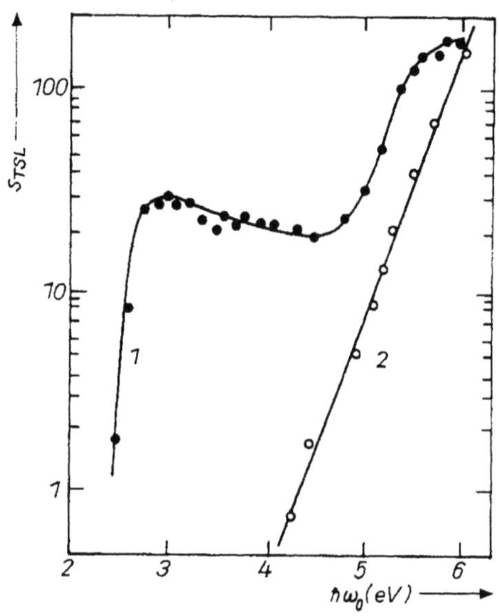

FIGURE 2.6. Spectral dependence of the efficiency of ionization S_{TSL} for the glass doped by 3 nm radius CdS NCs (curve 1) and for undoped glass (curve 2).

Figure 2.6 shows the spectral dependence of the nanocrystal ionization efficiency, S_{TSL}. It is seen that the ionization starts at the photon energy corresponding to the energy gap of the 3.0 nm radius CdS nanocrystals. The increase of the ionization efficiency at $\cong\omega_b$=4.5 eV corresponds to direct electron excitation over the semiconductor/glass band offset barrier, i.e., to the direct nanocrystal ionization. One can see in Fig. 2.6 that the electron ejection from the nanocrystal at excitation energies $\cong\omega < \cong\omega_b$, which from now on will be called nanocrystal autoionization, is significantly less efficient than the direct crystal ionization. The autoionization of the nanocrystals was also observed at room temperature.

2.3. AUTOIONIZATION OF THE NANOCRYSTALS AND AUGER QUENCHING OF THE PL IN THE NANOCRYSTAL WITH AN EXTRA CHARGE

The TSL and the PL time dependence measurements show clearly that the degradation of the PL is determined by the nanocrystal ionization. These experiments suggest the following model that describes light induced PL photodegradation and the single nanocrystal "blinking". The model is shown schematically in Fig. 2.7.

Optical excitation of an electron-hole pair in a neutral nanocrystal leads to the photon emission as a result of its annihilation. However, it is not always the case for ionized nanocrystals. Excitation of an electron-hole pair in such nanocrystals does not result in PL, because the annihilation energy is taken up not by a photon, but by the extra electron or hole. In a small quantum dot nanocrystal the Auger rate is much larger than the radiative recombination rate, and so the nanradiative Auger recombination quenches all the PL. The nonradiative decay time has been measured in semiconductor doped glasses to be on the order of 10-100 ps [17].

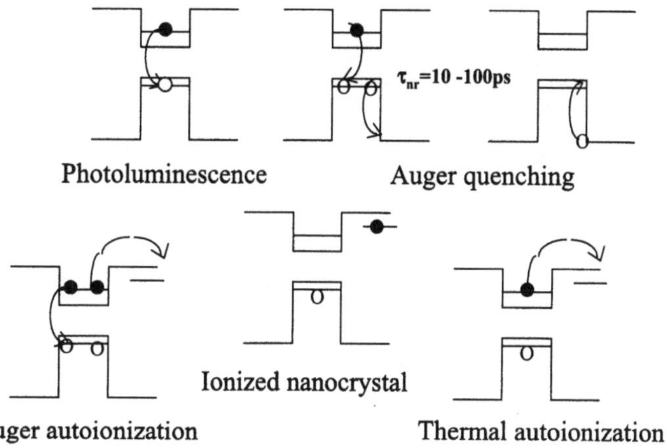

FIGURE 2.7. A scheme presenting various types of relaxation processes in a NC under steady state excitation by visible light. The ionized NC that quenches the PL can be created by visible light as result of Auger or thermal autoionization.

This extra charge can be created in a nanocrystal by optical excitation even if the photon energy is not enough for direct nanocrystal ionization. Two mechanisms of autoionization can be proposed, depending on the size of the nanocrystal, on temperature, on the band offset between the semiconductor nanocrystal and the surrounding matrix, and on the excitation intensity.

The first one is Auger autoionization of a nanocrystal when two electron-hole pairs are excited there simultaneously. In this case the annihilation energy of one of the electron-hole pairs goes into ejecting other electrons from the nanocrystal. The ejected electron (hole) is localized in a trap in the surrounding matrix (for example, at E_1 centers of the glass as was found in Ref. [19]) or at the nanocrystal surface.

The thermal or tunnel autoionization of a nanocrystal with a small band offset is another possible mechanism for forming a long-lived ionized state. An optically excited electron or hole in the nanocrystal has a finite probability of thermally exceeding the effective barrier height and then being localized at a deep trap in the matrix or tunneling directly to this trap.

The ejected electron (hole) returns to the quantum dot thermally after a time and restores the "on" periods of the PL. It is the length of time of this return process that determines the duration of the "off" period. The ratio of the "on" to the "off" periods determines the PL quantum efficiency.

The square dependence of the PL degradation rate on the excitation intensity (see Fig. 2.4) gives a clear indication that two photons are responsible for the nanocrystal ionization. This means that Auger autoionization is the major process leading to the PL degradation in CdS nanocrystals embedded in a glass matrix. Using this assumption we have described the time dependence of the PL degradation in these samples.

On can find four types of the nanocrystals under low excitation intensity of the glass sample: unexcited nanocrystals, nanocrystals with one electron–hole pair, nanocrystals with two electron–hole pairs, and ionized nanocrystals. The system of the rate equations describing the nanocrystal relative population can be written:

$$\frac{dN_0}{dt} = -W_1 N_0 + \frac{N_1}{\tau_1};$$

$$\frac{dN_1}{dt} = W_1 N_0 - (W_2 + \frac{1}{\tau_1})N_1 + \frac{N_2}{\tau_2} + \frac{N_+}{\tau};$$

$$\frac{dN_2}{dt} = W_2 N_1 - (\frac{1}{\tau_2} + \frac{1}{\tau_A})N_2;$$

$$\frac{dN_+}{dt} = \frac{N_2}{\tau_A} - \frac{N_+}{\tau}; \qquad (2.1)$$

where $N_{0,1,2}$ is the nanocrystal concentration with 0,1, and 2 electron-hole pairs, respectively, N_+ is the concentration of ionized nanocrystals, $N = N_0 + N_1 + N_2 + N_+$ is the total nanocrystal concentration, $\tau_{1,2}$ are, respectively, the radiative lifetime of nanocrystals with one and two electron-hole pairs, $1/\tau_A$ is the rate of Auger ionization of a nanocrystal with two electron-hole pairs, τ is the lifetime for a trapped electron (hole) to return to the nanocrystal, and $W_{1,2}=\sigma_{1,2}I_{exc}/\equiv\omega$, where $\sigma_{1,2}(\omega)$ are the 1 and 2 electron-hole pair excitation cross sections, respectively, of a NC for light of frequency ω, and I_{exc} is the excitation light intensity.

At low excitation intensity the luminescence intensity is proportional to the concentration of nanocrystals containing one electron-hole pair $I_{lum} \sim N_1(t)$. This is because $N_2 << N_1$ in our experiment and nanocrystals with two electron-hole pairs do not make a significant contribution to the PL. Therefore the PL degradation is the result of the slow decrease of N_1 due to ionization of the crystal. Using Eq. 2.1 we obtain the time dependence of the relative intensity of the PL, $J_{PL}(t)$:

$$J_{PL}(t) = \frac{N_1(t)}{N_1(0)} = \frac{\tau}{\tau + \tau_i} \exp\left[-t\left(\frac{1}{\tau} + \frac{1}{\tau_i}\right)\right] + \frac{\tau_i}{\tau + \tau_i}, \quad (2.2)$$

where τ_i is the typical time of the nanocrystal Auger autoionization:

$$\frac{1}{\tau_i} = \frac{W_1 W_2 \tau_1}{1 + W_1 \tau_1} \frac{\tau_2}{\tau_2 + \tau_A}. \quad (2.3)$$

One can see that at low excitation intensity ($W_1 \tau_1 << 1$) the rate of the NC autoionization is always proportional to $(I_{exc})^2$ since both W_1 and W_2 are proportional to I_{exc}. The biexciton lifetime $\tau_2 \sim 1$ ns and it is shorter than the time of Auger autoionization τ_A in practically all sizes of nanocrystals. This allows us to present the crystal ionization rate Eq.2.3 in a form:

$$\frac{1}{\tau_i} = \frac{(W_1 \tau_1)(W_2 \tau_2)}{\tau_A} \quad (2.4)$$

that has a very straightforward meaning. In this expression, $(W_1 \tau_1)$ is the steady state fraction of the NCs containing a single electron-hole pair, the product $(W_1 \tau_1)(W_2 \tau_2)$ is the steady state fraction of the crystals containing two electron-hole pairs, and $1/\tau_A$ is the probability of the Auger ionization of these crystals.

The time dependence $J_{PL}(t)$ completely describes the experimental data presented in Fig. 2.3, and allows us to measure the dependence of nanocrystal ionization rate τ_i as a function of the NC radius (see Fig. 2.8) and the time of the nanocrystal neutralization. The nanocrystal ionization rate τ_i in Fig. 2.8 increases strongly with the decrease of the nanocrystal radius. However, the time of the nanocrystal neutralization τ measured in our experiment at nitrogen temperature is within the experimental error of 25% independent of the crystal size and excitation intensity and turns out to be τ_{77K}=(1.3±0.25) h. From the size dependence of τ_i we have also calculated the size dependence of Auger ionization time τ_A with use of Eq. 2.4. The dependence (see Fig. 2.9) shows that τ_A decreases from 100 ns to 1 ns with nanocrystal size.

The reversible PL degradation observed in CdS nanocrystals is a very common property of different nanocrystal ensembles. It has been observed in porous Si at helium temperature [22]. Figure 2. 10 shows the PL spectra of porous Si at helium temperature excited by light with ω_{exc}=2.8 eV before and after degradation by a 20 W/cm^2 laser at an energy of 2.54 eV for 20 min. One can see that the PL intensity decreases considerably after sample illumination. Heating the sample up to room temperature and cooling it back to 5 K restores the initial intensity of the PL. Analysis of these measurements within the above model of Auger autoionization and Auger quenching of the PL [22] gives us τ_A = 17 ns and the time of the crystal neutralization at helium temperature τ_{5K}= 3 h. One can see that the time of Auger ionization measured in Si nanocrystals is consistent with those times measured in CdS nanocrystals.

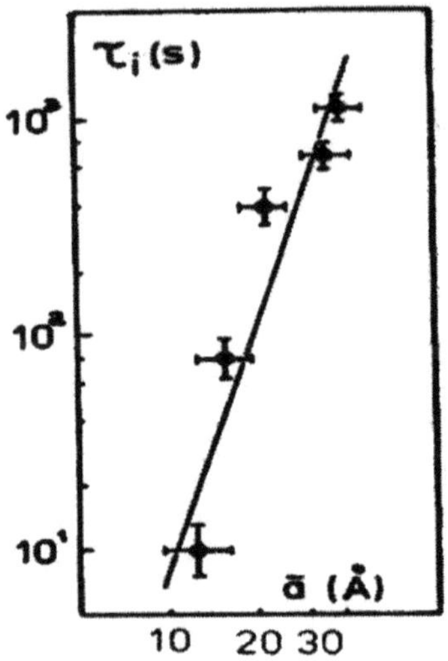

FIGURE 2.8. The dependence of ionization time τ_i on the average CdS NC size at fixed power of absorbed light P=1 mW (from Ref. [16]).

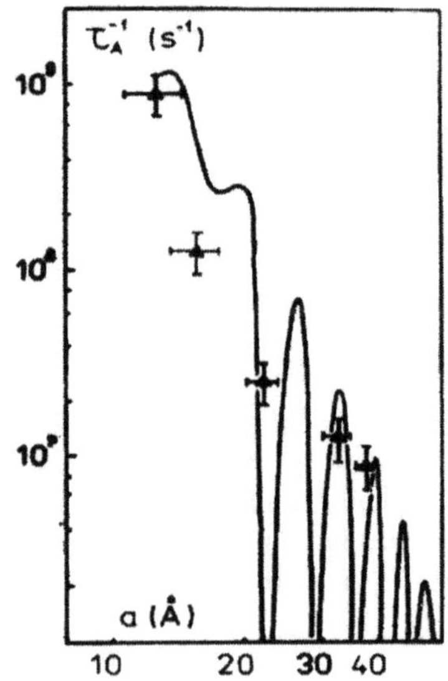

FIGURE 2.9. The size dependence of the Auger autoionization time of a NC with two electron-hole pairs, τ_A. The crosses show experimental data. The solid line is the theoretical dependence calculated with use of Eq. 2. 5 (from Ref. [16]).

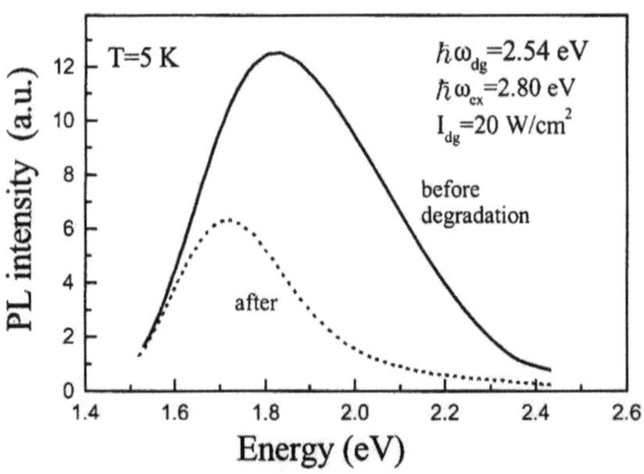

FIGURE 2.10. The effect of optical degradation by a 20 min. illumination by a $20 W/cm^2$ 488 nm line of a CW Ar^+ laser pump beam on the intensity and spectral position of the PL line in porous Si (from Ref. [22]).

All these results suggest that the random intermittence of the PL of a single NC and the light induced PL photodegradation of NC ensembles reflect essentially the same process: nanocrystal ionization and the quenching of the PL in auto-ionized nanocrystals. The only difference is the time in which it takes an electron or hole to return to the NC. For the carrier localized at a trap in the glass or at some surface states, the time should exponentially depend on the temperature $\tau \sim exp(\Delta E/kT)$, where T is the temperature and ΔE is the depth of the trap. In the case of photodegradation (low temperature and deep traps), this time is hours and in the random intermittence of the single dot PL (room temperatures and shallow traps) this time is seconds. However, by decreasing the temperature one can significantly increase the time of electron (hole) return to the nanocrystal and transfer the random intermittence into photodegradation.

2.4. RATE OF AUGER PROCESSES IN NANOCRYSTALS

In the model proposed above the rate of Auger processes is much faster than the rate of the carrier radiative recombination. This fact is also supported by experimental measurements of the time of the nonradiative Auger recombination, which was found to be on the order of 10-100 ps [17], while the time of the carrier radiative recombination is always on the order of several nanoseconds. The high rate of nonradiative Auger recombination is a surprising effect because in bulk wide-gap semiconductors such as CdS or CdSe the rate of Auger processes is considerably suppressed. In homogeneous bulk material both energy and momentum are conserved in the Auger process. This leads to a kinematic energy threshold for the reaction and a recombination rate that depends exponentially on the ratio Eg/T of the semiconductor energy gap Eg to the temperature T. Therefore, Auger processes are considerably suppressed in wide-band semiconductors.

There are two reasons why Auger processes are so efficient in the NC. First of all, the momentum is not a good quantum number for electron and hole motion in semiconductor nanocrystals. The second is an enhancement of Coulomb interactions between the carriers confined in a small volume of the NC. The rate of Auger processes in nanocrystals can be calculated using Fermi's golden rule:

$$\frac{1}{\tau_A} = \frac{2\pi}{\hbar} \sum_{k,l,m} |\langle \Psi^i | v(\vec{r}_1, \vec{r}_2) | \Psi^f_{k,l.m} \rangle|^2 \, \delta(E_i - E_f) \tag{2.5}$$

where E_i, E_f and Ψ^i, Ψ^f are the total energies and wave functions of the initial and final multi-electron state of the quantum dot, respectively. The sum goes over all final states of the system (k,l,m) and $v(\vec{r}_1, \vec{r}_2) = e^2/(\kappa |\vec{r}_1 - \vec{r}_2|)$ is the Coulomb potential, where κ is a dielectric constant.

Accurate calculation of the matrix element $M = \langle \Psi^i | v | \Psi^f \rangle$ in Eq. 2.5 is the major problem in Auger rate estimation. One can surmise that this matrix element evaluated for wave functions confined within the NC should be on the order of $M \sim e^2/\kappa a$, where a is the crystal radius. However, this estimation does not take into account the very large momentum that the Auger electron has in the final state. This momentum is $k_f \sim \sqrt{2m_e E_g}/\hbar$, where m_e is the electron effective mass, because practically the whole annihilation energy of the electron-hole pair ($\sim E_g$) is transferred into the electron kinetic energy. The electron wave function oscillates very rapidly inside the nanocrystal because of $k_f >> 1/a$. That drastically decreases the value of the matrix element M. The integrand in M can be written as a product of a rapidly oscillating function and a smooth function, $f(\mathbf{r})$. Integration over the internal volume of the nanocrystal gives essentially zero contribution to the integral. However, such integrals can be expanded as a power series in the small parameter $(a\, k_f)^{-1} << 1$, the coefficients of which are determined by the poles of the smooth function [23].

In the case of semiconductor nanocrystals the most important pole is at the nanocrystal surface [16]:

$$M = \langle f(\vec{r})\exp(-i\vec{k}\,\vec{r})\rangle \approx f'|_a\,(ak_f)^{-2} + f''|_a\,(ak_f)^{-3} + f'''|_a\,(ak_f)^{-4} \quad (2.6)$$

where the coefficients of $(a\ k_f)^{-2}$ and $(a\ k_f)^{-3}$ vanish because of continuity of the wave function and its derivative at the crystal surface. The first nonvanishing term is proportional to $f'''|_a$. A significant contribution to the matrix element M comes only from the surface of the nanocrystal. However, this matrix element does not vanish for any electron-hole pair state because, unlike the bulk, there is no momentum conservation in the nanocrystals that leads to a temperature threshold for Auger processes.

A semi-classical language is very useful for qualitatively understanding what stimulates the Auger process rate. The electron "likes" to take up the annihilation energy at the place where its kinetic energy, or its kinetic energy uncertainty, has a maximum value. For example, in the presence of an impurity, the Auger process takes place right at the center of the impurity, where the potential energy goes to minus infinity and the kinetic energy has a maximum.

In the case of Auger processes in NCs they take place right at the abrupt heterointerface, because of large uncertainty of the electron momentum. As a result electrons can get enough momentum at the interface. It is also directly seen in the quantum mechanical calculation of the matrix element M (see Eq. 2.6) that Auger processes take place right at the nanocrystal surface. The abrupt surface significantly accelerates the Auger rate in NCs with large surface to volume ratio.

Figure 2. 9 compares the result of the Auger rate calculations using Eq. 2. 5 with experimental data indicated by black triangles [16]. The theoretical dependence of Auger autoionization rate $1/\tau_A$ on size (solid line) has a strong oscillatory character. The maxima of $1/\tau_A$ correspond to the NCs for which the electron quantum size level with angular momentum 1 (P state) is very close the electron continuum or just becomes delocalized. The minimum of this dependence is due to the vanishing of the matrix element M that is caused by the quantum mechanical interference of the electron wave functions.

In real NCs these oscillations of $1/\tau_A$ are averaged out due to numerous reasons such as disordering of the glass matrix, dispersion of nanocrystal sizes and shapes, and fluctuations in the semiconductor glass band offset. The averaging of the theoretical size dependence in Fig. 2.9 gives the average rate:

$$\left\langle \frac{1}{\tau_A} \right\rangle \sim \frac{1}{a^5} \quad (2.7)$$

that increases significantly with decreasing NC size. The experimental size dependence measured in CdS doped glasses (see Fig. 2.9) can be described as $\langle 1/\tau_A \rangle_{\exp} \sim 1/a^{4.5}$, and that is slightly weaker than the theoretical one.

Actually, the calculation shows that the rate of Auger autoionization depends strongly on band offset and the size dependence may be even stronger. For the average size dependence Eq. 2. 5 gives:

$$\left\langle \frac{1}{\tau_A} \right\rangle \sim \frac{1}{a^{\nu}} \tag{2.8}$$

where $5 < \nu < 7$, depending on the band offset. The theoretical dependence shown in Fig. 2.9 and Eq. 2. 8 were calculated for the effective energy gap of the glass 7.02 eV obtained from direct photoionization data on CdS nanocrystals embedded in the glass [19].

The strong size dependence of an average Auger ionization rate described by Eqs. 7 and 8 also explains the shift of the PL line in porous-Si after the light induced degradation (see Fig. 2.10). In porous Si the size distribution of Si NCs is very broad. As a result the smallest NCs that before degradation contribute to the green part of the PL line are Auger ionized and do not contribute to the PL after degradation, while the largest Si crystals in the distribution stay neutral and contribute to the red edge of the PL line.

The oscillatory theoretical dependence of the rate of Auger ionization has not yet been observed. Still, this prediction shows that there are ways in principle to suppress considerably the rate of Auger processes by changing the shape of the nanocrystals and the value of the potential wall surrounding it by modifying the NC surface. This is a kind of Auger rate engineering.

2.5. RANDOM TELEGRAPH SIGNAL IN THE PL INTENSITY OF A SINGLE QUANTUM DOT

Now, let us return to a single nanocrystal experiment that shows the random intermittence of the PL that resembles a random telegraph signal (RTS). Although solutions of the kinetic state-filling equations (Eq. 2. 1) describe the time behavior of an ensemble of NCs very well [16], and also the time average behavior of a single dot, they do not describe a particular stochastic sequence of "on" and "off" periods occurring in a single NC.

The intermittence of the PL is a consequence of a particular sequence of events. Low intensity CW light creates electron-hole pairs which recombine radiatively emitting light ("on" period) until the quantum dot is ionized, either thermally or by Auger autoionization. The ejected electron (hole) is localized in a deep trap in the surrounding matrix, initiating the "off" period because nonradiative Auger processes in a charged crystal then quench all PL. The ejected electron (hole) returns to the NC and restores the "on" period of the PL.

This process is essentially a stochastic one. To simulate a particular sequence of "on" and "off" periods in a single quantum dot we construct a matrix of random transitions directly from Eq. 2. 1. For example, a NC containing one electron-hole pair is transformed into one with two pairs in a small time interval Δt with a probability $W_2\Delta t$, into one with no pairs with a probability $\Delta t/\tau_1$, and remains unchanged with a probability of $1- W_2\Delta t-\Delta t/\tau_1$ (the total probability is equal to one). The transition probabilities for the other states are obtained similarly. To determine which transition occurs in Δt we choose random numbers uniformly distributed in the interval [0,1] (a Monte Carlo procedure). We subdivide this interval into lengths equal to the appropriate probabilities described above. The subinterval in which a chosen random number falls determines which transition occurs. Thus we simulate the stochastic sequence of NC states as a function of time. Each step in this simulation depends only on the current NC state and not on the previous states.

Let us now discuss the parameters in Eq. 2. 1 that play an important role in the RTS and determine the conditions under which it can be observed.

The radiative lifetime of a single electron-hole pair, τ_1: In the case of strong size quantization, when the electron and hole motions are independent, the overlap integral of their wave functions does not depend on crystal size and is on the order of unity. [24] As a result, the radiative lifetime of a pair in a single dot, τ_1, should be on the order of several nanoseconds for almost all semiconductors.[25] However, formation of the optically forbidden ground exciton state (dark exciton) in the NC greatly increases the decay time, at low temperature, to as much as several microseconds (depending on the NC size [26]). Filling of the higher optically active states at higher temperature lowers this time to several nanoseconds again and makes it a function of the NC radius.

The radiative lifetime of nanocrystals with two *e-h* pairs, τ_2, is less than τ_1 because each electron (hole) can recombine with either of two holes (electrons), and formation of an optically forbidden biexciton state is impossible. As a result this time is always on the order of nanoseconds.

We have discussed the Auger autoionization time, τ_A, in a previous section. Generally, the NC can be Auger autoionized only if the band offset in the conduction or valence band is smaller than the NC energy gap. In CdS doped glasses, the measured conduction band offset satisfies this condition [19,20] and the ionized NC contains an unpaired hole. No measurements have been made of the band offset of CdSe NCs embedded in glass or polymers; however persistent [27] and dynamic [28] hole burning experiments performed on these samples clearly demonstrate the formation of an unpaired electron in the NCs. This electron occupies the first quantum size level of electrons and leads to the bleaching of corresponding optical transitions to this level. This is indirect proof of hole ejection from the NC. τ_A depends strongly on the NC radius, on the band offset at the boundary between the NC and the surrounding matrix (see Eqs. 7 and 8), and can be less than 0.1 ns in small NCs.

The time for the trapped electron (hole) to return to the NC, τ, can be phenomenologically written $\tau=\tau_{ph}\exp(\Delta E/kT)$, where ΔE is the depth of the trap in the matrix, T is the temperature, and τ_{ph} is a typical phonon scattering time in the matrix. At room temperature τ is about 10 min in semiconductor doped glasses [16]. For NCs embedded in polymers this time is much shorter (several seconds [10]), which probably reflects a smaller value of the trap depth there.

The single NC absorption cross section for excitation of the first electron-hole pair $\sigma_1(\omega)$, is on the order of $\sigma_1 \sim 10^{-16}\text{-}10^{-14}\text{cm}^{-2}$ and, far from the band edge, E_g, is proportional to the NC volume and to the electron and hole reduced density of states $\sigma_1(\omega) \sim \sqrt{\hbar\omega - E_g}$ in the parabolic band approximation [24] at the excitation energy $\hbar\omega$. The absorption cross section for excitation of a second electron-hole pair $\sigma_2(\omega)$ is approximately equal to $\sigma_1(\omega)$ far from the band edge, where the excitation energy is in the continuous spectrum. It can be much smaller than σ_1, however, for excitation near the band edge, where the discrete character of the excited electron-hole pair states plays an important role and excitation goes via phonon assisted transitions.

Figure 2. 11 from Ref. [29] shows the results of a simulation of the RTS using parameters appropriate for nanosize CdSe crystals at excitation frequencies far from the band edge. We choose an intensity of excitation close to 0.08 kW/cm^2 in order to

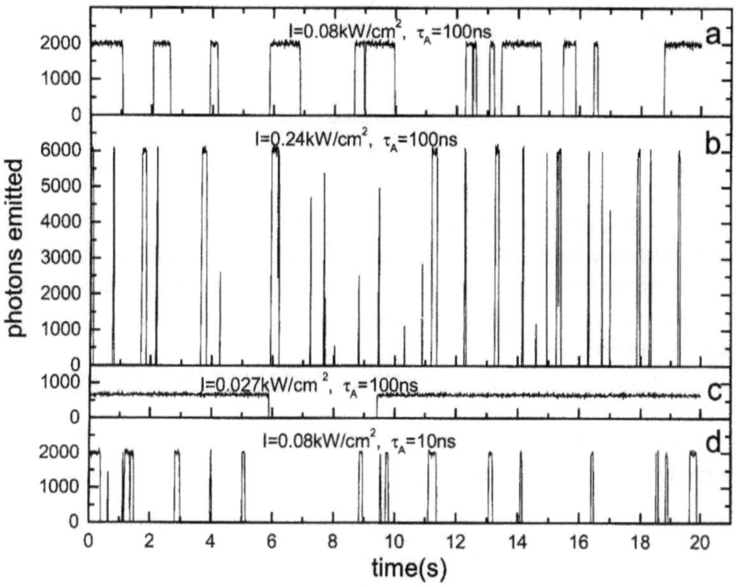

FIGURE 2.11. The number of phonons emitted in the simulation in 10 ms interval by a single NC quantum dot. We use τ_1=4 ns, τ_2=1 ns τ=0.8 s, $\sigma_1 = \sigma_2$= 1.0x10^{-15}cm^2, $\hbar\omega_{ex}$ =2.5 eV (from Ref. [29]).

observe quite long periods of brightness. On average, at this intensity, the time interval between photon absorption is $1/W_1$ = 5000 ns. All relaxation processes are much faster than this. The probability for exciting a two electron-hole pair state is very low, but when it occurs the NC is either almost instantly ionized or returns to a zero pair state. We integrate the intensity of the luminescence signal over a time interval Δt_m~10ms, as in the experiment of Ref. [10], i.e., we count the number of emitted photons in that time interval. One can see that the luminescence resembles a random telegraph signal (random pulses of equal magnitude) only in a narrow range of excitation intensity. Lowering the intensity by a factor of three greatly reduces the number of "off" periods seen, while a corresponding increase in the excitation intensity results in random luminescence impulses having random intensities. We also perform the calculation using a smaller value of τ_A in order to simulate the effect of a smaller effective band offset in bare NCs. The result, presented in panel d, clearly shows a shortening of the "on" period and a decrease of the PL quantum efficiency.

One can introduce the effective "on" and "off" states of quantum dots using the fact that the collecting time of the luminescence signal is much longer than all the dynamic processes taking place in the QD. The ratio of the probability of finding a QD in an "off" state to that for an "on" state determines the ratio of the lengths of the "off" and "on" average periods:

$$P_{off} / P_{on} = \tau / \tau_i \qquad (2.9)$$

where the nanocrystal ionization time τ_i in the case of Auger autionization is given by Eqs. 3 and 4. In the case of thermal autoionization of the QD with a small band offset, τ_i is given by the following rate of NC ionization [29]:

$$\frac{1}{\tau_i}=\frac{W_1\tau_1}{1+W_1\tau_1}\frac{1}{\tau_T} \qquad (2.10)$$

where $1/\tau_T \sim \exp(-\Delta E_{off}/kT)$ is the probability for electron or hole to leave the nanocrystal thermally and where ΔE_{off} is the band offset between the NC and the matrix. In this case the ionization rate depends linearly on excitation intensity in contrast to the Auger ionization rate (that has square dependence on excitation intensity) and now depends strongly on temperature.

The observation of the random intermittence of the PL intensity in a single QD changes completely our understanding of the PL quantum efficiency, η, of an ensemble of nanocrystals. The PL quantum efficiency of the nanocrystal ensemble has been suggested to be determined by the ratio

$$\eta=(N-N_{nr})/N \qquad (2.11)$$

of the number of "good dots" (N-N_{nr}) – that do not contained nonradiative channels - to the total number of the dots (N). However, the fact that the luminescence is completely quenched in ionized crystals and that we can ionize these crystals by Auger or thermal autoionization gives us a completely different picture.

The PL quantum efficiency is determined rather by the ratio of the duration of the "on" period to the duration of the sum of "on" and "off" periods or by the ratio of the ionization time to the sum of ionization and neutralization times. In ideal QDs that do not contain nonradiative channels:
h

$$\eta=\tau_i/(\tau_i+\tau) \qquad (2.12)$$

As can be seen from Eq. 2. 12 the quantum efficiency can be very low even at low excitation intensity if the time for the charge to return to the nanocrystal is long. The PL quantum efficiency even in an ensemble of ideal QDs depends on the temperature, size, and excitation intensity.

2.6. NONRADIATIVE AUGER RELAXATION IN NANOCRYSTALS WITH SEVERAL ELECTRON-HOLE PAIRS

As we mentioned in a previous section, the nonradiative Auger recombination of two electron-hole pairs leads to crystal ionization only if the band offset in the conduction or valence band is smaller than the NC energy gap. However, even if that is the case, the electron or the hole of the electron-hole pair that receives the annihilation energy of the first one does not always leave the NC. The Auger annihilation does not always lead to the NC ionization. This type of nonradiative Auger process, when the recombination energy of one pair is transferred to the other one, is an essential part of the stimulated emission process. This is because the lasing from direct-band semiconductor NCs requires more than one electron-hole pair per crystal on average that provides an inverse population at the NC PL band edge.

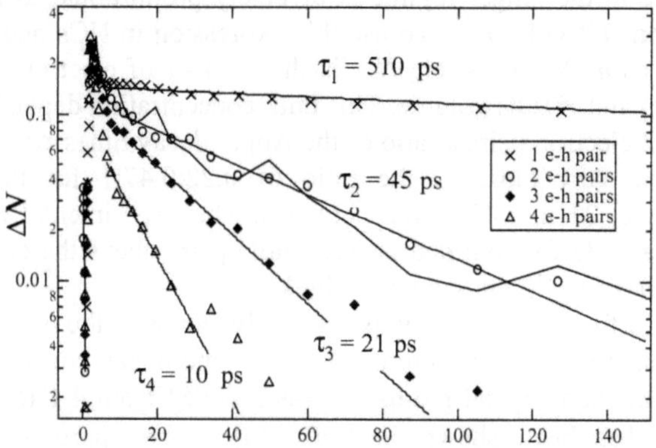

FIGURE 2.12. The decay dynamics of the 1-, 2-, 3-, and 4- electron-hole pair states in the 2.3 nm radius CdSe NCs. Experimental data extracted from transition absorption measurements are shown by symbols. The solid lines are their fit to a single exponential decay dependence (from Ref. [30]).

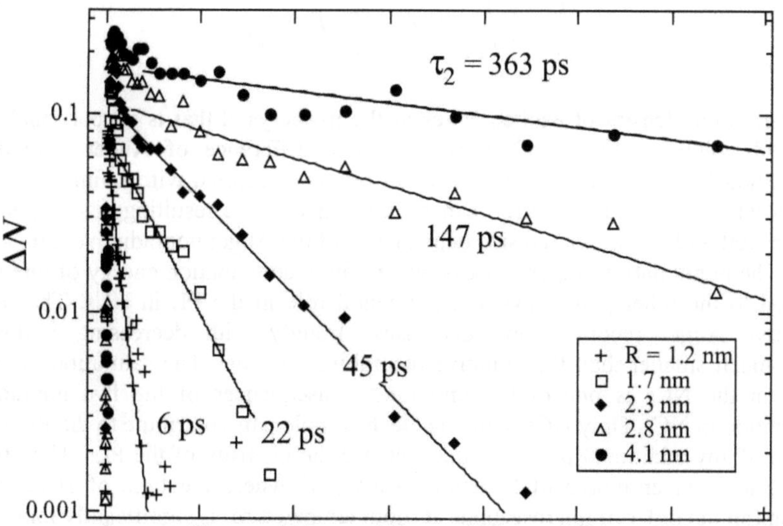

FIGURE 2.13. The time decay of the two electron-hole pair states measured in the NC quantum dots with radii 1.2, 1.7, 2.3, 2.8, and 4.1 nm (symbols are shown in the insert). The experimental data are fit by a single exponential dependencies (from Ref. [30]).

The rate of multiparticle Auger recombination has been measured in an ensemble of CdSe nanocrystals in Ref. [30] at different excitation levels. The dynamics of the 2-, 3-, 4-pair states decay in a 2.3 nm radius CdSe nanocrystal is shown in Fig. 2.12. One can see that the carrier decays faster as the number of electron-hole pairs in the NC increases, as expected for Auger recombination. In the bulk semiconductor the effective decay rate in the Auger regime is inversely proportional to the square of the carrier concentration: $1/\tau_A(n) \sim n^2$. To use this expression in NCs one has to determine the concentration as $n=N_{e\text{-}h}/V$, where $N_{e\text{-}h}$ is the number of electron-hole pairs in the nanocrystal ($N_{e\text{-}h}>1$) and V is its volume. This bulk concentration dependence predicts for NCs with 4, 3 and 2 electron pairs a ratio of the Auger decay times τ_4: τ_3: τ_2=0.25:0.44:1, respectively, that is very close to the ratio of 0.22:0.47:1 for the experimentally determined times (10, 21 and 45ps, respectively). The experimental results [30] show that the Auger decay rate for confined electron-hole pairs obeys the cubic concentration dependence: $dn_{e\text{-}h}/dt \sim (n_{e\text{-}h})^3$, the same as in bulk.

Figure 2. 13 from Ref. [30] compares the dynamics of the two electron-hole pair decay for CdSe nanocrystals of different size. The time decay constant decreases from 363 to 6 ps when the nanocrystal radius, a, was reduced from 4.1 to 1.2 nm. The size dependence of τ_2 has been shown in Ref. [30] to be approximated by a cubic dependence: $\tau_2(a) \sim a^3$.

There is no calculation of the decay time for the Auger nonradiative recombination of two electron-hole pairs. However, we can estimate this time using our calculations of the rate of Auger ionization (see Eq. 2. 8). This can be written:

$$\frac{1}{\tau_2} \sim \left\langle \frac{1}{\tau_A} \right\rangle \rho(a) \sim \frac{1}{a^{\nu}} a^3 = \frac{1}{a^{\beta}} \tag{2.13}$$

where $\rho(a)$ is the density of excited states in the nanocrystal that is proportional to the nanocrystal volume [24], and $2 < \beta < 4$. The difference of Auger nonradiative recombination from the Auger auto-ionization rate is connected with a finite density of the states where the Auger electron can be transferred. The resulting size dependence correlates well with the measured size dependence of the Auger nonradiative rate.

The nonradiative Auger process, where the recombination energy of one pair is transferred to the other pair, plays a fundamental role in the PL in NCs. The time of nonradiative Auger recombination decreases strongly with decreasing radius and becomes much smaller than the radiative recombination time. The saturation of the PL intensity in the NCs is one of the important consequences of the fast nonradiative recombination in NCs. Only NCs with one electron-hole pair contribute to the PL and the excitation of any additional pair does not increase the intensity of the PL. The effect is clearly seen in an ensemble of Si nanocrystals [31] where the time of electron-hole relaxation (on the order of microseconds at room temperature) is significantly longer than in CdSe nanocrystals. Figure 2. 14 shows the excitation intensity dependence of the PL intensity from porous Si excited well above the PL line. One can see that the PL intensity at the red edge of the PL line saturates earlier than at its blue edge, which is connected with differences of the electron-hole radiative decay times in small and large NCs. The radiative decay time in Si NCs decreases with the NC due to an enhance of the oscillator strength of the phononless transitions. This is because the electron states in the X and Γ points of the Brullouin zone are mixed in the small NCs. As a result the PL coming from the large NCs that contribute to the red edge of the PL line saturates earlier than the PL coming from the small NCs that contribute to the blue edge of the PL line.

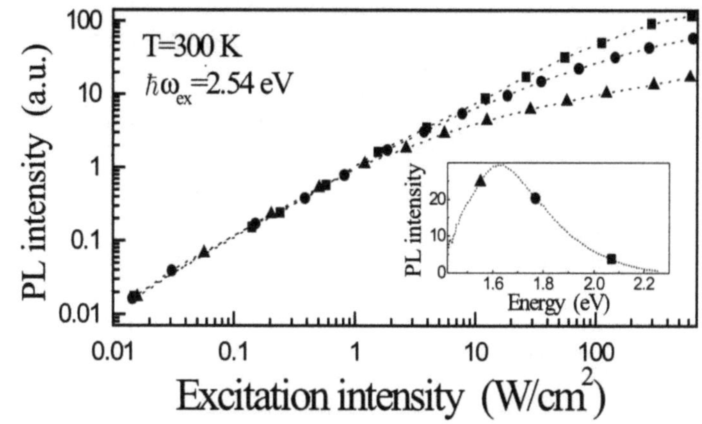

FIGURE 2.14. Dependence of the room temperature PL intensity in porous-Si on the excitation intensity for three different detection energies: 2.07 eV (squares), 1.77 eV (circles) and 1.55 eV (triangles). The excitation energy is 2.54 eV. The insert shows the PL spectra.

2.7. AUGER-LIKE THERMALIZATION IN NANOCRYSTALS

Not all of the Auger processes quench the PL in the NCs and decrease the PL quantum efficiency. For example, the Auger-like thermalization of carriers increases the PL quantum efficiency, because it is responsible for breaking a "phonon bottleneck" in small NCs. In small NCs the quantum size level spacing is much larger than the energy of an optical phonon. The direct transitions between levels have to be a multi-phonon process that has a very low probability. The standard phonon-assisted mechanism of carrier thermalization is then strongly suppressed: the effect is referred to as "phonon bottleneck." It was thought that this effect would make it impossible to use QDs for optical application.[32]

However, it was shown [33] that the rate of the Auger-like thermalization process is much faster than the radiative recombination time and it breaks the phonon bottleneck. This process is shown schematically in Fig. 2.15 from Ref. [34]. The spacing between the ground 1S and the first excited 1P electron levels is on the order of 200-300meV in small size NCs and is significantly larger than the optical phonon energy (20-30 meV) - this is the "phonon bottleneck." However, the electron-hole Coulomb interaction, although it is a small perturbation compared to the quantum size energy, becomes stronger in small NCs. In addition, hole level spacing are an order magnitude smaller than those of electrons, because of the higher hole effective mass and the degeneracy of the valence band. As a result the electron energy is transferred via the Coulomb interaction to the hole, which then relaxes rapidly through the almost continuous spectrum of valence band states. This process is called Auger-like thermalization.[1]

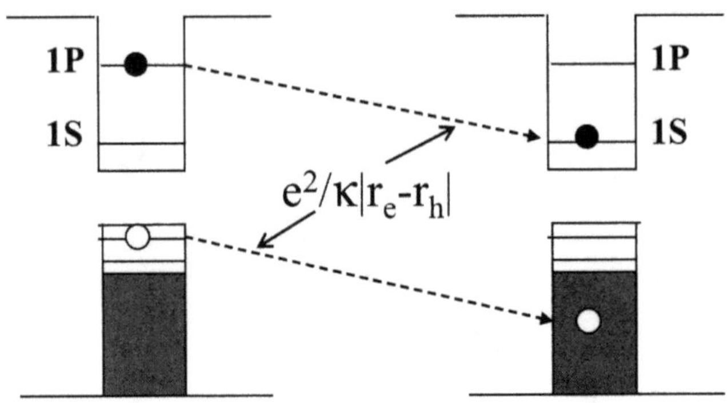

FIGURE 2.15. A scheme presenting an Auger-like thermalization process. The electron energy is transferred via the Coulomb interaction to the hole, which then relaxes rapidly through the almost continuous spectrum of valence band states.

The straightforward calculation of the size dependence of the rate of the electron transition from the 1P level to the 1S level in CdSe nanocrystals shows that the Auger-like thermalization time is on the order of 2 ps and almost constant over the range of radii from 2 to 4 nm. This time is significantly shorter than the electron-hole pair radiative decay time in this material and thus inelastic electron-hole collisions remove the phonon bottleneck problem in CdSe nanocrystals.

Indeed, this mechanism of fast electron relaxation was experimentally confirmed recently. The authors of Ref. [35] and [36] have studied the dynamics of the electron thermalization from the 1P state to the 1S state. Both experiments show that removing the photoexcited hole from the nanocrystal to the nanocrystal surface significantly suppresses the rate of electron thermalization. Removing the hole from the NC decreases the probability of energy transfer from the electron to the hole and can be interpreted as suppression of this Auger-like thermalization mechanism.

2.8. CONCLUDING REMARKS

The purpose of this review is to describe the role played by Auger processes in kinetics of the thermalization and interband relaxation of carriers. The Auger processes are considerably enhanced in NCs due to the abruptness of their interfaces and they affect all aspects of carrier relaxation and recombination: they strongly affect the PL quantum efficiency, and lead to PL blinking and reversible PL degradation. They require a significant shortening of the stimulated emission rate for lasing in order to overcome

---

[1] A similar type of the thermalization process when the electron energy is transfered to the hole via their Coulomb interaction has been studied in GaAs quantum wells. It is called inelastic scattering of the electron with heavy holes and is the next most efficient mechanism of the thermalization after thermalization via emission of the optical phonons.

the fast nonradiative decay in the NC with two electron–hole pairs. The misunderstanding of their role explains why quantum dot NC laser realization was delayed for almost 10 years. They also shorten the rate of carrier relaxation and allow one to use semiconductor doped glasses as a holographic material and as material for X-ray sensors.

Although the theory of Auger processes in a NC qualitatively describes many aspects of these processes, a quantitative description of their rates has yet to be developed. This is connected with the lack of a self-consistent theory that describes the electron-hole envelope functions at the NC surface. In our calculations we have used the envelope functions obtained with the help of the standard boundary conditions that assume the separate continuity of the envelope function and the normal envelope velocity at the interface. This assumption has never been justified and is a special case of the more general boundary conditions that conserve the total current across the interface. The discontinuities in the envelope wave functions at the heterointerface is one of the most important consequences of general boundary conditions.[37] They could strongly affect the calculation of the Auger processes rate, because they take place right at the nanocrystal surface and are affected by the surface property. The theoretical solution of this problem is important, because it may open the way to controlling the Auger processes rate through surface modifications.

2.9. ACKNOWLEDGEMENTS

I would like to thank my co-authors: V. A. Kharchenko and M. Rosen, who worked with me on the theory of Auger processes in nanocrystals. I also would like to thank my long–term collaborators: A. I. Ekimov, M. G. Bawendi, D. Norris, M. Nirmal, P. Alivisatos, as well as A. Nozik and L. Brus, whose challenging discussions stimulated most of these theoretical studies. I also thank M. Nirmal, V. Klimov and J. Cook for helping me with the figures. This work was supported by the Office of Naval Research.

REFERENCES

1. A. I Ekimov, A. A. Onushcenko, and V. A. Tzekhomskii, Sov. Phys. Chem. Glass **6**, 511 (1980).
2. A. Henglein, Ber Bunsenges, Phys. Chem. **88**, 301 (1982).
3. L. E. Brus, J. Chem. Phys. **79**, 5566 (1983).
4. *Spectroscopy of Isolated and Assembled Semiconductor Nanocrystals*, Eds. L.E. Brus, A. L. Efros and T. Itoh, Special Issue of J. of Luminescence **69**, North-Holland (1996).
5. L. E. Brus. Appl. Phys. A **53**, 465 (1991).
6. A. P. Alivisatos, Science **271**, 933 (1996).
7. Al. L. Efros and M. Rosen, Ann. Rev. Mater. Sci. **30**, 475 (2000).
8. V. I. Klimov, A. A. Mikhailovsky, Su. Xu, A. Malko, J. A. Hollingsworth, C. A. Leatherdale, H.-J. Eisler, and M. G. Bawendi, Science **290**, 314 (2000).
9. M. Bruchez Jr., M. Moronne, P. Gin, S. Weiss, and P. Alivisatos, Science **281**, 2013 (1998).
10. M. Nirmal, B. O. Dubbousi, M. G. Bawendi, J. J. Macklin, J. K. Trautman, T. D. Harris, and L. E. Brus. Nature, **383**, 802 (1996).
11. U. Banin, M. Bruchez, A.P. Alivisatos, T. Ha, S. Weiss, D. S. Chemla, J. Chem. Phys. **110**, 1195 (1999); M. Kuno, D. P. Fromm, H. F. Hamann, A. Gallagher, and D. J. Nesbitt, J. Chem. Phys. **112**, 3117 (2000); R. G. Neuhauser, K. Shimizu, W. K. Woo, S. A. Empedocles, and M. G. Bawendi, Phys. Rev. Lett. **85**, 3301 (2000).

12. J. Tittel, W. Gohde, F. Koberling, A. Mews, A. Kornowski, H. Weller, A. Eychmuller, Th. Basche, and Ber Bunsenges, Phys. Chem. **101**, 1626 (1997);_F. Koberling, A. Mews, Th. Basche, Phys. Rev. B **60**, 1921 (1999).
13. K. T. Shimizu, R. G. Neuhauser, C. A. Leatherdale, S. A. Empedocles, W. J. Woo, and M. G. Bawendi, Phys. Rev. B **63**, 5316 (2001); M. Kuno, D. P. Fromm, H. F. Hamann, A. Gallagher, and D. J. Nesbitt, J. Chem. Phys. **115**, 1028 (2001).
14. M. Kuno, D. P. Fromm, A. Gallagher, D. J. Nesbitt, O. I. Micic, and A. J. Nozik, Nano Lett. **1**, 557 (2001).
15. M. D. Mason, G. M. Credo, K. D. Weston, and S. K. Buratto, Phys. Rev. Lett. **80**, 5405 (1998).
16. D. I. Chepic, Al. L. Efros, A. I. Ekimov, M. G. Ivanov, I. A. Kudriavtsev, V. A. Kharchenco, and T. V. Yazeva, J. of Luminescece. **47**, 113 (1990).
17. P. Roussignol, D. Ricard, K.C. Rustagi, and C. Flytzanis, Optics Commun. **55**, 143 (1985).
18. A. I. Ekimov, I. A Kudriavtsev, M. G. Ivanov, and Al. L. Efros, J. of Luminescence **46**, 83 (1990).
19. V. J. Grabovskis, J. J. Dzenis, A. I. Ekimov, I. A. Kudriavtsev, and M. N. Tolstoii, Sov. Phys. Solid State **31**, 150 (1989).
20. A. I. Ekimov, and A. L. Efros, Phys. Stat. Sol. (b) **150**, 627 (1988)
21. J. H. MacKey, H. L. Smith, and A. I. Halperin, J. Phys. Chem. Solids **27**, 1759 (1966).
22. Al. L. Efros, M. Rosen, B. Averbuoukh, D. Kovalev, M. Ben-Chorim, and F. Koch, Phys. Rev. B **56**, 3875 (1997).
23. A. B. Migdal, *The qualitative methods of quantum theory,* Nauka, Moskva (1975),
24. Al. L. Efros and A. L. Efros, Sov. Phys.-Semicond. **16**, 772 (1982).
25. Al. L. Efros, Phys. Rev. B **46**, 7448 (1992).
26. M. Nirmal, D. J. Norris, M. Kuno, M. G. Bawendi, Al. L. Efros, and M. Rosen, Phys. Rev. Lett. **75**, 3728 (1995); Al. L. Efros, M. Rosen, M. Kuno, M. Nirmal, D. J. Norris, and M. Bawendi, Phys. Rev. **B 54**, 4843 (1996).
27. K. Naoe, L. G. Zimin, and Y. Masumoto, Phys. Rev. B **50**, 18200 (1994).
28. D. J. Norris, A. Sacra, C. B. Murray, and M. G. Bawendi, Phys. Rev. Lett. **72**, 2612 (1994).
29. Al. L Efros and M. Rosen, Phys. Rev. Lett. **78**, 1110 (1997).
30. V. I. Klimov, A. A. Mikhailovsky, D. W. McBranch, C. A. Leatherdale and M. G. Bawendi, Science **287**, 1011 (2000).
31. D. Kovalev, B. Averboukh, M. Ben-Chorin, F. Koch, Al. L Efros, and M. Rosen, Phys. Rev. Lett. **77**, 2089 (1996).
32. H. Benisty, C. M. Sotomayor Torres, and C. Weisbuch, Phys. Rev. B **44**, 10945 (1991).
33. Al. L. Efros, V. A. Kharchenko, and M. Rosen, Solid State Commun. **93**, 281 (1995).
34. Al. L. Efros, in: *Phonons in Semiconductor Nanostructures*, ed. J-P.Leburton, J. Pascual, C. Sotomayor Tores, KAP. Boston, London (NATO ASI-v.236), p.299, 1993.
35. P. Guyot-Sionnest, M. Shim, C. Matranga, and M. Hiens, Phys. Rev. B **60**, 2181 (1999).
36. V. I. Klimpov, A. A. Mikailovsky, D. W. McBranch, C. A. Leatherdale, and M. G. Bawendi, Phys. Rev. B **61**, R13349 (2000).
37. A. V. Rodina, A. Yu Alekseev, Al. L. Efros, M. Rosen, and B. K. Meyer, Phys. Rev. B **65**, 125302 (2002).

3

Carrier dynamics, optical nonlinearities, and optical gain in nanocrystal quantum dots

Victor I. Klimov
Chemistry Division, C-PCS, MS-J567, Los Alamos National Laboratory

3.1. INTRODUCTION

Nanoscale semiconductor particles in which carrier motion is restricted in all three dimensions are often referred to as semiconductor quantum dots (QDs). While the crystalline structure of the bulk solid is maintained in QDs, the three-dimensional (3D) quantum confinement imparted by the nanoscale size causes the bulk energy bands to collapse into discrete, atomic-like levels that exhibit strong size dependence.[1,2] One of the approaches to fabricating sub-10 nm semiconductor nanoparticles is through chemical synthesis. Chemically synthesized QDs are also called nanocrystals or nanocrystal QDs. Synthetic methods are particularly well developed for QDs of II-VI semiconductors. The two principle chemical routes for fabrication of these QDs are high-temperature precipitation in molten glasses[3,4] and colloidal synthesis using, e.g., organometallic reactions.[5] Glass samples provide rigidity and environmental stability; however, they have a broad QD size distribution (typically greater than 20%) and a large number of surface defects. A much higher level of synthetic flexibility and control is provided by colloidal QDs that can be chemically manipulated in a variety of ways including size-selective precipitation[5] (resulting in less than 5% size variations), surface modification by exchange of the passivation layer,[6,7] formation of layered core-shell heterostructures,[8,9] immobilization in sol-gel[10] and polymer[11] matrices, and self-assembly into 3D superlattices.[12,13]

The size-controlled tunability of QD optical properties combined with chemical flexibility provided by colloidal methods make nanocrystal QDs promising materials for applications in various optical technologies. This chapter reviews fundamental dynamical processes in chemically synthesized QDs that relate to their performance in nonlinear optical and light emitting (lasing) applications. Specifically, this review focuses on ultrafast carrier dynamics, band-edge optical nonlinearities, optical gain, and stimulated emission in CdSe nanocrystal QDs. Because of the highly-developed preparation techniques, CdSe QDs have been prototype systems for studying the effects of 3D confinement on electronic and optical properties of semiconductors. These dots have also been used to demonstrate the first technological applications of QD materials in such areas as biolabeling,[14,15] optical amplification and lasing,[16,17] photovoltaics,[11,18] light-emitting diodes,[19,20] and ultrafast dynamic holography.[21]

This chapter begins with an analysis of linear (section 3.2) and nonlinear (section 3.3) absorption spectra of CdSe QDs and mechanisms for band-edge optical nonlinearities. This analysis is followed by an overview of carrier intraband relaxation studies (section 3.4). Finally, we analyze the underlying physics of processes relevant to optical amplification and lasing in strongly confined QDs and, in particular, the effect of multiparticle Auger recombination on optical gain dynamics (section 3.5).

3.2. LINEAR ABSORPTION SPECTRA AND THE STRUCTURE OF OPTICAL TRANSITIONS

In this review, we concentrate on CdSe QDs fabricated by an organometallic synthesis.[5] This synthetic route provides routine preparation of dots with radii from 1 to 5 nm and with size dispersions as narrow as 5%. Since the exciton Bohr radius in CdSe is ~5 nm, the QDs in the sub-5 nm size range correspond to the regime of extremely strong quantum confinement for which the spacing between the lowest conduction-band levels is much greater than the bulk exciton binding energy (16 meV).

CdSe QDs prepared by the route described in Ref. 5 are passivated with organic molecules of trioctylphosphine oxide (TOPO) and trioctylphosphine (TOP). The capping layer prevents nanoparticles from aggregating and renders them soluble. It also provides a varying degree of "electronic" passivation of surface traps. The original passivation can be modified via surface exchange reactions,[5,7] which have been used to study the influence of surface treatment on the electronic and optical properties of QDs.[7,22,23] In addition to passivation with organic molecules, CdSe QDs can be overcoated with a layer of a wide-gap inorganic semiconductor such as ZnS.[9] This procedure allows one to greatly improve "electronic" passivation of QD surfaces, resulting, in particular, in a significant increase in the photoluminescence (PL) quantum yield.

Detailed PL-excitation studies described in Ref. 24 demonstrate that the structure of optical transitions in CdSe QDs can be explained by the effective-mass model which takes into account mixing between three valence sub-bands.[25] In this model, conduction-band states are characterized by two quantum numbers: l, the angular momentum of the envelope wave function, and n, the number of the state within a series of states of the same symmetry. States with $l = 0, 1, 2, \ldots$ are labeled as S, P, D ... states. In QDs, the three lowest electron states are 1S, 1P, and 1D.

A description of QD hole states requires considering an additional quantum number, F, which is the total hole angular momentum calculated as a sum of the Bloch function and envelope function momenta.[25,26] In the notation of hole states, the F number is usually shown as a subscript. For example, the notation for the lowest hole state is $1S_{3/2}$. The hole states are $(2F + 1)$- fold degenerate. This degeneracy is lifted if the effects of the crystal field in a hexagonal lattice,[27] the QD nonspherical shape,[28] and the electron-

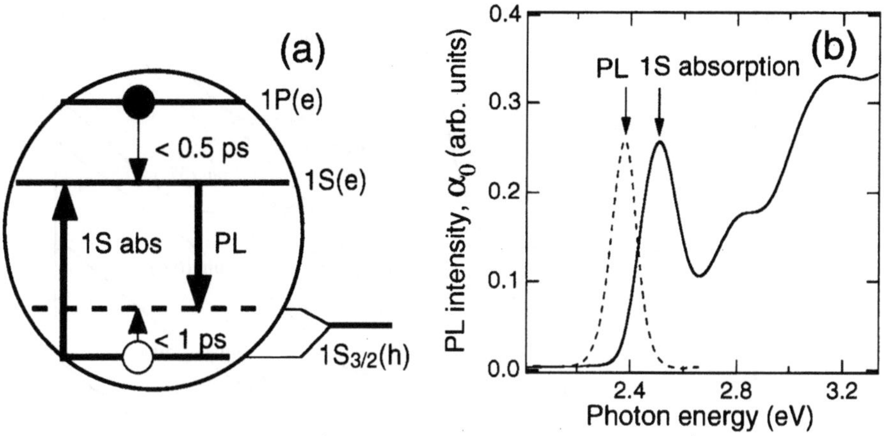

FIGURE 3.1. (a) Schematics of "absorbing" and "emitting" transitions in CdSe NQDs along with intraband relaxation processes leading to a population buildup of the "emitting" transition (these processes are discussed in section 3.4). (b) Absorption and PL spectra of CdSe QDs with R = 1.2 nm (T = 300 K) illustrating a large "global" Stokes shift between the 1S absorption peak and the PL maximum.

hole (e-h) exchange interactions[29,30] are taken into account, leading to a fine structure of the lowest exciton state.[30]

The fine-structure hole states form two groups of closely spaced levels that are separated by a gap that can be as large as ~100 meV in small size dots [Fig. 3.1(a)]. This gap is observed in optical spectra as a large "global" Stokes shift between the lowest 1S absorption maximum and a band-edge PL detected under nonresonant excitation [Fig. 3.1(b)]. The intense 1S absorption peak results from a strong ("absorbing") optical transition that couples high-energy, fine-structure hole states to the lowest 1S electron state. In contrast, the band-edge PL band is due to a weaker ("emitting") transition involving the lower-energy manifold of hole states.

Fig. 3.2 displays linear absorption spectra (α_0) of five colloidal samples with QD radii R = 1.2, 1.7, 2.3, 2.8, and 4.1 nm. These spectra show well-resolved features that correspond to interband optical transitions coupling different electron and hole quantized states. The two low-energy bands can be assigned to transitions involving the lowest electron state (1S) and two different hole states ($1S_{3/2}$ and $2S_{3/2}$). In the following text, we will refer to these transitions as 1S [1S(e) – $1S_{3/2}$ (h)] and 2S [1S(e) – $2S_{3/2}$ (h)]. The linear absorption spectra of QDs with $R > 2.5$ nm also exhibit a well-resolved band associated with the transition coupling the 1P electron state to the $1P_{3/2}$ hole state (1P transition) and a higher-lying band associated with the transition involving the 1S electron state and a hole state originating from the spin-orbit split-off band ($3S_{1/2}$ or $2S_{1/2}$, depending on the QD radius).[24]

The broadening of optical transitions observed in absorption spectra of colloidal QDs is primarily due to inhomogeneity arising from size dispersion. In the strong-confinement regime, the size dependence of energy transitions in QDs is dominated by a term proportional to $1/R^2$. In this case, the inhomogeneous broadening (Γ_i) of the transition with energy $\hbar\omega_i$ is proportional to the shift of this transition with respect to the

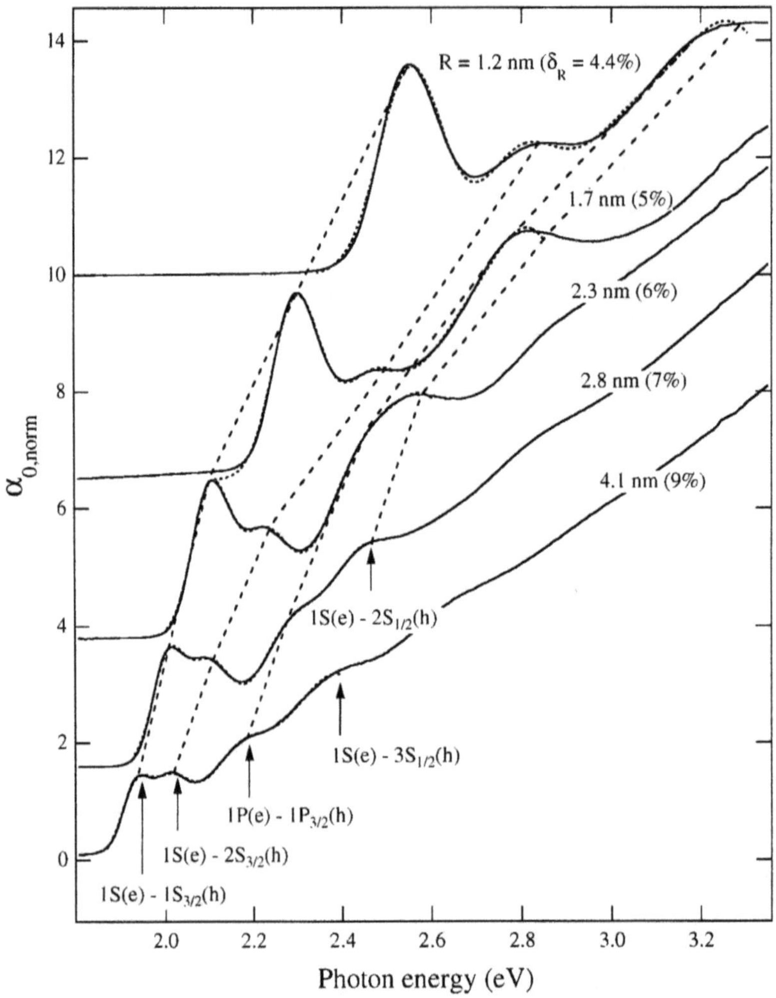

FIGURE 3.2. Linear absorption spectra of CdSe QDs with radii from 1.2 to 4.1 nm (solid lines) fit to a sum of four Gaussian bands (dotted lines). QD size dispersions derived from the fits are indicated in parentheses. Arrows mark the positions of low-energy optical transitions involving the 1S and 1P electron states.

bulk material energy gap (E_g). For a Gaussian size distribution with a standard deviation Δ_R,

$$\Gamma_i = \frac{2\Delta_R}{R}(\hbar\omega_i - E_g) = 2\delta_R\Delta_i, \tag{3.1}$$

where $\Delta_i = \hbar\omega_i - E_g$ and $\delta_R = \Delta_R/R$.

Using the transition-energy dependent broadening given by Eq. (3.1), one can accurately fit band-edge absorption spectra of QD samples to a sum of four Gaussian bands (dotted lines in Fig. 3.2):

$$\alpha_0(\hbar\omega)=\sum_{i=1}^{4}\frac{a_i}{\sqrt{2\pi}\Gamma_i}\exp\left[-\frac{(\hbar\omega-\hbar\omega_i)^2}{2\Gamma_i^2}\right], \tag{3.2}$$

where a_i is the area of the band centered at $\hbar\omega_i$. Three low-energy bands in Eq. (3.2) account for the 1S, 2S, and 1P transitions (in the order of increasing energy); while the fourth high-energy band is likely due to a transition involving the 1S electron state and one of the $S_{1/2}$ hole states originating from the spin-orbit split-off band (see above). In addition to providing energies for optical transitions, this multi-Gaussian fit allows one to derive the size dispersion of QD samples (indicated in Fig. 3.2).

3.3. TRANSIENT ABSORPTION SPECTRA AND RESONANT OPTICAL NONLINEARITIES

3.3.1. Mechanisms for resonant optical nonlinearites

This section concentrates on resonant optical nonlinearities, i.e., on nonlinearities that are due to absorption/refraction changes associated with photoexcited carriers. In strongly-confined colloidal QDs, resonant optical nonlinearites are primarily due to state filling and Coulomb multiparticle interactions (the carrier-induced Stark effect).[31]

State filling arises from Pauli exclusion and leads to the bleaching of the interband optical transitions that involve populated quantized states. The absorption changes resulting from this effect are proportional to the sum of the electron and hole occupation numbers. If we present the linear absorption spectrum of QDs as a sum of separate absorption bands corresponding to different "quantized" optical transitions, the state-filling induced absorption changes ($\Delta\alpha$) can be calculated using the expression:

$$\Delta\alpha(\hbar\omega)=-\sum_i a_i G_i(\hbar\omega-\hbar\omega_i)\left(n_i^e+n_i^h\right), \tag{3.3}$$

where $G_i(\hbar\omega-\hbar\omega_i)$ is the unit-area absorption profile of the $\hbar\omega_i$ transition, a_i is its area, and n_i^e and n_i^h are occupation numbers of the electron and hole states involved in the transition. Under thermal quasi-equilibrium (after the intraband relaxation is finished), the occupation numbers can be found using the Fermi distribution function.

Because of the degeneracy of the valence band and the large difference between electron and hole masses (in CdSe, $m_h/m_e \approx 6$), at spectral energies above the 1S transition, hole states form a very dense spectrum [quasicontinuum; see section 3.4.2 below]. Therefore, room temperature occupation probabilities of single electron states are much greater than occupation probabilities of the coupled hole states (hole populations are spread over many adjacent levels by the thermal distribution). As a result, the state-filling-induced absorption changes are dominated by electrons[32,33] (with the exception of spectral energies below the 1S transition; see the discussion concerning the contribution of hole populations to state filling effect and optical gain in section 3.5).

In addition to state filling, the photoexcitation of carrier populations leads to the generation of local fields that modify the optical spectra of QDs due to the Stark effect. The Stark effect leads to a shift of optical transitions and changes in transition oscillator strengths as a result of modifications in the selection rules. In terms of traditional pump-probe spectroscopy [see section 3.3.2 below], the carrier-induced Stark effect can be interpreted as arising from Coulomb interactions between one or several e-h pairs excited by the pump pulse and the e-h pair generated by the probe pulse. In contrast to state filling that *selectively* affects only transitions involving populated states, the carrier-induced Stark effect does not have this selectivity and modifies all QD transitions, with a stronger effect on transitions involving *unpopulated* states.[34] The Stark effect is particularly important in the case of charge separated e-h pairs, which can be formed, for example, as a result of the surface trapping of one type of carriers (the trapped-carrier induced Stark effect).[35] In the case of two interacting e-h pairs, the Stark effect can be considered as arising from biexciton Coulomb interactions,[34,36,37] and the transition shift can be used as a measure of the binding energy of the 3D-confined biexciton. Spatial confinement leads to an enhancement in the biexciton binding energy in comparison with that in bulk materials.[36] In bulk CdSe, the biexciton binding energy is only about 4 meV,[38] whereas in small-size CdSe QDs it is in the tens-of-meV range.[34,37]

In the case of an isolated optical transition, and in the limit of small fields (the Stark shift Δ_S is small in comparison with the transition broadening Γ), absorption changes due to the Stark effect can be calculated as proportional to the first derivative of the linear absorption spectrum: $\Delta\alpha(\hbar\omega) \approx \Delta_S \alpha'_0(\hbar\omega)$. In the presence of two closely spaced broad transitions (energy spacing $\Delta_{12} < \Delta_S$, and $\Delta_{12} < \Gamma$), the Stark effect leads to a repulsion between the transitions, with an associated absorption change that can be approximated by a second derivative of the absorption profile: $\Delta\alpha(\hbar\omega) \propto \alpha''_0(\hbar\omega)$. This second limit likely applies to QDs because large, carrier-induced Stark shifts should lead to strong interactions between closely-spaced, QD valence band states.[35]

3.3.2. Transient absorption: Femtosecond dynamics and pump dependences

Time-resolved nonlinear optical spectroscopy, specifically transient absorption (TA) spectroscopy, is a powerful tool for studying the electronic processes in different classes of bulk and low-dimensional materials.[39–41] Recently, TA spectroscopy was extensively applied in studies of energy structures, mechanisms for optical nonlinearities, and carrier dynamics in colloidal QDs.[22,23,42,43] TA, or the pump-probe technique, is based on monitoring absorption changes induced by e-h pairs generated by a short picosecond (ps) or femtosecond (fs) pump pulse. One of the configurations of a high-sensitivity, fs TA experiment is described in Ref. 44. Briefly, samples are pumped at 3.1 eV by frequency doubled 100 fs pulses from an amplified Ti-sapphire laser. The pump-induced absorption changes are probed over the range from 1.3 to 3 eV with a fs white-light continuum generated in a sapphire plate. In the case of time-resolved, sub-ps measurements, it is important to account for the temporal dispersion of the frequency (chirp) of the probe pulse.[45] In order to perform TA measurements not affected by chirp, the experiment in Ref. 44 utilizes single wavelength, phase-sensitive detection and spectral scanning, with simultaneous adjustment of the time delay between pump and probe pulses. This technique allows chirp-free, high-sensitivity detection of TA signals with an accuracy up to 10^{-5} in differential transmission.

In Fig. 3.3, we show early time TA spectra (a pump-probe delay time $\Delta t = 100$ fs) detected with the technique described above for a series of CdSe QDs with radii from 1.7 to 4.1 nm. These spectra were taken at low pump intensities corresponding to excitation of less than one e-h pair per dot on average. Initial QD populations, $\langle N_0 \rangle$, can be calculated as follows: $\langle N_0 \rangle = \Phi_p \sigma_0$, where Φ_p is the pump fluence (presented in photons per cm^2) and σ_0 is the QD absorption cross-section that can be estimated using a procedure from Ref. 41.

The shapes of the early-time TA spectra (solid lines in Fig. 3.3) closely match those generated by calculating second derivatives of the absorption spectra (dotted lines in Fig. 3.3). This similarity can be explained in terms of the carrier-induced Stark effect in the system of closely-spaced, broad transitions. This effect leads to absorption bleaching ($\Delta\alpha < 0$) at the positions of the "original" transitions (i.e., transitions seen in linear absorption) and to photoinduced absorption (PA) ($\Delta\alpha > 0$) at the positions of the "new" (shifted) transitions. For QDs with $R = 4.1$ nm, the bleaching bands B_1, B_2, B_3, and B_4 (Fig. 3.3) can be assigned to the 1S, 2S, 1P, and 1S(e) – $3S_{1/2}$(h) transitions, respectively, consistent with a previous assignment of optical transitions in linear

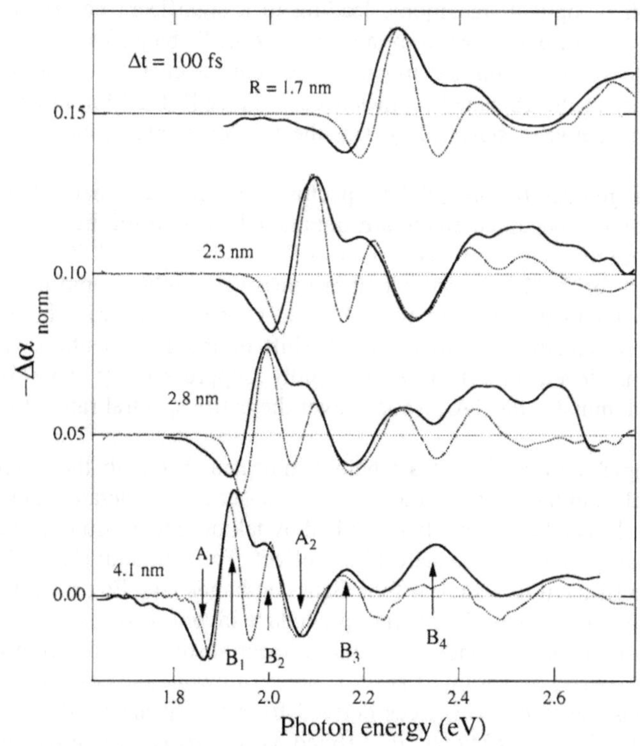

FIGURE 3.3. Early-time TA spectra (Δt = 100 fs) of CdSe QDs with radii from 1.7 to 4.1 nm (solid lines) compared to the second derivative of the absorption spectra (dotted lines).

absorption spectra in Fig. 3.2. In Fig. 3.3, one can also resolve PA features A_1 (due to a low-energy-shift of the 1S transition) and A_2 (generated as a result of a low-energy shift of the 1P transition and/or a high-energy shift of the 2S transition).

Initial TA dynamics provide information on re-distribution of carrier populations between different quantized states during the energy relaxation. Fig. 3.4(a) displays time-resolved TA spectra of 4.1-nm CdSe QDs detected between 0.1 and 2 ps after excitation. At Δt = 0.1 ps (almost immediately after the completion of the pump pulse), TA signals are due to both the Stark effect and state filling induced by nonequilibrium carriers generated directly in various electron states. As Δt is increased up to 0.5 ps, we observe a growth of B_1, B_2, and B_3 bleaching features, which is due to an increased contribution from filling of the 1S and 1P electron states as a result of carrier intraband relaxation. In the 2-ps spectrum, while the B_1 and B_2 features are further increased, the amplitude of the B_3 feature is decreased, indicating relaxation of the 1P electrons into the lowest 1S state.

The complementary population/depopulation dynamics of the 1S and 1P electron levels are well pronounced in TA time transients recorded at the positions of the 1S and 1P transitions [Fig. 3.4(b)]. These time transients indicate an extremely fast relaxation process with a time constant of about 540 fs. Interestingly, the build-up of the B_1 band is accompanied by the decay of the PA feature (A_1) below the 1S transition [circles in Fig. 3.4(b)]. This feature is associated with the 1S transition Stark shift. The magnitude of this feature is strongly affected by intraband relaxation that is a result of a large bandwidth of optical transitions, leading to a significant contribution of the 1S bleaching even at large detunings from the center of the B_1 band. Therefore, the increased bleaching of the 1S transition accompanying the carrier energy relaxation overwhelms the A_1 PA feature as the 1S state becomes occupied (see Ref. 34 for a numerical analysis of a competition between state filling and the Coulomb-interaction-induced transition shifts).

It is a general feature of TA spectra of CdSe QDs that TA signals within transitions involving occupied states are dominated by state-filling with only a minor contribution from the Stark shift. This point is illustrated in Fig. 3.5(a) which shows TA spectra recorded for the 4.1-nm sample at progressively higher pump intensities (Δt = 2 ps). In the low-intensity spectrum ($\langle N_0 \rangle$ = 0.9), one can clearly discern the A_2 PA feature that is presumably due to the Stark shift of the 1P transition. However, with increasing pump level, this feature is rapidly suppressed by state-filling induced bleaching that dominates the TA response over the entire spectral range from B_1 to B_4 at $\langle N_0 \rangle > 2$.

A simple numerical analysis below indicates that both the magnitude of the bleaching signals and the overall structure of TA spectra at moderate and high excitation levels ($\langle N_0 \rangle \geq 1$) can be accurately described by taking into account a single source for optical nonlinearity, namely, the state-filling effect. To quantitatively analyze spectra in Fig. 3. 5(a), we fit them to a sum of four Gaussian bands [see Eq.(3.2)] that account for the B_1–B_4 bleaching features. The fitting procedure involves two steps. First, we derive positions and widths of the bands by fitting a linear absorption spectrum in Fig. 3.5(a). Then we fit TA spectra by varying transition areas (a_i) while the positions and the widths of the transitions are fixed. The areas derived from the fit are used to calculate pump dependent "saturation" coefficients, p_i, defined as $p_i = a_i/a_{0i}$, where a_{0i} are the areas of transitions in the linear absorption spectrum. In terms of the state filling model, the

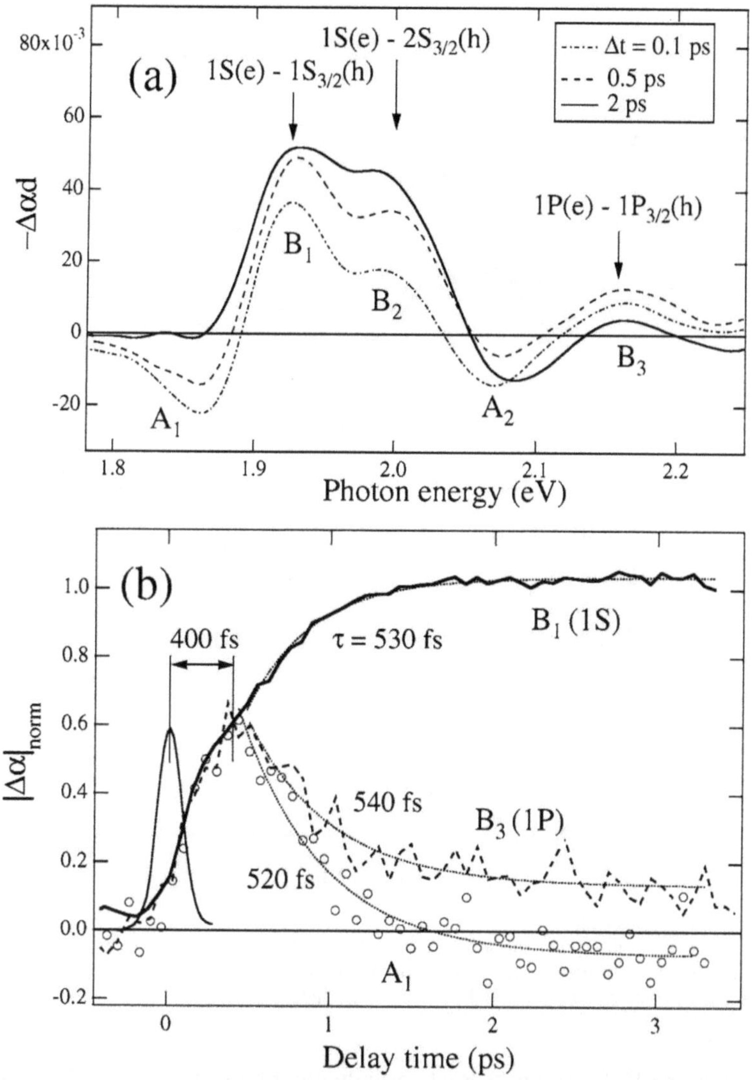

FIGURE 3.4. (a) TA spectra of 4.1-nm CdSe QDs detected at 0.1, 0.5, and 2 ps after excitation. (b) TA dynamics recorded for the same sample at the positions of the B_1 (thick solid line), B_3 (thick dashed line), and A_1 (circles) features, fit to a single-exponential growth/decay (thin dotted lines). The thin solid line is the pump-probe cross-correlation.

coefficients p_i account for the population factors in Eq. (3.3): $p_i = n_i^e + n_i^h$. For the 1S and higher lying transitions, $n_i^e \gg n_i^h$ and, therefore, $p_i \approx n_i^e$.

The pump-density dependence of p_i is shown in Fig. 3.5(b). This dependence is almost identical for p_1 and p_2, as expected for the 1S and 2S transitions that are bleached due to a population increase of the same 1S electron state. Coefficient p_3 shows a delayed growth with the onset at $\langle N_0 \rangle \approx 2$, which is consistent with the threshold expected for the filling of the 1P electron level. The growth of p_4 is further delayed (the onset at ~5),

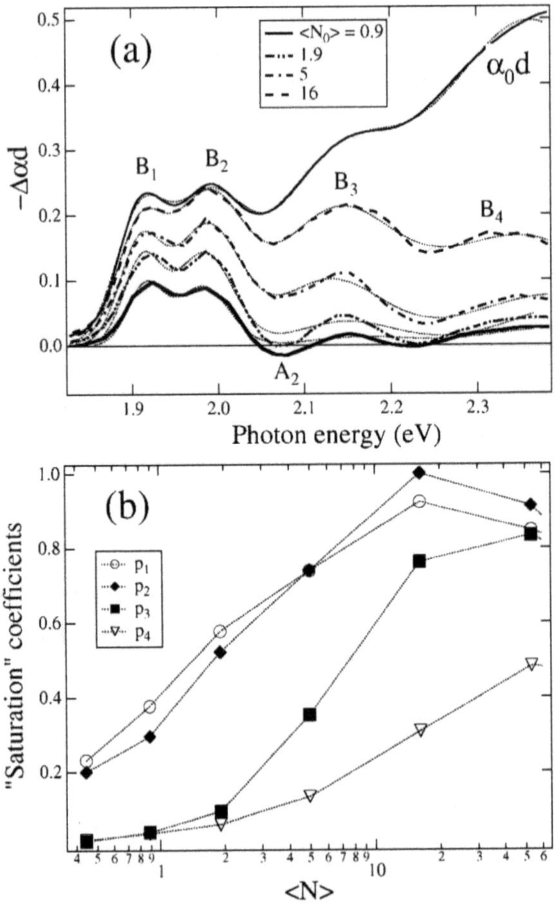

FIGURE 3.5. (a) Pump-dependent TA spectra (thick lines) of CdSe QDs with R = 4.1 nm taken at $\Delta t = 2$ ps in comparison to the linear absorption spectrum (thin solid line). The thin dotted lines are fits to a sum of four Gaussian bands (see text for details). (b) Pump dependence of the p_1–p_4 "saturation" coefficients (normalized B_1–B_4 bleaching areas) derived from fits to the spectra in the upper panel.

indicating that the B_4 feature is at least partially contributed by the population of the 1D electron level.

In Fig. 3.6(a), we compare the pump dependence of the normalized "saturation" coefficients p_1 (circles) and p_3 (squares) with pump dependent occupation numbers of the 1S and 1P electron states calculated within the state-filling model (lines). In the case for which the pump photon energy is much greater than the QD energy gap (this case corresponds to experimental conditions used to obtain data in Fig. 3.5), carrier-induced absorption saturation at the pump wavelength is insignificant, and the probability of generating an e-h pair in a dot is independent of the number of pairs already existing in it. This assumption results in the Poisson distribution of QD populations: $P(N) = \langle N \rangle^N e^{-\langle N \rangle} / N!$, where $P(N)$ is the probability of having N e-h pairs in a selected QD when the average population of QDs is $\langle N \rangle$. The occupation numbers of

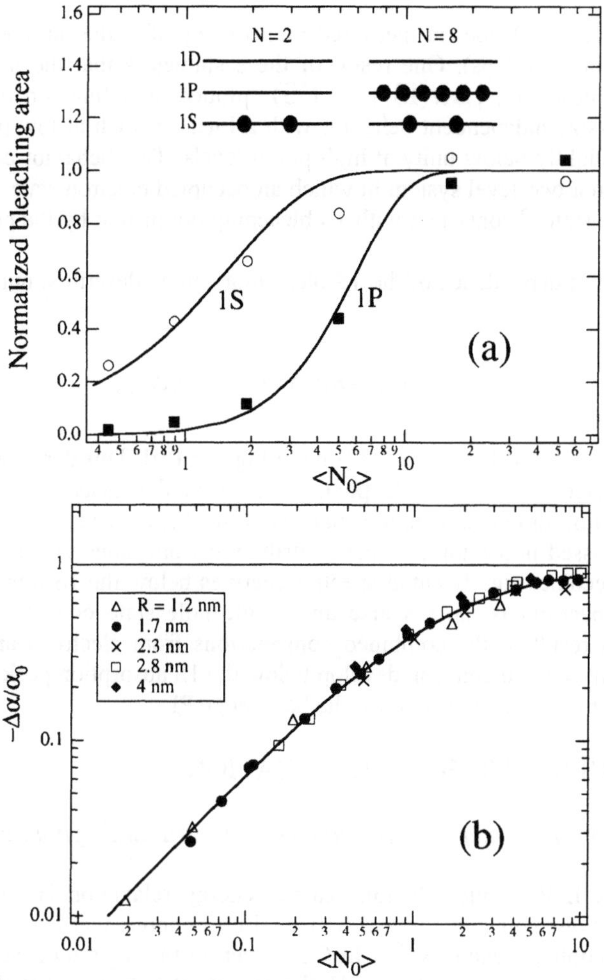

FIGURE 3.6. (a) Pump dependence of the normalized 1S and 1P bleaching areas (symbols) in comparison to the pump-dependent average occupation numbers of the two-fold-degenerate 1S and the six-fold-degenerate 1P electron states calculated assuming a "Poisson" type population growth (lines). Inset: distribution of electron populations between different QD states in the case of 2 and 8 e-h pairs excited per dot. (b) Pump dependence of normalized 1S absorption changes for QDs of different sizes (symbols) fit to the "universal" pump dependence $-\Delta\alpha/\alpha_0 = k_1\langle N_0\rangle/(k_2+\langle N_0\rangle)$ (line).

the 1S and 1P states in this model can be calculated as follows: $n^e_{1S} = 0.5P(1)+\sum_{i=2}^{\infty}P(i)$, $n^e_{1P} = \sum_{i=3}^{7}P(i)(i-2)/6+\sum_{i=8}^{\infty}P(i)$.

The pump dependence of the calculated occupation numbers [lines in Fig. 3.6(a)] closely matches the dependence measured for "saturation" coefficients [symbols in Fig. 3.6(a)] providing strong evidence that bleaching of QD transitions that couple occupied quantized states is primarily due to state filling.

Experiments in Refs. 46, 47 establish an accurate relationship between the magnitude of band-edge bleaching signals and QD average populations by studying the pump dependence of TA spectra measured immediately after the intraband relaxation is finished (Δt from 1 to 2 ps). One result of these studies is that the magnitude of the normalized 1S bleaching, $|\Delta\alpha(1S)|/\alpha_0(1S)$, plotted as a function of $\langle N_0 \rangle$ shows "universal" QD-size independent behavior with a linear growth at low pump intensities and saturation slightly below unity at high pump levels. This behavior can be explained by state filling in a two-level system in which an occupied electron state is coupled to an unoccupied hole state,[46] consistent with 1S bleaching dominated by the population of the 1S electron level.

The pump dependence of the 1S bleaching can be described numerically by the expression:

$$-\Delta\alpha/\alpha_0 = k_1\langle N_0 \rangle/(k_2 + \langle N_0 \rangle) \tag{3.4}$$

with $k_1 = 1.05 \pm 0.15$ and $k_2 = 1.6 \pm 0.3$ [line in Fig. 3.6 (b)]. This dependence allows one to accurately determine instant QD populations from the measured $\Delta\alpha$ signals. This capability is useful, for example, in the analysis of the Auger effects [see section 3.5.2]. As will be discussed in section 3.5, the contribution from holes to state-filling-induced nonlinearities becomes significant at spectral energies below the 1S transition where the valence band spectrum is more sparse and single hole state occupation numbers are increased. As a result of the combined contributions from electrons and holes, under certain conditions, optical gain can develop below the 1S absorption peak in the region of the "emitting transition" responsible for the band-edge PL.

3.4. CARRIER INTRABAND RELAXATION

3.4.1. Electron-hole interactions and electron intraband dynamics

In bulk II-VI semiconductors, carrier energy relaxation is dominated by the Fröhlich interactions with longitudinal optical (LO) phonons that lead to fast (typically sub-ps) carrier cooling dynamics.[48–50] In QDs, even in the regime of weak confinement when the level spacing is only a few meV, the carrier relaxation mediated by interactions with phonons is hindered dramatically because of restrictions imposed by energy and momentum conservation, leading to a phenomenon called a "phonon bottleneck."[51,52] Further reduction in the energy loss-rate is expected in the regime of strong confinement, for which the level spacing can be much greater than LO phonon energies, and hence carrier-phonon scattering can only occur via weak multiphonon processes.

Despite these theoretical predictions, a number of experimental results indicate that carrier relaxation in II-VI QDs is not significantly slower than in bulk materials.[42,43,53] Strong evidence against the existence of the phonon bottleneck in 3D-confined systems has been obtained in recent studies of intraband relaxation in strongly-confined colloidal CdSe QDs.[33,43] These studies indicate sub-ps 1P-to-1S electron relaxation even for very small QDs for which the 1S–1P energy separation is greater than ten LO phonon energies. The results of these studies are reviewed below.

As pointed out in the discussion of fs TA signals in section 3.3.2, the decay of the 1P bleaching [B_3 in Fig. 3.4(a)] and the complementary growth of the 1S bleaching [B_1 in Fig. 3.4(a)] provide a measure of the rate of electron energy relaxation. As indicated by dynamics in Fig. 3.4(b), the 1P-to-1S relaxation is extremely fast. For 4.1-nm QDs, this relaxation has a time constant of 540 fs despite the very large 1S–1P energy separation of about eight LO phonon energies. Interestingly, intraband relaxation becomes even faster with decreasing QD size, as evident from a comparison of the 1S state population dynamics for QDs of different radii shown in Fig. 3.7(a). These data indicate a decrease in the 1S build-up time (τ_{1S}) with decreasing QD radius: τ_{1S} is 530 fs for R = 4.1 nm and shortens down to 120 fs for R = 1.7 nm, roughly following a linear size dependence.

In Fig. 3.7(b), we compare size-dependent rates of energy relaxation (dE_e/dt) observed in QDs with the energy-relaxation rate characteristic of the electron–LO-phonon scattering in bulk CdSe [shaded region in Fig. 3.7(b)].[33] In 4.1-nm QDs, the

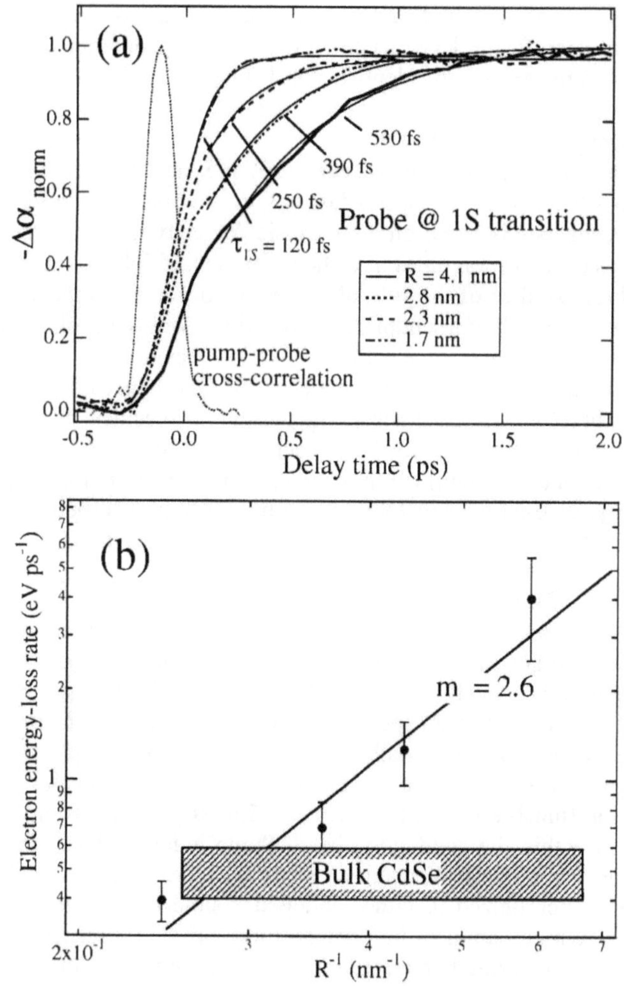

FIGURE 3.7. (a) The 1S-bleaching build-up dynamics in CdSe QDs with radii 1.7, 2.3, 2.8, and 4.1 nm (thick lines), fit to a single-exponential growth (thin solid lines). The thin dotted line is the pump-probe cross-correlation. (b) Size dependence of the electron energy-loss rate in CdSe QDs (symbols) in comparison to the $R^{-2.6}$ fit (line). The shaded region shows the range of energy-loss rates characteristic of the electron-LO-phonon scattering in bulk CdSe.

energy-loss rate is ~0.44 eV ps^{-1}. While this rate is close to rates expected for a bulk material, it is many orders of magnitude greater than rates estimated for a multiphonon emission.[54] The energy-loss rate rapidly increases (roughly as $R^{-2.6}$) with decreasing QD size [Fig. 3.7(b)] and is ~4 eV ps^{-1} for 1.7-nm dots, which is a rate almost an order of magnitude greater than for LO phonon emission in bulk materials.

Extremely fast electron relaxation, as well as a confinement-induced enhancement in the relaxation process, clearly indicate that energy relaxation in QDs is dominated by *nonphonon* energy-loss mechanisms. Recent works have suggested that coupling to defects,[55] Auger interactions with carriers outside the QD,[56] or Auger-type e-h energy transfer[57] can lead to fast energy relaxation not limited by a phonon bottleneck. The first two mechanisms are not intrinsic to QDs and cannot explain relaxation data for colloidal samples.[33] These data indicate that in colloidal nanoparticles, energy relaxation does not show any significant dependence on QD surface properties (i.e., the number of surface defects) and remain almost identical for different liquid- and solid-state matrices, including transparent optically-passive glasses, polymers, and organic solvents, for which no carriers are generated outside the QD.

The energy-loss mechanism proposed in Ref. 57 involves transfer of the electron excess energy to a hole, with subsequent fast hole relaxation through its dense spectrum of states. This mechanism is based on the intrinsic Auger-type e-h interactions and leads to significantly faster relaxation times than those for the multiphonon emission.

Most direct studies of the role of e-h interactions in intraband relaxation have used CdSe QDs in which e-h coupling (separation) was controlled by the surface ligand.[23,58] In Ref. 58, the rate of electron relaxation was studied as a function of e-h separation using either core/shell TOPO-capped CdSe QDs overcoated with ZnS or CdSe QDs passivated directly with hole-accepting pyridine molecules. Since a ZnS surface layer creates a confining potential for both electrons and holes, in ZnS-overcoated dots, the e-h coupling should not significantly change during electron/hole relaxation. On the other hand, in pyridine-passivated QDs, the e-h coupling is strong immediately after photoexcitation but is reduced dramatically after the hole is transferred to a capping group. This change in the coupling strength should have a strong effect on electron intraband relaxation, if it is indeed due to the e-h interactions.

In order to monitor electron intraband dynamics at different stages of hole relaxation/transfer, studies in Ref. 58 used a three-pulse fs TA experiment [Fig. 3.8(a)] in which the sample was excited by a sequence of two ultrashort pulses [one in the visible and the second in the infrared (IR) spectral range] and was probed by broad-band pulses of a fs white-light continuum. The visible interband pump (3.1 eV photon energy; 100 fs pulse duration) was used to create an e-h pair in the QD, whereas a time-delayed, intraband IR pump (tunable within 0.49–1.1 eV; 250 fs pulse duration) was used to re-excite an electron within the conduction band. Pump-induced absorption changes were detected over the range 1.2–3 eV using a chirp-free phase-sensitive technique[44] with a lock-in amplifier synchronized to either chopped visible or chopped IR pump beams. Signals obtained by modulation of the visible pump provide information on absorption changes induced by either interband excitation alone (IR pump is "off"), or by combined inter- and intraband excitations (IR pump is "on"). Modulating the IR pump beam, one can directly detect changes in the *interband* TA due to the *intraband* re-excitation.

In Fig. 3.8(b) we show 1S bleaching dynamics in ZnS-capped QDs (R = 1.15 nm) recorded with a modulation of either the visible pump [dashed line (IR pump is off)

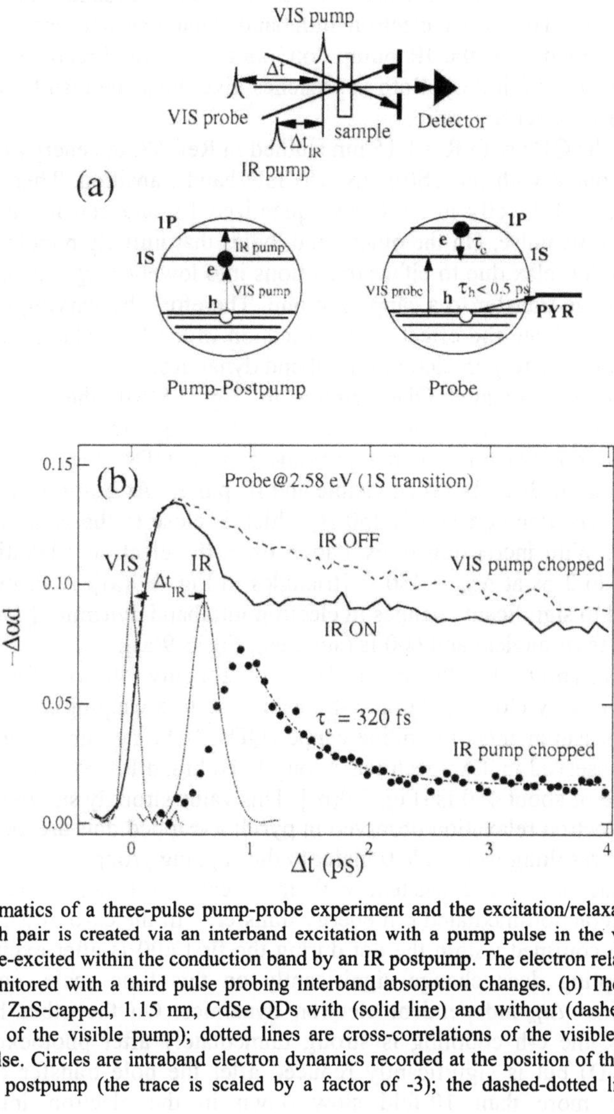

FIGURE 3.8. (a) Schematics of a three-pulse pump-probe experiment and the excitation/relaxation processes monitored in it. An e-h pair is created via an interband excitation with a pump pulse in the visible spectral range. The electron is re-excited within the conduction band by an IR postpump. The electron relaxation back to the ground state is monitored with a third pulse probing interband absorption changes. (b) The 1S bleaching dynamics recorded for ZnS-capped, 1.15 nm, CdSe QDs with (solid line) and without (dashed line) the IR postpump (modulation of the visible pump); dotted lines are cross-correlations of the visible and IR pump pulses with a probe pulse. Circles are intraband electron dynamics recorded at the position of the 1S bleaching by modulating the IR postpump (the trace is scaled by a factor of -3); the dashed-dotted line is a single exponential fit.

and solid line (IR pump is on)] or the IR pump (circles). The delay between the visible and the IR pump pulses (Δt_{IR}) is 0.6 ps, and the IR postpump photon energy is 0.55 eV, which corresponds to the intra-conduction-band 1S–1P resonance. In the absence of the IR pump, the 1S bleaching shows a fast, ~150 fs growth followed by a monotonic, nonexponential decay due to depopulation of the 1S electron state (dashed line). Application of the IR pulse leads to an abrupt decrease in the 1S bleaching due to re-excitation of the 1S electrons into the 1P state. A decay of perturbations associated with the IR pump provides a direct measure of the electron intraband relaxation back into the 1S ground state. To extract electron intraband dynamics one can either subtract time transients recorded with the IR pump "on" and "off", or directly record them using a modulated IR beam (circles). Both approaches give the same result; however the second method yields a better S/N ratio.

For the QDs with R = 1.15 nm studied in Ref. 58, the energy of the visible pump is almost resonant with the 1S(e)–$2S_{3/2}$(h) interband transition. Therefore, the electrons that are generated directly in the lowest quantized 1S state remain in this state until re-excited by the IR pulse. On the other hand, holes that initially populate the excited $2S_{3/2}$ state can rapidly relax due to either transitions into lower-energy, quantized states inside the dot or charge-transfer to a capping group. Therefore, by varying the IR re-excitation time one can evaluate the effect of modification in the hole charge distribution (during hole relaxation/transfer) on electron intraband dynamics.

In ZnS-overcoated QDs [circles in Fig. 3.8(b)], the 1P-to-1S relaxation is extremely fast (320 fs time constant) for all Δt_{IR} delays from 100 fs to 1 ps. In sharp contrast, electron dynamics in the pyridine-capped QDs [Fig. 3.9(a)] show a strong dependence on the delay between visible and IR pulses. At Δt_{IR} = 70 fs [diamonds in Fig. 3.9(a)], the relaxation constant is 250 fs, which is close to the value measured for ZnS-capped QDs. With increasing re-excitation time, the electron relaxation time gradually increases up to 3 ps at Δt_{IR} = 430 fs [triangles in Fig. 3.9(a)]. A further increase in Δt_{IR} does not lead to significant changes in electron intraband dynamics [compare traces taken at Δt_{IR} = 430 fs (triangles) and 600 fs (squares); Fig. 3.9(a)].

Importantly, the threshold delay of roughly 400 fs found in three-pulse experiments is very close to the hole-transfer time to a capping molecule inferred from visible TA measurements of pyridine-capped QDs.[58] The pyridine-cation (formed by hole transfer) is observed in TA spectra as a long-lived broad PA band below ~2.5 eV, and it has a risetime of about 450 fs [Fig. 3.9(b)]. This value strongly suggests that the dramatic changes in electron relaxation observed in pyridine-capped dots are due to changes in the e-h coupling, resulting from hole transfer to the capping group.

These data are consistent with the electron-relaxation mechanism involving Coulomb-interaction-mediated e-h energy transfer.[57] In the case of ZnS-capped samples, the holes are confined within the dot during the first 600 fs after excitation.[46] Therefore the e-h coupling does change significantly on this time scale, consistent with our observations of Δt_{IR}-independent electron dynamics. On the other hand, in pyridine-capped QDs, the e-h coupling is strong immediately after photoexcitation (holes are inside the QD) but is significantly reduced after the hole transfer to pyridine, which explains the more than 10-fold slow down in the electron intraband relaxation [Fig. 3.9(a)]. However, even in the charge-separated system, the electron relaxes back to the ground state on a ps time scale that is much faster than expected for multiphonon processes. This relatively fast relaxation indicates that even in the case of a significant spatial separation between an electron and a hole, Coulomb e-h interactions can be strong enough to provide an efficient channel for electron energy losses [the Coulomb coupling does not require a direct overlap between electron and hole wave functions and scales relatively slowly (inversely) with the e-h separation].

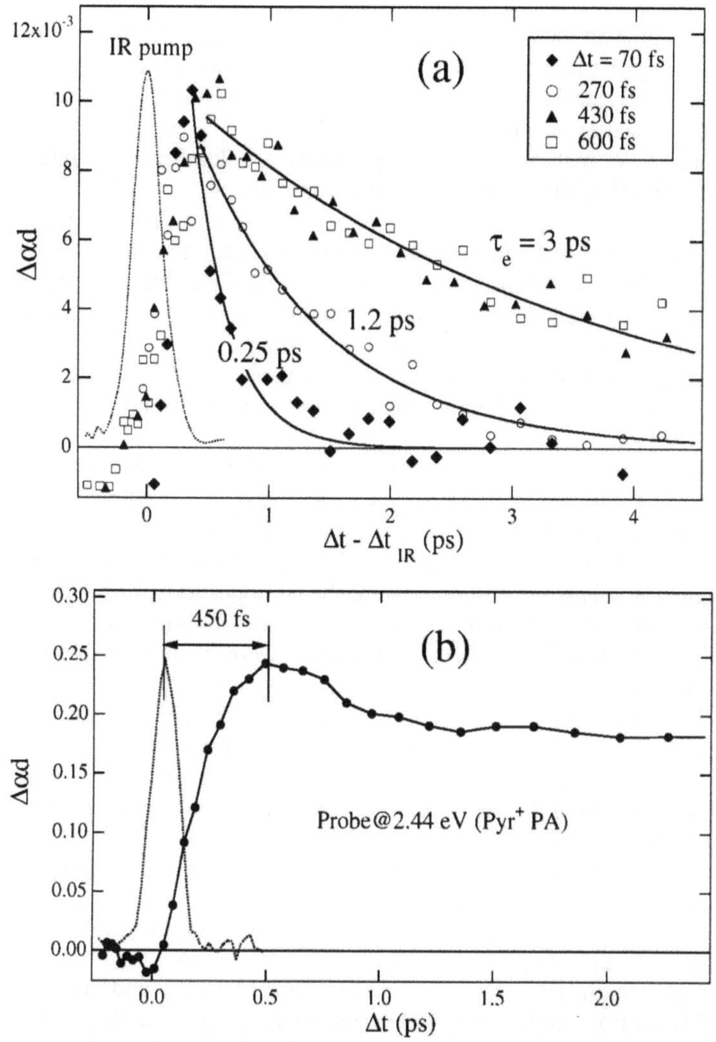

FIGURE 3.9. (a) Dynamics of the IR-postpump-induced 1S bleaching changes (electron intra-band relaxation) in pyridine-capped 1.15-nm CdSe QDs detected at different delay times (Δt_{IR}) between the visible and IR pump which correspond to different stages of hole transfer to a capping molecule (symbols are experimental data points, lines are single-exponential fits). (b) Dynamics of the pyridine-cation-related PA (circles) detected in the traditional two-pulse pump-probe experiment. The dotted line is the pump-probe cross-correlation.

3.4.2. Intraband hole relaxation and the structure of valence-band states

The data discussed in the previous section provide strong evidence that electron energy losses in small-size QDs are dominated by Auger-type energy transfer from electrons to holes. In contrast, the reverse process (energy transfer from holes to electrons) is not efficient due to small inter-level separations in the valence band. The energy that would be released during a hole "jump" to a lower energy state is not sufficient to re-excite an electron to a higher lying state. Because of this "unidirectional" nature of the energy transfer, energy dissipation from a photoexcited e-h system in QDs should ultimately rely on hole interactions with QD lattice vibrations (phonons) or with matrix/solvent molecules outside the dot.

In bulk II-VI materials, hole energy losses are dominated by interactions with LO phonons.[48] In QDs, the efficiency of these interactions is predicted to be strongly modified due to the discrete structure of energy states (which makes it more difficult to satisfy energy conservation) and to changes in the strength of carrier-phonon coupling. For example, carrier polar coupling to LO phonons (the Fröhlich mechanism) should be strongly reduced compared to bulk systems,[59,60] while coupling via deformation potential is predicted to be enhanced by a factor that is proportional to R^{-2}.[61,62]

Experimental studies of hole intraband dynamics are complicated by difficulties in detecting valence-band populations. Most carrier intraband relaxation studies in QDs have been performed using femtosecond (fs) TA experiments.[23,42,43] However, holes are only weakly pronounced in TA spectra because of a high spectral density of valence band states resulting in a significant spread of hole populations over multiple levels. One of the approaches to detecting hole intraband relaxation is by monitoring ultra-fast dynamics of "hot" PL.[63] In contrast to state-filling-induced TA signals that are proportional to the sum of the electron and hole occupation numbers, PL is proportional to the product of these numbers. Therefore, electron and hole dynamics can, in principle, be decoupled by performing both PL and TA studies. Such studies were conducted in Ref. 63. In this work, sub-ps time resolution in PL measurements was achieved using a fs up-conversion experiment.[64] In this experiment, a sample emission is frequency mixed with an ultra-short gating pulse in a nonlinear optical crystal. The sum-frequency signal is proportional to an instant PL intensity at the moment defined by an arrival time of the gating pulse. By scanning the gating pulse with respect to the PL pulse, one can obtain information about a PL temporal profile.

The pump photon energy used in Ref. 63 was close to the energy of the 1S(e)–$3S_{3/2}$(h) transition (~3.2 eV). Therefore, for a significant number of dots in the sample, electrons were generated directly in the lowest 1S state. For such dots, photogenerated holes are very "hot" and carry a large excess energy of ~0.9 eV. "Hot" hole dynamics are well pronounced in time-resolved PL spectra [Fig. 3.10(a)] and indicate a gradual relaxation of holes from high-energy states (manifested, e.g., as a "hot" PL peak at ~2.8 eV) to the lowest "emitting" state giving rise to the band-edge PL band.

A descent of holes through the ladder of the valence band states is easily seen in PL time transients recorded at different spectral energies [Fig. 3.10(b)]. These transients show a progressive increase in the delay (t_d) of the PL maximum with reduction in the detection energy (E) as indicated by the plot in Fig. 3.11 (solid circles). The derivative, dE/dt_d, provides a measure of the hole energy-loss rate. For 1.8-nm dots, the energy relaxation rate is nearly constant (1.5 eV ps^{-1}) between 2.3 eV and 3.1 eV, but reduces dramatically (down to ~0.26 eV ps^{-1}) below the 1S absorption peak, i.e., at the stage of final hole relaxation between "absorbing" and "emitting" fine-structure states [see schematics of hole energy levels in Fig. 3.1(a)]. Two stages in hole relaxation are also observed for dots with radii 1.2 and 3 nm (Fig. 3.11, open squares and open triangles,

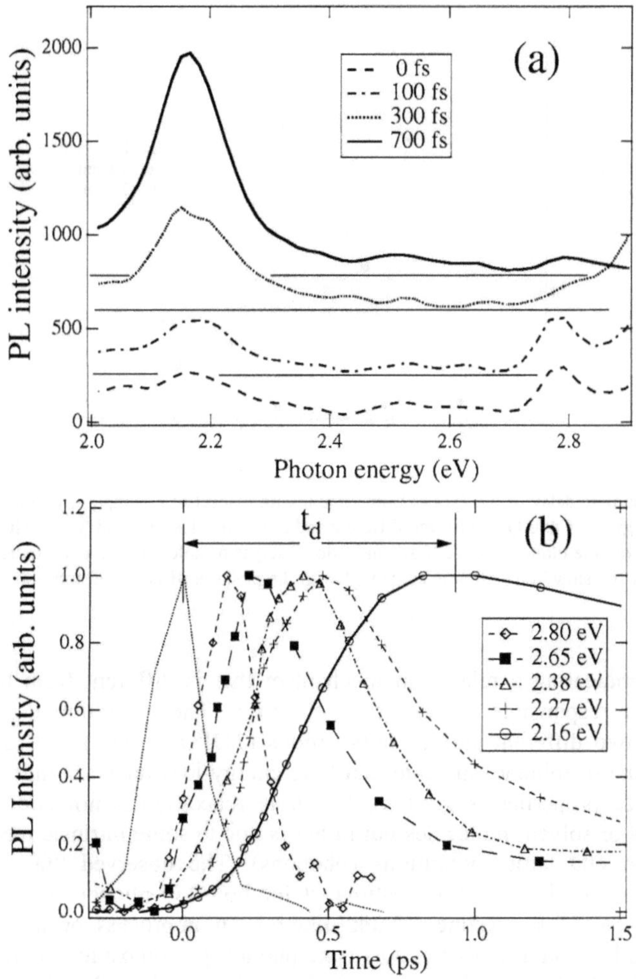

FIGURE 3.10. (a) Time-resolved PL spectra of CdSe QDs (R = 1.8 nm) detected at 0, 100, 300, and 700 fs after excitation (spectra arbitrarily offset for clarity). (b) Dynamics of "hot" PL detected at different spectral energies (symbols) along with a pump-pulse autocorrelation (dotted line).

respectively). The spectral onset of the "slow" relaxation region is size dependent, closely following the position of the 1S absorption peak. The "fast" and "slow" relaxation rates were in the range 1.3 to 1.8 eV ps^{-1} and 0.19 to 0.3 eV ps^{-1}, respectively.

In Fig. 3.12(a) we compare size-dependent relaxation rates observed for hole and electrons [electron relaxation data are from Ref. 33; see also Fig. 3.7(b)]. In contrast to electron relaxation rates that increase with decreasing dot size, hole rates show an opposite trend. In this case, both the "fast" and the "slow" rates decrease as the dot radius

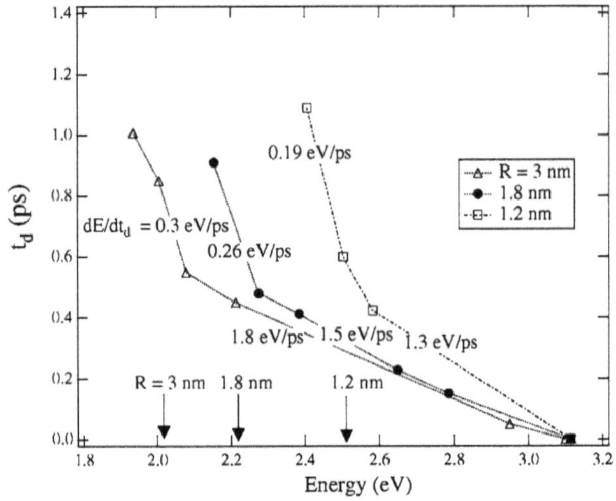

FIGURE 3.11. The time delay of the "hot" PL maximum with respect to a pump pulse plotted as a function of the detection energy for CdSe QDs of three different radii: 1.2 nm, 1.8 nm, and 3 nm. The energy-loss rates (dE/dt_d) derived from the plotted data at "fast" and "slow" stages of hole relaxation are indicated in the figure. The spectral onset of "slow" relaxation is compared with the position of the 1S absorption resonance (marked by arrows).

is decreased, indicating a relaxation mechanism that is different from the Auger-type energy transfer responsible for electron relaxation. The relaxation data obtained for samples with two different surface passivations (TOPO or ZnS) and in two different solvents (hexane or toluene) indicate that hole energy-loss rates are independent of QD surface/interface properties, suggesting that hole relaxation is not due to coupling to surface defects or solvent molecules but rather is due to some intrinsic mechanisms such as coupling to QD lattice vibrations (phonons). The observed "fast" hole energy-relaxation rates are close to those estimated for hole-LO phonon interactions in bulk CdSe (~1.4 eV ps^{-1}),[49] suggesting a "bulk"-like relaxation process, or more specifically, a cascade of single-phonon emission acts (one phonon per relaxation step). The fact that phonon emission by "hot" holes is apparently not hindered by the discrete character of the energy levels implies that the valence band states form a very dense spectrum (*quasi-continuum*) at spectral energies above the 1S absorption peak. Large effective hole masses, the existence of three valence subbands strongly intermixed by quantum confinement,[25] and the fine-structure splitting of valence band states[30] are all factors that can lead to a high density of hole states [the effect of fine-structure splitting is qualitatively illustrated in Fig. 3.12(b) taking into account S- and P-type hole states; an even denser spectrum is expected if the states of other symmetries (D, F, etc.) are also taken into account]. The hole energy structures are further smeared out by broadening due to, e.g., dephasing induced by elastic carrier-phonon scattering.[65] Another factor that simplifies the process of meeting energy conservation requirements in QDs is the relaxation of momentum conservation which allows for a wider spread of *k*-vectors (i.e., phonon energies) in phonon-assisted processes.

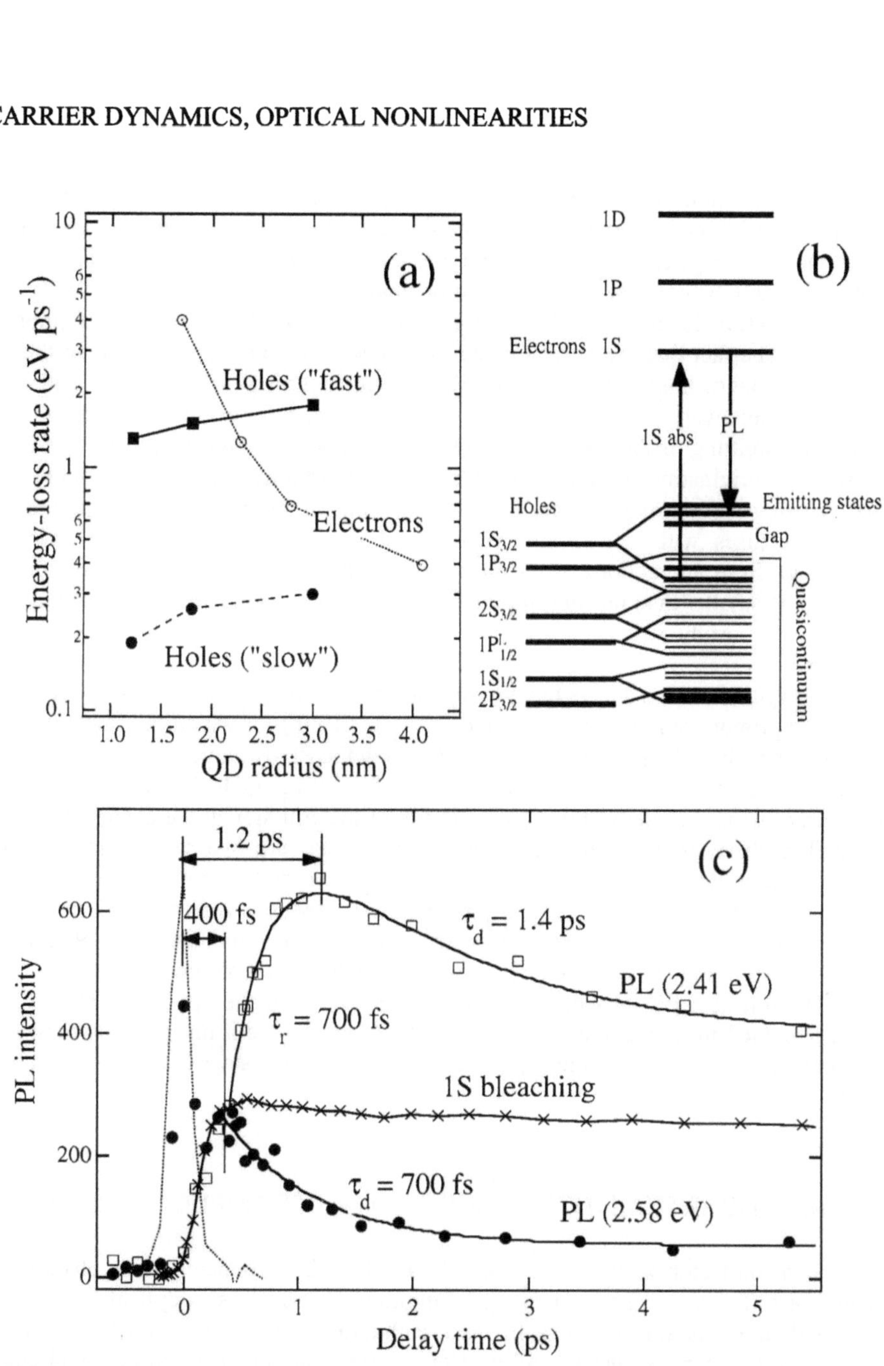

FIGURE 3.12. (a) Energy-loss rates as a function of QD radius for "fast" (solid squares) and "slow" (solid circles) stages of hole relaxation, and for the 1P-to-1S electron relaxation (open circles; from Ref. 33). (b) Schematic illustration of the formation of the valence-band quasicontinuum separated by an energy gap from the lowest "emitting" hole states. Arrows show the transitions observed in optical spectra as the 1S absorption peak and the band-edge PL. (c) Complementary PL dynamics detected at the positions of the "absorbing" (solid circles) and "emitting" (open squares) transitions for 1.2 nm CdSe QDs (dotted line is a pump-pulse autocorrelation). The PL traces are compared to the 1S bleaching dynamics (crosses) detected using a TA experiment.

The reduction in the energy-loss rate in the final stage of hole relaxation (solid circles in Fig. 3.11) can be explained by the fact that the hole spectrum becomes more sparse at the band-edge energies, i.e., at energies close to the 1S absorption resonance. In the case of large inter-level separations, two or more phonons are required to satisfy the energy conservation during hole relaxation, leading to a reduced relaxation rate ("phonon bottleneck"). The fact that the onset for reduced rates occurs at energies close to the 1S absorption resonance implies that the "absorbing" and "emitting" valence-band states are separated by a relatively wide energy gap that can be formed, for example, as a result of a fine-structure splitting of the $1S_{3/2}$ level.[30] One may argue that the superposition of fine-structure states originating from different "unperturbed" valence-band levels should close this gap. However, the strength of the e-h exchange (one of the factors contributing to the splitting) decreases with decreasing wave-function overlap. Therefore, the splitting resulting, e.g., from interactions of the first excited $1P_{3/2}$ hole state with the electron 1S state should be smaller than the splitting of the S-symmetry ground hole state, leaving the gap at the bottom of the valence band at least partially open [see schematic illustration in Fig. 3.12(b)].

Slowed relaxation across the gap separating the "emitting" and "absorbing" hole states is well pronounced in very small-size QDs ($R \approx 1$ nm) for which the fine structure splitting is particularly large. In Fig. 3.12(c), we show the PL dynamics at the positions of the 1S absorption and the PL maximum recorded for 1.2-nm QDs. The 1S PL shows a 700-fs decay that is complementary to the growth of the emission at the center of the cw PL band. This decay is consistent with a hole relaxing between two, well-defined levels. In the same plot we show 1S bleaching dynamics recorded using a TA experiment (crosses). A nearly flat 1S bleaching signal at times >500 fs is indicative of an almost constant population of the 1S electron states, which further indicates that initial PL dynamics shown in Fig. 3.12(c) are due entirely to a hole relaxation process.

The ultrafast PL data discussed above indicate that the "phonon bottleneck," which is bypassed in the conduction band due to Auger-type e-h interactions, may still significantly affect hole dynamics, particularly in the final stage of hole relaxation between the manifolds of "absorbing" and "emitting" states.

3.5. MULTIPARTICLE DYNAMICS AND LIGHT AMPLIFICATION

Semiconductor materials are widely used in both optically and electrically pumped lasers. The use of semiconductor quantum well structures as optical gain media has resulted in important advances in laser technology. Quantum wells have a two-dimensional (2D), step-like density of electronic states that is nonzero at the band-edge, enabling a higher concentration of carriers to contribute to the band-edge emission and leading to a reduced lasing threshold, improved temperature stability, and a narrower emission line. A further enhancement in the density of the band-edge states and an associated reduction in the lasing threshold are, in principle, possible using quantum wires and QDs in which the confinement is in two and three dimensions, respectively. In very small dots, the spacing of the electronic states is much greater than the available thermal energy (strong confinement), inhibiting thermal depopulation of the lowest electronic states. This effect should result in a lasing threshold that is temperature insensitive at an excitation level of only one e-h pair per dot on average.[66] Additionally, QDs in the strong confinement regime have an emission wavelength that is a pronounced function of size, adding the advantage of continuous spectral tunability over a wide energy range simply by changing the size of the dots.

Lasing has previously been demonstrated for epitaxially grown III-V QDs.[67,68] These dots have relatively large lateral sizes (typically greater than 10 nm); therefore, the spacing between their electronic states is smaller than room-temperature carrier energies (weak confinement), and the lasing threshold is temperature sensitive. In addition, large lateral dimensions and difficulties in size control restrict the spectral tunability of these dots using quantum confinement effects. As a result, emission wavelengths in epitaxial dots are typically controlled by a material's composition rather than QD size.

On the other hand, direct colloidal syntheses provide routine preparations of nanocrystal QDs with sub-10 nm sizes that correspond to the regime of extremely strong confinement. For this regime, size-controlled spectral tunability over an energy range as wide as 1 eV can be achieved. The prospect of realizing lasers for which the output color can be controlled by facile manipulation of the dot size has been a strong driving force in colloidal QD research for more than a decade.

Despite providing indications of optical gain performance,[69,70] strongly confined QDs have failed to yield lasing in numerous efforts. Difficulties in achieving lasing have often been attributed to high nonradiative carrier losses due to trapping at surface defects, a direct consequence of the large surface-to-volume ratio characteristic of sub-10 nm particles. However, as was shown in Ref. 17, carrier surface trapping turns out to be of much lesser importance compared to such largely unforeseen problems as nonradiative, multiparticle Auger recombination[47] and interference from excited state absorption (PA) due to carriers trapped at QD interfaces. These two effects are analyzed below. This analysis indicates that although complicating realization of lasing, both of the problems can be circumvented by fabricating films of close-packed QDs (QD solids). These films do not show PA at the position of the emitting transition and exhibit large optical gain with a magnitude that is sufficiently high to successfully compete with multiparticle Auger decay. QD films show narrow-band amplified spontaneous emission (ASE) at both cryogenic and room temperatures, and the emission color is tunable with dot size. Moreover, the QD films can be incorporated into microcavities such as micro-capillary tubes that produce lasing into whispering gallery modes.

3.5.1. *Optical gain and excited state absorption*

In order to detect light amplification/attenuation, one can use a TA experiment [see section 3.3.2] that allows one to monitor the absorption of the sample with (α) and without (α_0) pump. In absorption spectra, optical gain corresponds to $\alpha < 0$, i.e., to pump-induced absorption bleaching ($\Delta\alpha = \alpha - \alpha_0 < 0$) that is greater than α_0 ($-\Delta\alpha/\alpha_0 > 1$).

Many initial attempts to observe QD optical gain (such as one in Ref. 46) used solution samples. In Fig. 3.13(a) we show absorption spectra of a hexane solution of CdSe QDs (R = 1.2 nm) recorded at 2 ps after excitation for progressively higher pump intensities (the pump photon energy is 3.1 eV). These spectra do not show any evidence for gain ($\alpha < 0$) even at the highest pump density of $\langle N_0 \rangle = 6$. Instead, the 1S absorption bleaching saturates slightly below a level $|\Delta\alpha|/\alpha_0 \approx 1$, i.e., right before a crossover to optical gain [Fig. 3.13(b), open squares], consistent with previously-discussed data in Fig. 3.6(b). As analyzed in section 3.3.2, this behavior indicates that the 1S transition is bleached by only one type of carriers (electrons), which is consistent with fs PL data [Fig. 3.12(c)] indicating very fast hole relaxation between the "absorbing" (responsible for the 1S absorption) to the lower energy "emitting" (involved in the PL transition) fine-structure states.

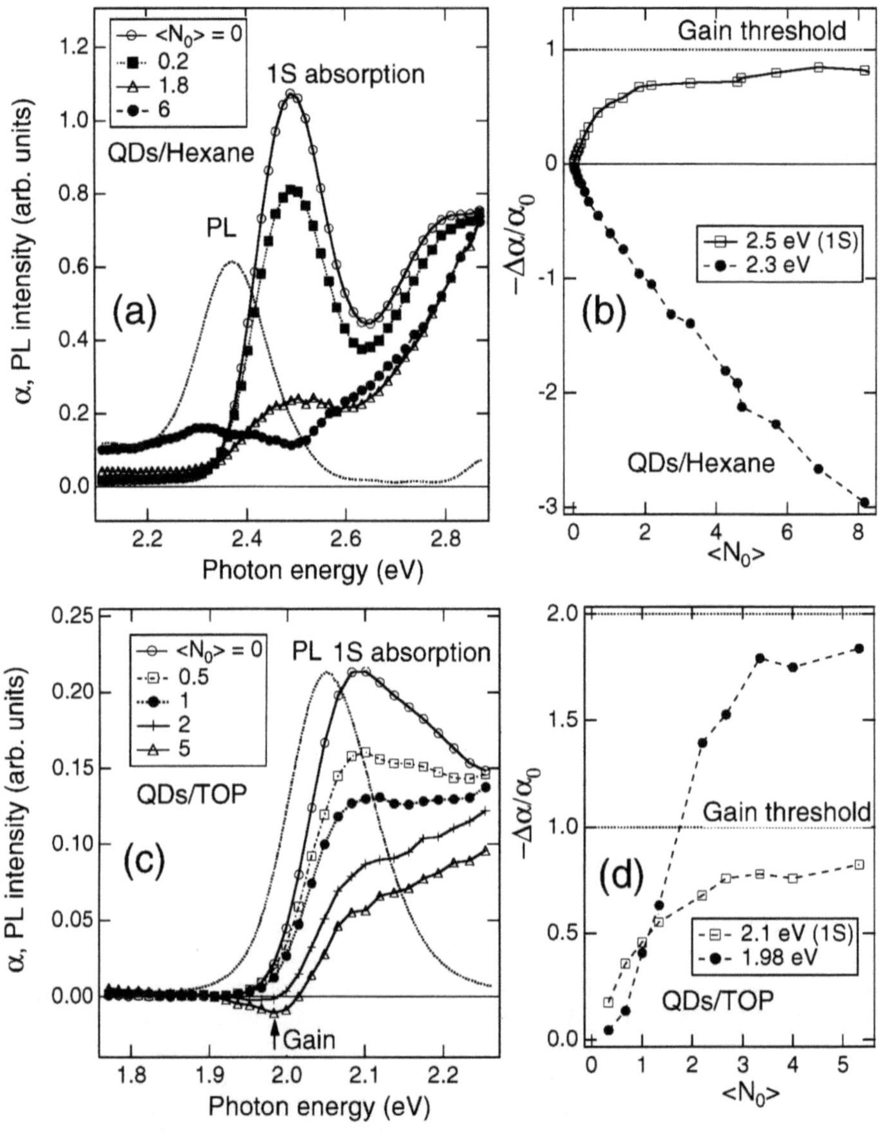

FIGURE 3.13. Pump-intensity dependent absorption spectra (symbols) of CdSe QDs in (a) hexane and (c) TOP solutions in comparison to the emission spectra (dotted line). Pump-intensity dependence of normalized absorption changes at the positions of the 1S bleaching (squares) and PL (circles) for (b) hexane and (d) TOP solutions. QD mean radii R = 1.2 nm (a, b) and 2.3 nm (c, d).

However, for QD/hexane samples, gain at the position of the "emitting" transition is not detected either. In the region of this transition, CdSe QD/hexane samples show *increased* absorption [Fig. 3.13(a)]. In contrast to 1S bleaching, which saturates at high pump intensities, PA does not show saturation [circles in Fig. 3.13(b)] and, therefore, cannot be circumvented by simply increasing the excitation density.

Analysis *of* data for QDs in different liquid and solid-state matrices indicates that the PA is dependent on the matrix material, suggesting that it is due to excited-state absorption involving carriers trapped at matrix/solvent-related interface states. Such commonly used solvents as toluene, chloroform and heptamethylnonane show a PA band comparable in intensity with that in hexane. However, in the case of QDs dispersed in polyvinyl butyral, the PA is reduced in magnitude and is red shifted with respect to the PL band. The strongest PA suppression was observed for TOP, one of the QD growth solvents. A CdSe QD/TOP sample ($R = 2.3$ nm) in Fig. 3.13(c) shows gain at the position of the PL band. The transition from absorption to gain occurs at carrier densities between 1 and 2 e-h pairs per dot on average [circles in Fig. 3.13(d)], consistent with the 1 e-h pair theoretical threshold expected for 3D strongly confined systems.[66] The gain saturates at a level of $\sim 0.8\alpha_0$, corresponding to ~90% population inversion in terms of a simple two-level model. The maximum gain values detected in this experiment correspond to a gain cross section (σ_g) of $\sim 5 \times 10^{-17}$ cm^2 (the gain cross section is defined as $\sigma_g = G/n_0$, where G is the gain magnitude and n_0 is the density of QDs in the sample).

Recent studies indicate that in addition to the type of solvent material, optical gain performance is very sensitive to the excitation wavelenghth.[71] For example, for CdSe QD/hexane solutions, the PA below the 1S absorption peak can be suppressed by choosing the pump-photon energy in the vicinity of the 1S absorption resonance. In contrast to pumping at 3.1 eV which results in a strong PA band, the quasiresonant pumping into the 1S absorption peak produces optical gain. This result suggests that trapping at interface states that gives rise to excited state absorption occurs preferentially from higher lying (excited) QD states.

Despite the fact that CdSe QDs solution samples under certain conditions show optical gain, these samples do not exhibit ASE. In this case, as well as in other cases of relatively dilute QD systems (e.g., QD-doped glasses fabricated by high-temperature precipitation), the development of ASE is inhibited by ultrafast decay of the optical gain due to Auger effects as analyzed below.

3.5.2. Multiparticle Auger recombination and optical gain dynamics

If we model the band-edge emission in QDs using a two level system with two-fold spin-degenerate states, we find that optical gain (i.e., population inversion) begins at a carrier density of $\langle N \rangle = 1$, with gain saturation (i.e., complete population inversion) at $\langle N \rangle = 2$. These values imply that the QD band-edge gain should be primarily due to two e-h pair states (i.e., due to quantum confined biexcitons).

Studies of multiparticle dynamics[47] show that the intrinsic decay of multiple e-h pair states (including biexciton states) in CdSe QDs is dominated by Auger recombination. Auger recombination is a nonradiative process that leads to the recombination of e-h pairs via energy transfer to a third particle (an electron or a hole) that is re-excited to a higher energy state within the dot or outside it (Auger ionization).

Auger recombination has a relatively low efficiency in bulk semiconductors, for which significant thermal energies are required to activate the effect.[72,73] However, Auger decay is greatly enhanced in quantum confined systems, in which the relaxation in momentum conservation removes the activation barrier.[74,75]

In bulk semiconductors, Auger decay is characterized by a cubic carrier decay rate $C_A n_{eh}^3$, where C_A is the Auger constant, and n_{eh} the carrier density. In this case, one can also introduce an instant time constant, $\tau_A = (C_A n_{eh}^2)^{-1}$, that changes continuously as the carrier density is reduced during recombination. In contrast to this continuous decay, Auger recombination in QDs occurs via a sequence of "quantized" steps from N, to $N-1$, $N-2$, ... and finally to the 1 e-h pair state, with each step characterized by a discrete exponential decay constant [Fig. 3.14(a), inset]. The decay of multiple-pair states in QDs via a sequence of quantized steps can be described by the following set of coupled rate equations:

$$\begin{aligned} \frac{dn_N}{dt} &= -\frac{n_N}{\tau_N} \\ \frac{dn_{N-1}}{dt} &= \frac{n_N}{\tau_N} - \frac{n_{N-1}}{\tau_{N-1}} \\ &\cdots \\ \frac{dn_1}{dt} &= \frac{n_2}{\tau_2} - \frac{n_1}{\tau_1}, \end{aligned} \tag{3.5}$$

where n_i ($i = 1, 2\ldots N$) is the concentration of dots in the sample that contain i e-h pairs. If one neglects recombination during the pump pulse, these equations have a straightforward solution for time dependent QD average populations: $\langle N(t)\rangle = \sum_{i=1}^{N} A_i \exp(-t/\tau_i)$, where A_i are coefficients determined by initial carrier densities and τ_i is the lifetime of the i-pair QD state. In the case for which absorption saturation at the pump wavelength is small, the initial conditions for Eq. (3.5) can be calculated using the Poisson distribution:

$$n_m(t=0) = n_0 \frac{<N_0>}{m!} e^{-<N_0>}, \quad m=1,2,\ldots N. \tag{3.6}$$

Quantized steps during Auger decay in strongly confined CdSe QDs were experimently resolved in Ref. 47 using a fs TA experiment. In this experiment, dynamics of QD populations, $\langle N(t)\rangle$, were derived from time-resolved 1S bleaching signals using the "universal" relationship between $|\Delta\alpha|/\alpha_0$ and $\langle N\rangle$ given by Eq. (3.4) (section 3.3.2). Fig. 3.14(a) displays these dynamics measured for a 1.7-nm sample at carrier injection levels, $\langle N_0\rangle$, from 0.2 to 3.2. In Ref. 47, a simple subtractive procedure was used to extract *single* exponential dynamics that are characteristic of the decay of different multiple-pair QD states from the measured $\langle N(t)\rangle$ time transients. This procedure is based on the fact that at long times after photoexcitation, the decay is governed by singly-excited QDs, independent of the initial carrier density. This allows one to normalize time transients taken at different pump levels to match their long-time-delay "tails" as shown in Fig. 3.14(a). By subtracting the low-pump intensity trace ($\langle N \rangle < 1$; single e–h pair decay) from traces recorded at $\langle N_0 \rangle > 1$, it was possible to derive the dynamics, $\Delta N(t)$, due to relaxation of multiple-pair states with $N \geq 2$. For initial carrier densities $1 < \langle N_0 \rangle < 2$, this procedure yielded the two-pair state dynamics [open circles in Fig. 3.14(b)], with a faster initial component due to contribution from states with a larger number of excited pairs. By further subtracting two-pair decay from the $\Delta N(t)$ transients detected at $\langle N_0 \rangle > 2$, one could derive dynamics of states with $N \geq 3$. This procedure could be repeated to extract dynamics of QD states with $N = 4$, 5, etc.

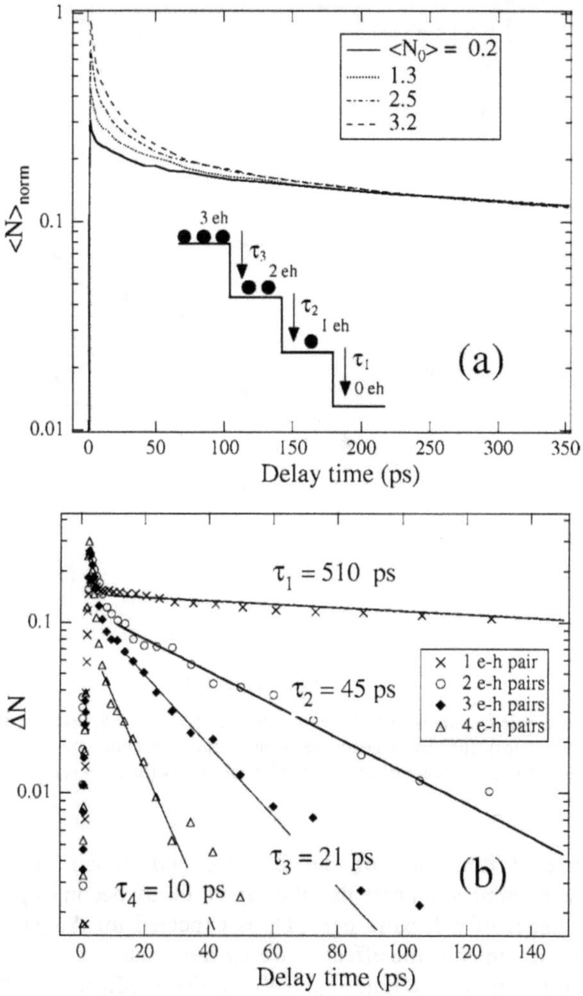

FIGURE 3.14. (a) Pump-dependent dynamics of QD average populations in the 1.7-nm sample normalized to match their "tails." Inset: Quantized steps in quantum-confined Auger recombination in QDs. (b) Dynamics of the 1–, 2–, 3–, and 4–e–h pair states (symbols) extracted from the TA data for the 1.7-nm sample (see text), fit to a single exponential decay (solid lines).

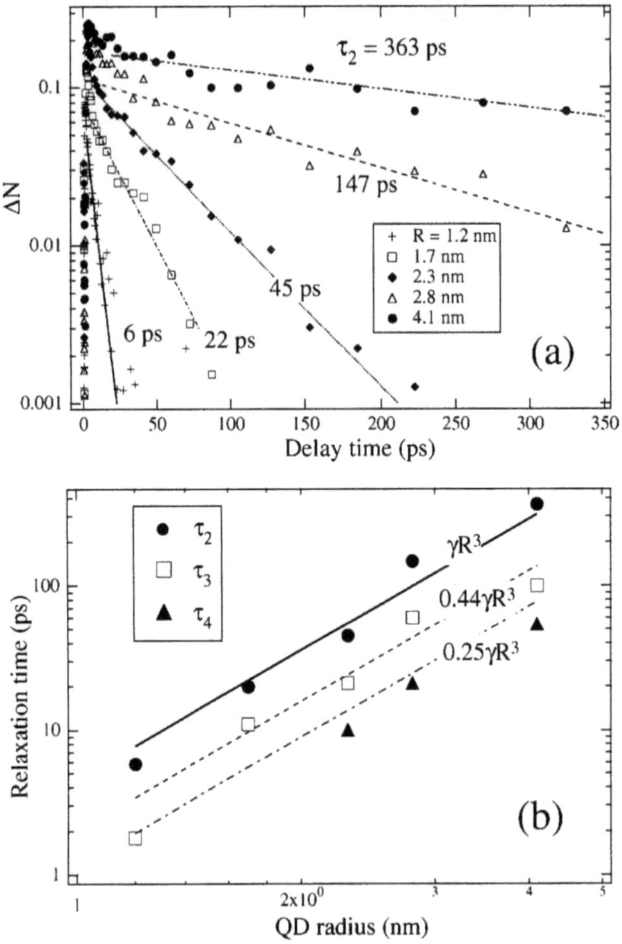

FIGURE 3.15. (a) 2–e–h pair dynamics in QDs with radii 1.2, 1.7, 2.3, 2.8, and 4.1 nm (symbols), fit to a single-exponential decay (lines). (b) Size-dependence of relaxation constants of 2–, 3–, and 4–e–h pair states (symbols), fit to the dependence γR^3, $0.44\gamma R^3$, and $0.25\gamma R^3$, respectively (lines); γ is a size-independent constant.

The extracted dynamics of the 2–, 3–, and 4–pair states are shown in Fig. 3.14(b). These dynamics indicate that the carrier decay becomes progressively faster with increasing number of e–h pairs per QD, as expected for Auger recombination. In bulk semiconductor arguments, the effective decay time constant, τ_N, in the Auger regime is given by the expression $\tau_N^{-1} = C_A(N/V_0)^2$ ($N \geq 2$) that predicts the following scaling of times for 4–, 3–, and 2–pair relaxation: τ_4: τ_3: τ_2 = 0.25:0.44:1. This scaling is very close to one observed experimentally (0.22:0.47:1), indicating that the decay rates for quantum-confined Auger recombination are cubic with respect to the carrier density ($dn_{eh}/dt \propto -n_{eh}^3$), just as in bulk materials.

TA data obtained for QDs of different sizes [Fig. 3.15(a)] indicate that the τ_2 time constant (the biexciton life time) rapidly decreases with decreasing QD size following a cubic size dependence ($\tau_2 \propto R^3$), as obvious from the plot in Fig. 3.15(b) (solid circles). The two-pair lifetime shortens from 363 ps to only 6 ps as the dot radius is decreased from 4.1 to 1.2 nm. Interestingly, the time constants measured for the 3– and 4–pair decay [open squares and solid triangles in Fig. 3.15(b), respectively] follow the same size dependence as that of the 2–pair state such that the time-constant ratios predicted by the bulk-semiconductor model hold for all QD sizes. Simple bulk-material reasoning would suggest that the enhancement in the Auger decay in smaller particles is caused by an effective increase in carrier concentrations resulting from the increased spatial confinement. This reasoning would predict the R^6 scaling for Auger times ($\tau_N \propto (N/V_0)^{-2} \propto R^6$). However, the experimentally-measured size dependence is R^3, indicating that in 3D-confined systems the Auger "constant" depends on the particle size. For CdSe QDs, C_A scales approximately as R^3 and decreases from ~7×10^{-29} cm^6 s^{-1} to ~2×10^{-30} cm^6 s^{-1} as the dot radius is reduced from 4.1 to 1.2 nm (Fig. 3.16). For all QD sizes, the C_A values calculated by using lifetimes of 2–, 3–, and 4–e–h pair states are close to

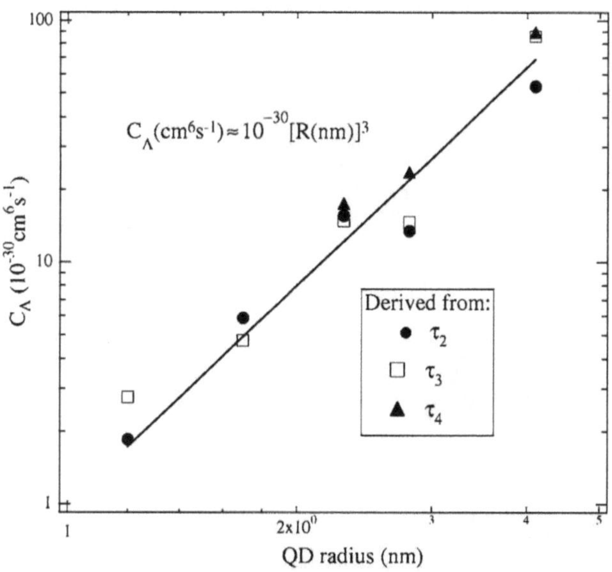

FIGURE 3.16. Size dependence of the Auger constant (symbols), fit to the dependence $C_A \propto R^3$ (line); circles, squares, and triangles correspond to data derived from the lifetimes of the 2–, 3–, and 4–pair states, respectively.

each other (compare data shown by circles, squares and triangles in Fig. 3.16), indicating that the cubic density dependence of the Auger rates holds for all sizes between 1 and 4 nm.

The effect of QD surface/interface properties on Auger rates was studied in Ref. 47 using colloidal samples with different types of surface passivation. Fig. 3.17 displays 1–pair (a) and 2–pair (b) dynamics for 1.7-nm CdSe QDs either capped with TOPO [squares (freshly-prepared sample) and circles (aged sample)] or overcoated with four monolayers of ZnS (crosses). As indicated by photoluminescence quantum-yield measurements for these three samples, the best surface passivation was provided by the epitaxial ZnS layer. Freshly prepared TOPO-capped samples had a lower degree of surface passivation, which was further reduced upon the sample aging. TA data taken at $\langle N_0 \rangle = 0.11$ [Fig. 3.17(a)] show that the reduction in the degree of surface passivation leads to faster initial dynamics of the single e-h pair state, indicating that these dynamics are affected by trapping at surface defects. In contrast, the 2 e–h pair decay [Fig. 3.17(b] is not significantly altered by changing surface/interface properties which strongly suggests that Auger recombination in CdSe NQDs is dominated by processes involving states inside the dot, without a significant contribution from the Auger ionization.

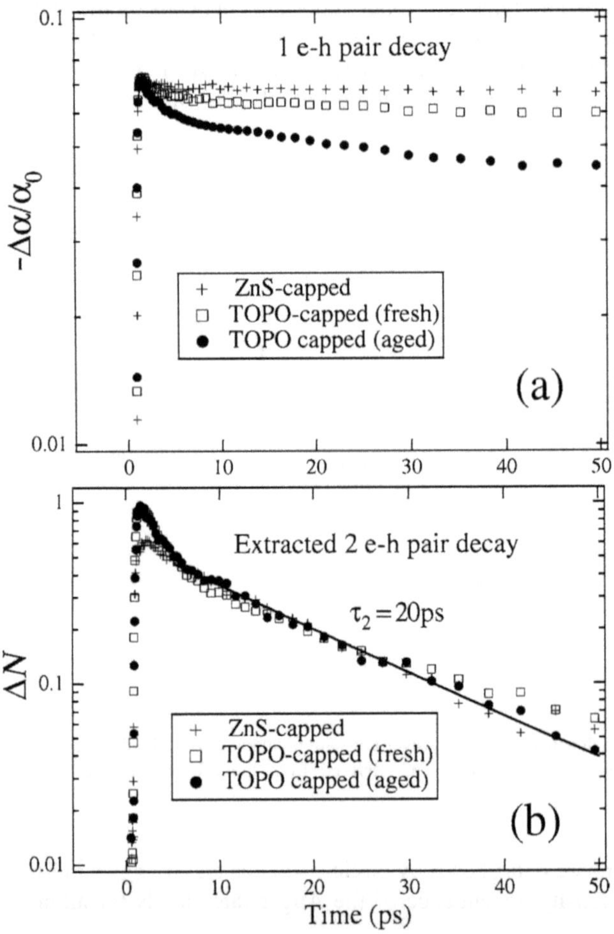

Figure 3.17. (a) The low-pump-intensity 1S-bleaching dynamics ($\langle N_0 \rangle = 0.11$) for CdSe QDs (R = 1.7 nm) with differently prepared surfaces: fresh (squares) and aged (circles) TOPO-capped QD samples, and a sample of QDs overcoated with a ZnS layer (crosses). (b) 2-e–h pair decay for the same set of samples.

Because of their strong size dependence, Auger effects, which play a minor role in large epitaxial dots, become significant in strongly confined colloidal nanoparticles.

For moderately well-passivated dots, nonradiative Auger relaxation of doubly excited dots (QD biexcitons) is more efficient than the surface trapping that imposes an *intrinsic* limit on the lifetime of the optical gain in strongly confined QDs. The dominance of the Auger effect in QDs also implies that even relatively poorly passivated samples could behave as well as the best samples with respect to gain and stimulated emission because as long as trapping rates are lower than Auger rates, the two e-h pair life time is determined by the intrinsic Auger decay.

3.5.3. *Amplified spontaneous emission and lasing*

To overcome the problem of intrinsic ultrafast gain decay, development of amplified emission must occur on time scales that are shorter than those for the Auger decay. The ASE buildup time (τ_{ASE}) is inversely proportional to the gain magnitude: $\tau_{ASE} = n_r/(G_C)$, where c is the light velocity and n_r is the sample refractive index. One can further rewrite this expression in terms of the gain cross section and the volume fraction (filling factor) of semiconductor material in the sample ($\xi = 4\pi n_0 R^3/3$):

$$\tau_{ASE} = \frac{4\pi R^3}{3}\frac{n_r}{\xi\sigma_g c}. \tag{3.7}$$

In the case for which the gain decay is dominated by the intrinsic two-pair Auger recombination (decay time τ_2), ASE can only be observed if $\tau_{ASE} < \tau_2$. The Auger data from Ref. 47 (see also previous section) indicate that τ_2 scales with dot size as R^3 ($\tau_2 \approx \beta R^3$, where $\beta \approx 5$ ps nm^{-3}) which allows one to derive the following critical condition for the development of ASE:

$$\xi > \frac{4\pi n_r}{3c\beta\sigma_g}. \tag{3.8}$$

For $\sigma_g = 10^{-17}$ cm^2 (a lower estimate for the gain cross section) and $n_r = 2$ (an upper estimate for the refractive index), Eq. (3.8) indicates that minimal filling factors required for achieving ASE are of the order of 0.005 (0.5%). Such semiconductor volume fractions are difficult to obtain in colloidal solutions or in solid-state, semiconductor-doped glasses that may explain numerous unsuccessful attempts to realize ASE and lasing in these types of samples.

Very high dot densities are readily achieved in close-packed QD films (QD solids). For example, films made of 1.3-nm CdSe QDs capped with TOPO have filling factors as high as ca. 20% (1.1 nm length of a capping molecule is assumed) for which the condition given by Eq. (3.8) is easily satisfied. Solid state films of close-packed CdSe QDs were used in Ref. 17 to demonstrate ASE. These films were fabricated by drop casting from hexane/octane solutions using either TOPO-capped or ZnS-overcoated dots. Transmission electron microscopy (TEM) indicated random close-packing of dots in the samples [Fig. 3.18(a)]. The films' emission color was tunable over an entire visible range by changing the dot size [Fig. 3.18(b)].

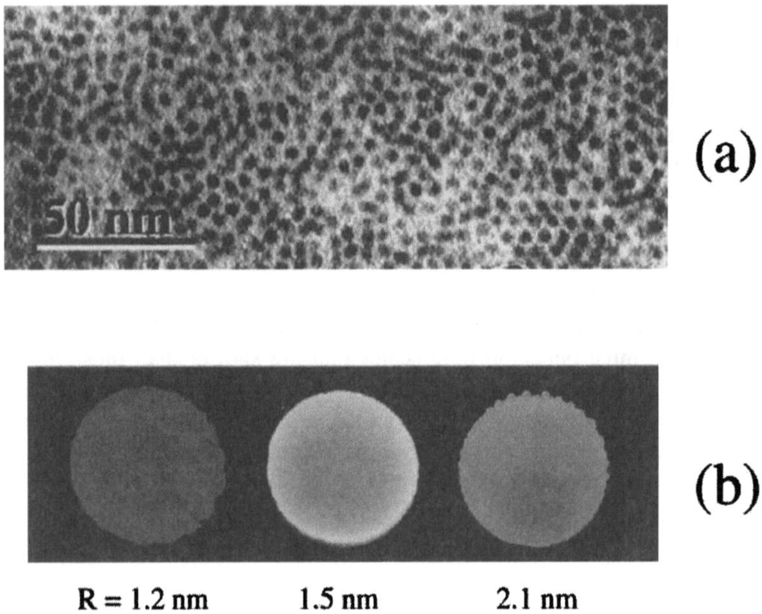

FIGURE 3.18. (a) A TEM image of a close-packed CdSe QD film. (b) Optical images of three CdSe QD films taken under the ultra-violet (UV) illumination; the films fabricated from dots with radii 1.2, 1.5, and 2.1 nm produce emission of green, yellow, and orange color, respectively.

TA measurements indicate that QD films do not exhibit PA at the position of the emitting transition but instead show optical gain at both liquid nitrogen and room temperature [Fig. 3.19(a)]. The gain threshold is close to the theoretical limit of one e–h pair per dot on average, independent of dot size. In terms of the excitation fluence, the gain threshold scales roughly as R^{-3} (the pump photon energy is ~3 eV), as expected from the R^3 scaling of QD absorption cross sections at spectral energies well above the energy gap.[41]

The QD films also show ASE as illustrated by pump intensity dependent emission spectra in Fig. 3.19(b) (TOPO-capped QDs with R = 2.1 nm; sample temperature 80 K). These spectra were taken using a frequency-doubled output of an amplified Ti-sapphire laser as a pump (the photon energy is 3.1 eV, the pulse duration is 100 fs); the ASE was detected at the edge of the film, which acts as an optical waveguide. At pump levels of approximately 1.5 e-h per dot, a sharp, ASE peak develops on the low energy side of the spontaneous emission band. The pump intensity dependence of this peak [Fig. 3.19(b), inset] shows a threshold behavior that is a clear signature of optical amplification. The ASE band is spectrally tunable over a range of ~300 meV by changing the dot size from 2.2 to 1.3 nm [Fig. 3.19(c)]. For relatively large dots (R > 2 nm), the observation of ASE does not require the use of ZnS-overcoated dots. TOPO-capped QDs have poorer surface passivation than CdSe/ZnS core-shell structures and hence substantially smaller PL quantum yields. However, because in relatively large dots the Auger

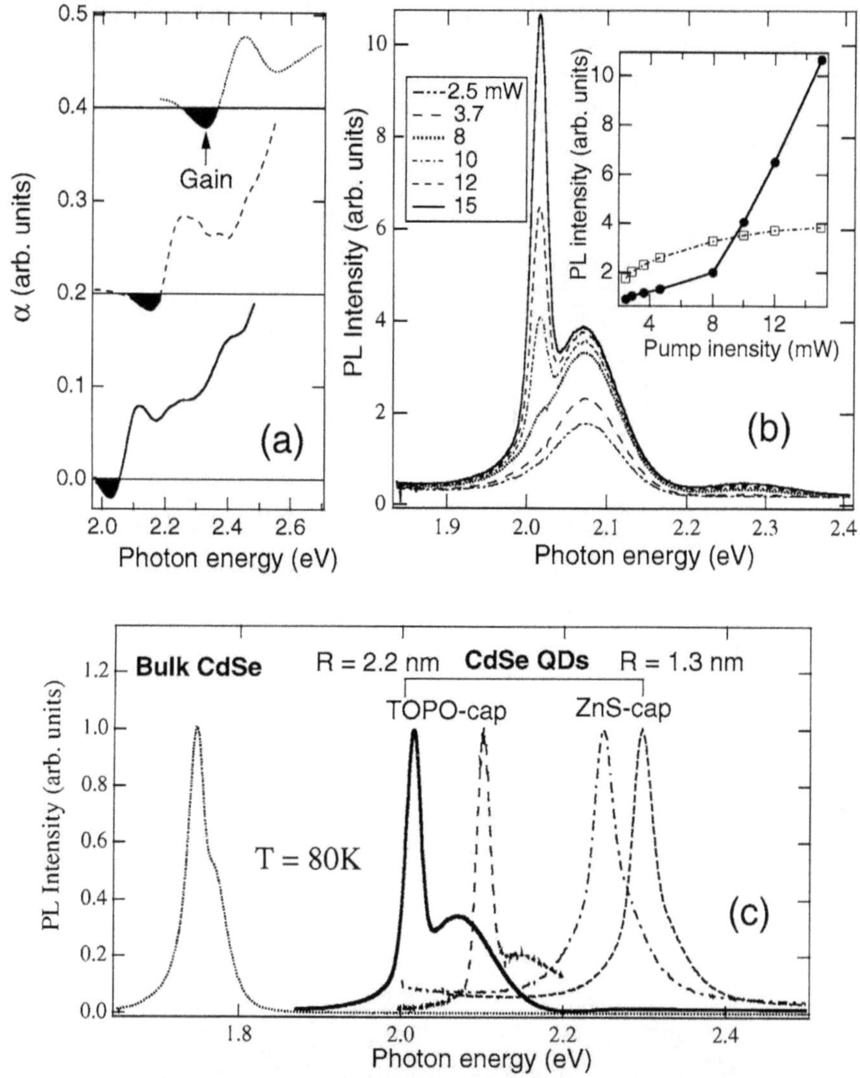

FIGURE 3.19. (a) Absorption/gain spectra of CdSe QD films fabricated from dots with R = 1.3, 1.7, and 2.1 nm (T = 300 K). (b) Development of a sharp ASE band as a function of pump intensity in PL spectra of the film fabricated from TOPO-capped CdSe QDs with R = 2.1 nm (T = 80 K). Inset: Superlinear intensity dependence of the ASE peak (circles) compared to the sublinear dependence of the PL intensity outside this peak. (c) Size-controlled ASE spectra of films fabricated from both TOPO- and ZnS-capped CdSe QDs in comparison with ASE in bulk CdSe (T = 80 K). All data shown in the figure were taken using 100-fs pump pulses at 3.1 eV.

timescale is faster than surface trapping, both types of samples (TOPO- and ZnS-capped) show similar properties in the ASE regime. Femtosecond PL data indicate that in TOPO-capped dots with $R < 2$ nm, hole surface trapping becomes extremely fast (time scales < 5 ps). Therefore, for these "ultrasmall" sizes, overcoating with ZnS is essential for achieving optical amplification.

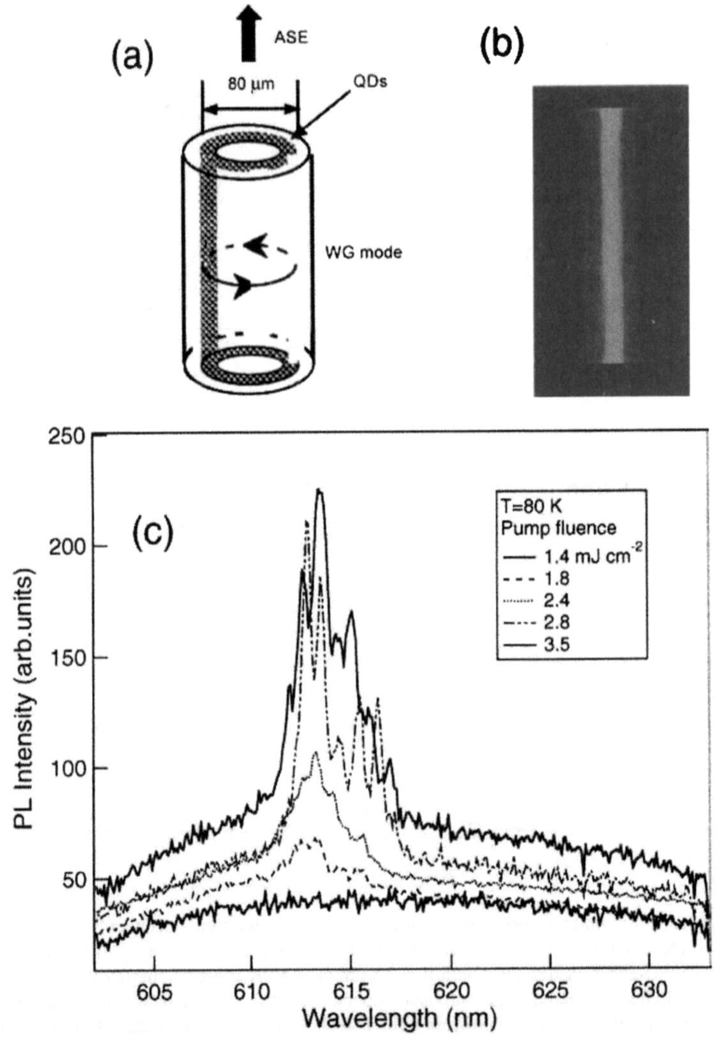

FIGURE 3.20. (a) A schematic and (b) a microphotograph of a QD layer deposited on the inner side of a glass microcapillary tube (the microphotograph was taken under the UV illumination). (c) Emission spectra of QDs incorporated into a microcapillary tube taken at progressively higher pump intensities (100-fs pump pulses with a photon energy of 3.1 eV).

QD films also show ASE at room temperature. Interestingly, the same pump fluences that are used to excite room temperature ASE in CdSe QDs are not sufficient to produce light amplification in bulk CdSe samples. Optical gain in bulk CdSe can be due to both low-threshold excitonic and high-threshold e-h plasma mechanisms. Thermal dissociation of excitons at room temperature (the exciton binding energy in CdSe is 16 meV) results in a significantly increased optical gain threshold. Because of a large interlevel spacing, "quantum confined" excitons in QDs are more robust than bulk excitons, allowing one to excite room temperature ASE at pump levels comparable to those at cryogenic temperatures. This is an illustrative example of enhanced temperature stability in lasing applications expected for strongly confined QDs.

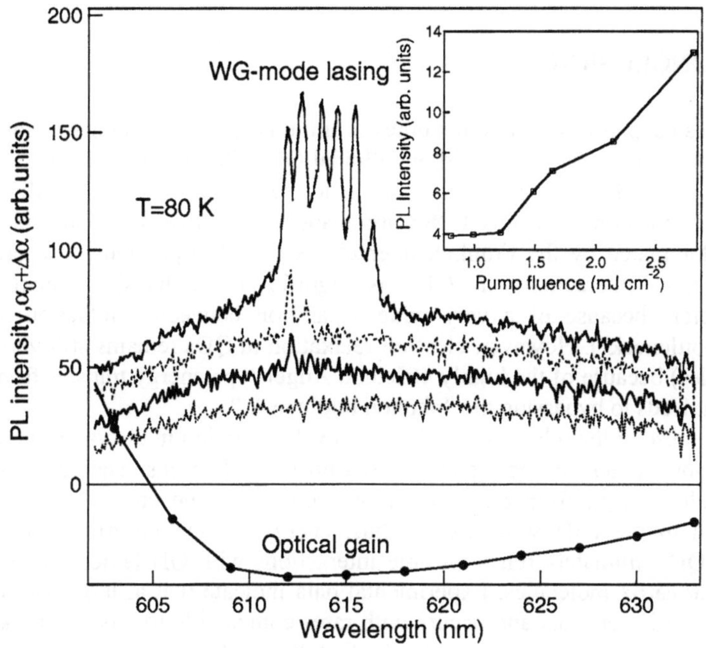

FIGURE 3.21. Pump-dependent emission spectra of QDs solids incorporated into a microcapillary tube that demonstrate the regime of pure microring lasing (the emission spectra are shown by lines, while the gain spectrum is shown by solid circles). The emission spectra are taken at progressively higher fluences of 1, 1.4, 1.6 and 2.8 mJ cm^{-2}. Inset: Pump-fluence dependence of the intensity of the 612.0 nm mode demonstrating a sharp transition to the lasing regime.

QDs can be coupled to microcavities to produce lasing into cavity modes. In Ref. 76, microring lasing in CdSe QDs was demonstrated using QD films deposited on inner walls of microcapillary tubes [see Fig. 3.20(a, b)]. In such a configuration, light amplification can be promoted both along the length of the tube (waveguide modes) and around the inner circumference of the tube [whispering gallery (WG) modes]. However, while for modes propagating along the tube, one can only achieve the ASE regime (no optical feedback is present), the WG modes can support a true lasing action (microring lasing[77]). In PL spectra recorded at progressively higher pump levels [Fig. 3.20(c)], one first observes the development of stimulated emission due to the propagation along the tube (band at 613 nm) and then the development of sharp oscillatory structures due to lasing into WG modes. The data in Fig. 3.21, which were taken for a different sample, provide an example of pure micro-ring lasing. Lasing into a single, sharp WG mode develops at ~1.25 mJ cm^{-2} (the lasing threshold is clearly seen in the pump-dependent mode intensity in the inset to Fig. 3.21). The position of this mode (612.0 nm) corresponds to the optical gain maximum (solid circles in Fig. 3.21). As the pump fluence is increased, additional WG modes develop on the low-energy side of the 612.0 nm mode, which is consistent with an asymmetric shape of the gain band. From the intermode spacing (~0.75 nm), one can derive an effective index of the gain medium of 1.8, indicating that WG modes are confined primarily within the QD layer.

3.6. CONCLUSION

This chapter has reviewed studies of ultrafast carrier dynamics, resonant optical nonlinearities and optical gain in CdSe colloidal QDs. Because of a discrete structure of energy states, mechanisms for intraband carrier relaxation in QDs are different from those in bulk materials. In bulk CdSe, intraband energy losses for both electrons and holes are dominated by the Fröhlich interactions with LO phonons. The efficiency of electron-phonon interactions in QDs is significantly reduced compared to bulk semiconductors because of a very wide separation between conduction band states ("phonon bottleneck"). However, electron relaxation in QDs remains still very fast (sub-ps time scales) because of the highly efficient, Auger-type energy transfer from electrons to holes that leads to bypassing the "phonon bottleneck."

Because of the relatively small inter-level separation in the valence band, energy transfer is not efficient in the direction from holes to electrons. Therefore, all accesses energy resulting, e.g., from photoexcitation well above the energy gap is eventually accumulated in the hole sub-system. Thus, energy dissipation from non-equilibrium carriers in QDs ultimately relies on hole interactions with QD lattice vibrations and/or with solvent/matrix molecules. Experimental data indicate that at large access energies, hole relaxation is very fast and occurs with rates comparable to those in bulk materials. This fast relaxation can be explained by the fact that well above the gap, hole states form a very dense spectrum (quasicontinuum) in which relaxation occurs in the same way as in bulk materials, i.e., via a cascade of single-phonon-emission acts. However, hole relaxation rates reduce dramatically (up to a factor of ten) at the band-edge spectral energies. The rate reduction occurs likely due to the existence of a relatively wide gap at the bottom of the valence band that can only be overcome via low-efficiency, multi-phonon processes. This gap can be formed due to a fine-structure splitting of the lowest $1S_{3/2}$ hole state manifested as a large "global" Stokes shift between the lowest 1S absorption peak and the band-edge PL maximum. Thus, the "phonon bottleneck," which is bypassed in the conduction band due to highly efficient Auger-type interactions, may still play a significant role in the final stage of hole relaxation.

Three-dimensional quantum confinement has a strong effect on mechanisms for band-edge optical nonlinearities. A large interlevel separation inhibits thermal depopulation of the lowest quantized states and enhances the role of state filling in QDs as compared with bulk materials. For example, excitation of a single e-h pair per dot is sufficient to bleach the strong 1S absorption resonance by almost 50%. Quantum confinement also enhances optical nonlinearities due to Coulomb interactions (carrier-induced Stark effect) leading, in particular, to a large shift of optical transitions that is particularly well-pronounced in early stages of carrier relaxation before the population of quantized states is established. Under thermal quasi-equilibrium, optical nonlinearities in QDs are primarily due to bleaching of optical transitions which is dominated by the filling of electron quantized states. The contribution of holes to state-filling induced nonlinearites is reduced compared to that of electrons because of a high density of valence-band states, which results in a wide spread of hole populations over multiple levels.

Band-edge optical gain in strongly confined QDs is dominated by two-e-h-pair states (quantum confined biexcitons). For moderately well-passivated dots, nonradiative Auger recombination of two-e-h-pair states is more efficient than both radiative decay and surface trapping that imposes an intrinsic limit on the lifetime of the optical gain. The fast Auger relaxation of optical gain (ps time constants) inhibits the development of ASE in dilute QD systems such as colloidal solutions and semiconductor-doped glasses fabricated by high temperature precipitation. Another factor that hinders development of optical gain and ASE in QDs is the interference from excited-state absorption features associated with carriers trapped at QD interfaces. Such features develop in most of the commonly used solvents such as hexane and toluene. The excited state absorption can be eliminated by an appropriate choice of the solvent/matrix and/or by tuning the excitation wavelength in quasiresonance with the 1S absorption maximum. Excited state absorption is also suppressed in matrix-free solid-state QD films. Due to high dot packing densities, QD solid-state films show optical gain of sufficient magnitude to successfully compete with multiparticle Auger recombination. These films exhibit narrowband ASE at both cryogenic and room temperatures, and the emission color is tunable with dot size as expected from confinement effects. Moreover, the QD solids can be incorporated into microcavities of different geometries (e.g., microspheres and microcapillary tubes) that produce lasing into whispering gallery modes.

I would like to acknowledge contributions of postdoctoral fellows and students to the work reviewed here, especially A. A. Mikhailovsky, J. A. Hollingsworth, A. Malko, S. Xu, B. Kraabel, C. Schwarz, and M. Petruska. I am also greatful to Al. Efros and M. Bawendi for numerous discussions on the photophysics of quantum dots. This work was supported by Los Alamos Directed Research and Development Funds, under the auspices of the U. S. Department of Energy.

REFERENCES

1. A. L. Efros and A. Efros, Sov. Phys. Sem., **16**, 772 (1982).
2. L. Brus, J. Chem. Phys., **80**, 4473 (1984).
3. A. Ekimov, J. Lumin., **70**, 1 (1996).
4. N. Borrelli, D. Hall, H. Holland, and D. Smith, J. Appl. Phys., **61**, 5399 (1987).
5. C. Murray, D. Norris, and M. Bawendi, J. Am. Chem. Soc., **115**, 8706 (1993).
6. S. Majetich and A. Carter, J. Phys. Chem., **97**, 8727 (1993).
7. M. Kuno, J. K. Lee, B. O. Dabbousi, F. V. Miulec, and M. G. Bawendi, J. Phys. Chem., **106**, 9869 (1997).
8. A. Kortan, R. Hull, R. Opila, M. Bawendi, M. Steigerwald, P. Carroll, and L. E. Brus, J. Am. Chem. Soc. **112**, 1327 (1990).
9. M. Hines and P. Guyot-Sionnest, J. Phys. Chem., **100**, 468 (1996).
10. V. C. Sundar, H.-J. Eisler, and M. G. Bawendi, Adv. Mater. **14**, 739 (2002).
11. N. Greenham, X. Peng, and A. Alivisatos, Phys. Rev. B, **54**, 17628 (1996).
12. C. Murray, C. Kagan, and M. Bawendi, Science, **270**, 1335 (1995).
13. O. Micic, K. Jones, A. Cahill, and A. Nozik, J. Phys. Chem. B, **102**, 9791 (1998).

14. M. Bruchez, M. Moronne, P. Gin, S. Weiss, and P. Alivisatos, Science, **281**, 2013 (1998).
15. W. C. W. Chan and S. Nie, Science, **281**, 2016 (1998).
16. Y. Vandyshev, V. Dneprovskii, V. Klimov, and D. Okorokov, JETP Lett., **54**, 442 (1991).
17. V. I. Klimov, A. A. Mikhailovsky, S. Xu, A. Malko, J. A. Hollingsworth, C. A. Leatherdale, H. J. Eisler, and M. G. Bawendi, Science, **290**, 314 (2000).
18. A. Hagfeldt and M. Graetzel, Chem. Rev., **95**, 49 (1995).
19. V. Colvin, M. Schlamp, and A. Alivisatos, Nature, **370**, 354 (1994).
20. B. O. Dabbousi, M. Bawendi, O. Onitsuka, and M. F. Rubner, Appl. Phys. Lett., **66**, 1316 (1995).
21. B. Kraabel, A. Malko, J. A. Hollingsworth, and V. I. Klimov, Appl. Phys. Lett., **78**, 1814 (2001).
22. S. Logunov, T. C. Green, S. Marquet, and M. A. El-Sayed, J. Phys. Chem. A, **102**, 5658 (1998).
23. P. Guyot-Sionnest, M. Shim, C. Matranga, and M. Hines, Phys. Rev. B, **60**, 2181 (1999).
24. D. Norris and M. Bawendi, Phys. Rev. B, **53**, 16338 (1996).
25. A. I. Ekimov, F. Hache, M. C. Schanne-Klein, D. Ricard, C. Flytzanis, I. A. Kudryavtsev, T.V. Yazeva, A. V. Rodina, and A. L. Efros, J. Opt. Soc. Am. B, **10**, 100 (1993).
26. J.-B. Xia, Phys. Rev. B, **40**, 8500 (1989).
27. A. L. Efros, Phys. Rev. B, **46**, 7448 (1992).
28. L. Efros and A. Rodina, Phys. Rev. B, **47**, 10005 (1993).
29. T. Takagahara, Phys Rev. B, **47**, 4569 (1993).
30. M. Nirmal, D. Norris, M. Kuno, M. Bawendi, A. L. Efros, and M. Rosen, Phys Rev. Lett., **75**, 3728 (1995).
31. V. Klimov, in *Handbook on Nanostructured Materials and Nanotechnology*, edited by H. Nalwa (Academic Press, San Diego, 1999).
32. S. Hunsche, T. Dekorsy, V. Klimov, and H. Kurz, Appl. Phys. B, **62**, 3 (1996).
33. V. I. Klimov, D. W. McBranch, C. A. Leatherdale, and M. G. Bawendi, Phys. Rev. B, **60**, 13740 (2000).
34. V. Klimov, S. Hunsche, and H. Kurz, Phys. Rev. B, **50**, 8110 (1994).
35. D. Norris, A. Sacra, C. Murray, and M. Bawendi, Phys. Rev. Lett., **72**, 2612 (1994).
36. Y. Hu, S. Koch, M. Linberg, N. Peyghambarian, E. Pollock, and F. Abraham, Phys. Rev. Lett., **64**, 1805 (1990).
37. K. Kang, A. Kepner, S. Gaponenko, S. Koch, Y. Hu, and N. Peyghambarian, Phys. Rev. B, **48**, 15449 (1993).
38. K. H. Pantke, J. Erland, and J. M. Hvam, J. Crystal Growth, **117**, 763 (1992).
39. J. Shah, *Ultrafast Spectroscopy of Semiconductors and Semiconductor Nanostructures* (Springer, Berlin, 1999).
40. T. Elsaesser and M.Woerner, Phys. Rep., **321**, 253 (1999).
41. V. I. Klimov, J. Phys. Chem. B, **104**, 6112 (2000).
42. U. Woggon, H. Giessen, F. Gindele, O. Wind, B. Fluegel, and N. Peyghambarian, Phys. Rev. B, **54**, 17681 (1996).
43. V. I. Klimov and D. W. McBranch, Phys. Rev. Lett., **80**, 4028 (1998).
44. V. I. Klimov and D. W. McBranch, Opt. Lett., **23**, 277 (1998).
45. R. Fork, C. V. Shank, C. Hirlimann, and R. Yen, Opt. Lett., **8**, 1 (1983).
46. V. I. Klimov, C. J. Schwarz, D. W. McBranch, C. A. Leatherdale, and M. G. Bawendi, Phys. Rev. B, **60**, 2177 (1999).
47. V. I. Klimov, A. A. Mikhailovsky, D. W. McBranch, C. A. Leatherdale, and M. G. Bawendi, Science, **287**, 1011 (2000).
48. E. Conwell, *High Field Transport in Semiconductors* (Academic Press, New York, 1967).
49. S. Prabhu, A. Vengurlekar, and J. Shah, Phys. Rev. B, **51**, 14233 (1995).
50. V. Klimov, P. H. Bolivar, and H. Kurz, Phys. Rev. B, **52**, 4728 (1995).
51. H. Benisty, C. Sotomayor-Torres, and C. Weisbuch, Phys. Rev. B, **44**, 10945 (1991).
52. U. Bockelmann and G. Bastard, Phys. Rev. B, **42**, 8947 (1990).
53. K. Shum, W. B. Wang, R. Alfano, and K. Jones, Phys. Rev. Lett., **68**, 3904 (1992).
54. T. Inoshita and H. Sakaki, Phys. Rev. B, **46**, 7260 (1992).
55. P. Sercel, Phys. Rev. B, **51**, 14532 (1995).
56. U. Bockelmann and T. Egler, Phys. Rev. B, **46**, 15574 (1992).
57. A. L. Efros, V. A. Kharchenko, and M. Rosen, Solid State Commun., **93**, 281 (1995).
58. V. I. Klimov, A. A. Mikhailovsky, D. W. McBranch, C. A. Leatherdale, and M. G. Bawendi, Phys. Rev. B, **61**, 13349 (2000).
59. S. Schnitt-Rink, D. Miller, and D. Chemla, Phys. Rev. B, **35**, 8113 (1987).
60. A. L. Efros, A. Ekimov, F. Kozlovski, V. Petrova-Koch, H. Schmidbaur, and S. Shumilov, Solid State Commun., **78**, 853 (1991).
61. T. Takagahara, Phys. Rev. Lett., **71**, 3577 (1993).

62. T. Takagahara, J. Luminescence, **70**, 129 (1996).
63. S. Xu, A. Mikhailovsky, J. A. Hollingsworth, and V. I. Klimov, Phys. Rev. B **65**, 53191 (2002).
64. J. Shah, IEEE J. Quantum Electron., **24**, 276 (1988).
65. A. V. Uskov, Phys. Rev. Lett., **85**, 1516 (2000).
66. M. Asada, Y. Miyamoto, and Y. Suematsu, IEEE J. Quantum Electron., **22**, 1915 (1986).
67. N. N. Ledentsov, V. M. Ustinov, A. Y. Egorov, A. E. Zhukov, M. V. Maksimov, I. G. Tabatadze, and P. S. Kopev, Semiconductors, **28**, 832 (1994).
68. N. Kistaedter, N. N. Ledentsov, M. Grundmann, D. Bimberg, V. M. Ustinov, S. S. Ruvimov, M. V. Maximov, P. S. Kopev, Z. I. Alferov, and U. Richter, Electron. Lett., **30**, 1416 (1994).
69. J. Butty, Y. Z. Hu, N. Peyghambarian, Y. H. Kao, and J. D. Mackenzie, Appl. Phys. Lett., **67**, 2672 (1995).
70. F. Gindele, R. Westphaeling, U. Woggon, L. Spanhel, and V. Ptatschek, Appl. Phys. Lett., **71**, 2181 (1997).
71. A. V. Malko, J. A. Hollingsworth, A. A. Mikhailovsky, and V. I. Klimov, unpublished.
72. D. Chatterji, *The Theory of Auger Transitions* (Academic Press, London, 1976).
73. P. Landsberg, *Recombination in Semiconductors* (University Press, Cambridge, 1991).
74. D. Chepic, A. L. Efros, A. Ekimov, M. Ivanov, V. A. Kharchenko, and I. Kudriavtsev, J. Luminescence, **47**, 113 (1990).
75. V. A. Kharchenko and M. Rosen, J. Luminescence, **70**, 158 (1996).
76. A. V. Malko, A. A. Mikhailovsky, M. A. Petruska, J. A. Hollingsworth, H. Htoon, M. G. Bawendi, and V. I. Klimov, Appl. Phys. Lett. **81**, 1303 (2002).
77. S. L. McCall, A. F. Levi, R. E. Slusher, S. J. Pearton, and R. A. Logan, Appl. Phys. Lett. **60**, 289 (1992).

4

Novel Device Applications of Stranski-Krastanov Quantum Dots

Karl Brunner[*] and Artur Zrenner[†]

** Walter Schottky Institut, Technische Universität München, Am Coulombwall, D-85748 Garching, Germany; New address: Universität Würzburg, Am Hubland, D-97074 Würzburg, Germany*

† Universität Paderborn, Warburger Str. 100, D-33098 Paderborn, Germany

4.1. INTRODUCTION

The development of information technology is certainly one of the most important achievements of applied solid state physics. Most of the corresponding key components are based on semiconductor devices with electronic or opto-electronic functions[1]. After decades of intense research, the performance of those devices is excellent. But information technology has an enormous impact on human society and hence there is an increasing demand for further developments.

In search for new concepts, the field of semiconductor science and technology moved therefore more and more towards quantum dot (QD) structures[2]. From technological point of view the most direct and flexible way to produce QDs is to laterally pattern existing two-dimensional (2D) heterostructures[3]. This concept turned out to be specifically successful in the field of single electron devices[4]. For optical applications the available methods of lateral patterning turned out to introduce too many defects which dramatically enhance nonradiative recombination processes. In many ways more advantageous are here self assembled QDs formed by Stranski-Krastanow growth[5,6]. With typical dimensions of the order of 10 nm, they exhibit dominant quantization effects with energy scales, which basically allow for room temperature device applications. Stranski-Krastanow growth is a general concept for the self assembly of QD structures. It can be realized in a variety of different material systems and it is a cheap and fast process, compared to routes of lithographic patterning. The resulting QDs can be regarded as defect free, ensembles of QDs are however subjected to inhomogeneous broadening effects.

In our current article we will present a variety of novel device applications based on Stranski-Krastanow QDs. The most advanced application of self-assembled Stranski-Krastanow islands are laser diodes with In(Ga)As QDs as an active medium. Quantum dots mainly improve the thermal stability and extend the spectral range of laser emission to longer wavelengths compared to quantum well laser structures. Some aspects of lasing from chemically synthesized QDs are discussed in chapter 3 of this book. We will concentrate here on other classes of optoelectronic devices. After a general discussion of Stranski-Krastanov growth mechanisms and properties of GaAs/InGaAs and Si/Ge islands, we will cover QD intraband photodetectors, optical memory device structures based on QDs, and interband photodetectors including single QD photodiodes.

4.2. STRANSKI-KRASTANOV GROWTH AND PROPERTIES OF GaAs/InGaAs AND Si/Ge ISLANDS

4.2.1. Structural Properties

Early optical studies of low dimensional semiconductor structures were performed with laterally patterned, etched QW structures, with locally interdiffused structures or with natural quantum dots formed by local fluctuations of the composition and width of narrow QWs.[7,8,9,10] Etched structures and 2D systems with split-gate electrodes are very powerful for unipolar structures like electron gases in low-dimensional systems or for quantum electronic structures like point contacts and single electron tunnelling transistors (SET) studied at low temperature.[11] A strong confinement of electron as well as hole states, however, can be reached only by a modulation of composition at a length scale of about 10 nm. For fabricating such structures with discrete excitonic transitions epitaxial deposition techniques like molecular beam epitaxy (MBE) in the Stranski-Krastanov growth mode (S-K) were established in the last decade to be the most powerful method. The S-K growth mode of the material systems Ge on Si and InAs on GaAs was intensively studied.[12,13] It was also demonstrated in many other strained III-V semiconductor heterosystems like InP/InAs,[14] GaAs/InGaAlSb,[15] GaSb/InSb[16] as well as for II-VI semiconductors.[17] We will focus on the heterosystems Si/Ge and GaAs/InAs that have a lattice mismatch of about 4% and 7.2%, respectively. (Si)Ge and In(Ga)As can be deposited in atomically smooth, compressively strained layers up to a certain thickness of a few monolayers (ML). At adequate growth conditions, of which a substrate temperature of about 500°C required for a high surface mobility of adatoms and a typical deposition rate of 0.02 nm/s are the most important, island nucleation takes place at a coverage of about 6 ML Ge or 1.8 ML InAs, respectively. This amount is less than required for inelastic strain relaxation and no dislocations nucleate in that stage. The morphological phase transition to islands on a wetting layer rapidly takes place when the elastic energy gained by elastic strain relaxation within the islands exceeds the additional surface energy of the island structure. A huge number of islands that are free of defects evolves with areal densities up to about 10^{11} cm^{-2}, heights in the nm-range, and a lateral size of typically 20 nm or above, depending on growth conditions. Figure 4.1 illustrates the different morphologies of S-K islands in two extreme cases. Fig. 4.1(a) shows an atomic force microscopy (AFM) image of a dense ensemble of Ge islands on a (001)Si substrate formed from 8 ML Ge at T=510°C. The

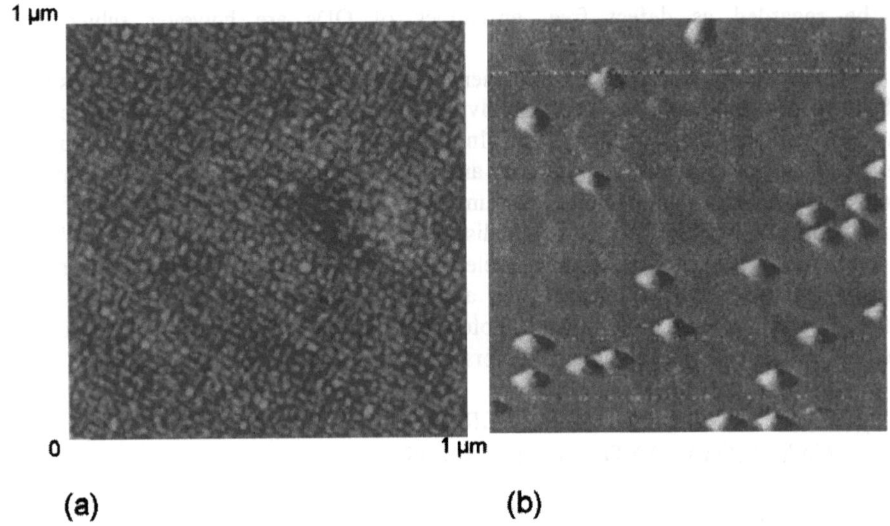

FIGURE 4.1. AFM images of (a) Ge QDs formed by the Stranski-Krastanov growth mode by deposition of 8 monolayers Ge on a (001)Si substrate at a temperature of 510°C and (b) of a dilute $In_{0.4}Ga_{0.6}As$ QD layer at about 7.5 ML coverage, slightly above the critical value for island nucleation. The height scale is 3 nm and 6 nm, respectively.

island height is about 2 nm and the lateral width about 20 nm. These Ge quantum dots represent hut-like island structures with {105}-facetted surfaces and are in part spontaneously elongated in [100]- or [010]-direction. Such ensembles with a high density of islands are required for applications in which the efficiency of optical absorption, for example, scales with the number of active quantum dots. Fig. 4.1(b) shows an AFM image of self-assembled $In_{0.4}Ga_{0.6}As$ quantum dots formed from about 7.5 ML that is just slightly above the critical thickness for nucleation of such S-K islands. At adequate coverage, the areal density of islands can be reduced down to a few islands per μm^2. Such low densities allow spatial selection of individual QDs by optical spectroscopy with sub-µm lateral resolution described in section 4.5.2. With increasing temperature, coverage, and time duration of growth, islands generally increase in size and decrease in areal density due to the kinetics and energetics of island formation and Ostwald ripening.[18] Thus, the structural parameters of islands depend very critically on the actual growth parameters. The formation of a rather homogeneous ensemble of islands with an energetically favoured island size and the suppression of ripening was observed for certain growth conditions and is theoretically understood by taking into account the energy contributions of the island edges.[19,20] Elastic relaxation of compressive lattice strain preferentially in the island apex drives the island nucleation. During subsequent deposition of host material, first the valleys between islands are epitaxially filled and finally the island apex is capped. Embedding islands in the semiconductor host material changes their shape from a facetted pyramid to a flattened, more lens-shaped structure and also the local strain distribution is modified. Model calculations based on finite element or valence force field approaches indicate strong relaxation along the short dimension of islands, i.e. in growth direction and just slight lateral relaxation.[21,22,23] As an example, the lateral lattice constant at the centre of an embedded Si/Ge island with 2 nm height and 20 nm diameter is just about 0.2% larger than a_{Si} in bulk

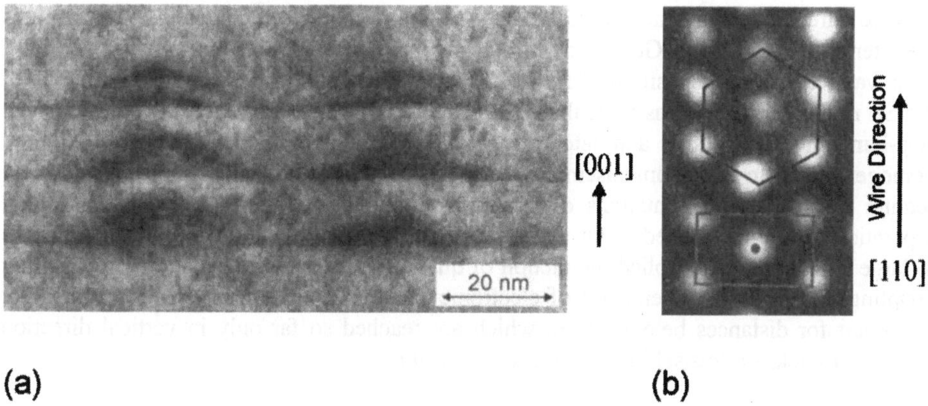

FIGURE 4.2. (a) High resolution TEM cross-section of GaAs/InAs QD stacks self-ordered in the modified S-K growth mode and (b) AFM image of a hexagonal array of self-ordered Si/Ge islands on a Si/SiGe wire multiple layer structure.

Si. In growth direction, the lattice constant is about $1.035a_{Si}$. This is close to the value of bulk Ge, but is considerably smaller than 1.07aSi in a biaxially strained Ge layer on (001)Si due to the Poisson effect. The local strain fields also extend into the host material with a length scale that is comparable to the lateral island size that is about 20 nm for flat S-K islands discussed here.[24] Such strain fields affect the confining potentials for carriers described in the next section.

Strain fields reaching the surface during heteroepitaxy also influence the nucleation of islands in subsequently deposited layers within multilayer structures. Relaxation of compressive island strain causes a slight increase of the lateral lattice constant in the host crystal on top of the islands. In(Ga)As or (Si)Ge material deposited accumulates preferentially in that regions, as the mismatch strain energy is lower there. This local material accumulation causes a reduction of the nominal critical coverage for island nucleation and islands mainly form right on top of underlying islands.[25,26] Figure 4.2(a) shows a cross-sectional transmission electron microscopy (TEM) image of stacked GaAs/InAs quantum dots fabricated by this modified S-K growth mode. The InAs islands are perfectly stacked with a vertical separation of a few nm, given by the nominal width of GaAs separation layers of 10 nm minus the island height of about 5 nm. The island size in these stacks is remarkably homogeneous. A significant increase of island size was observed within subsequent layers. This size inhomogenity is attributed to the reduced critical thickness for island nucleation and it may appear inevitable. It could be strongly reduced, however, by a properly adjusted coverage in subsequent layers.[24,27] The formation of vertically correlated S-K islands in multiple layer structures is a fundamental and very efficient self-ordering process. The spatial separation of correlated islands is well controlled by the epitaxial growth parameters. Vertical self-alignment by the local strain fields extending from buried islands to the surface additionally causes a gradual self-ordering within the layer plane and a homogenisation of island size in structures with a large number of layers.[28] Very regular 3-dimensional arrays of islands were synthesized in material systems with strong elastic anisotropy.[29,30] Lateral self-ordering of quantum dots was reached by aligning islands by natural surface steps,[31] by mesa edges on prepatterned substrates[32] and by stripe-like strain fields originating in buried regular arrays of strained

Si/SiGe wire structures that self-assembled on vicinal surfaces.[33] Figure 4.2(b) illustrates the lateral arrangement of Ge islands in a hexagonal densely packed 2-dimensional array. This was reached by deposition of 5 ML Ge on top of a 20-period (2.5 nm $Si_{0.55}Ge_{0.45}$/10 nm Si) multilayer that forms a regular SiGe wire array of 120 nm period by strain-induced step-bunching.[34] The size and lateral separation of such self-ordered islands is more homogeneous than in random ensembles. Such regular cells of laterally ordered islands may remind to some implementations of quantum computers, although the size and lateral separation of islands realised so far in the range of 100 nm is about one order of magnitude to large for enabling controlled interaction of quantum dots at finite temperature. Quantum coupling and Coulomb interaction of electronic states in neighbouring nanostructures get important for distances below 10 nm which are reached so far only in vertical direction within dot stacks or dots self-aligned on wire structures.

4.2.2. Band Structure Properties

The band structure realised in GaAs/InAs quantum dots is dominated by the strong reduction of the direct band gap energy at the Γ-point in bulk InGaAs alloys with increasing In content x, given by $E_G(In_xGa_{1-x}As) = (1.519-1.584x+0.475x^2)$ eV at low temperature.[35] The conduction band (CB) offset is about 67% of the band gap reduction and decreases just slightly for increasing In content in strained InGaAs QWs on GaAs, according to model calculations of Van de Walle.[36] Thus, a type I band alignment is realised with electrons and (heavy) holes localised within In(Ga)As. The band offsets and effective masses in the CB and especially in the valence band (VB) of pseudomorphic GaAs/InGaAs QW structures are affected by biaxial lattice strain.[37,38] The actual values and spatial distributions of In content and strain in GaAs/InAs QDs are difficult to analyse in detail. Nevertheless, a good agreement of model calculations using adequate parameters and experimental results from optical and structural studies can be reached. The QDs are often described by a strong, QW-like confinement in vertical (growth) direction and a parabolic lateral confinement potential. In a QD of cylindrical symmetry, electron or hole states are quantized with a ground state n=1 of degeneracy 2 (due to spin polarisation). The higher excited states with index n=2,3,... of degeneracy 2n are related to lateral localisation of carriers. These levels are also called *s*-, *p*-, *d*-shell in analogy to the atomic level scheme. Excited QD states related with confinement in growth direction are at much higher energy or even unbound due to the small effective dot height. Other QD shapes like truncated pyramids or slightly asymmetric potentials were also assumed and verified by spectroscopy.[22] The details in island shape and in In composition profile depend strongly on growth conditions due to intermixing, segregation, island ripening. They appear to differ for QD structures fabricated by different groups, but the main features are very consistent. Experimental results on the complex level structure in GaAs/InAs S-K quantum dots formed from 2.5 ML InAs deposited at 530°C at an As-pressure of 1.5×10^{-5} Torr are schematically summarized in Figure 4.3. The QD parameters are 27 nm in lateral size, 9 nm in height, and an areal density 1×10^{10} cm^{-2}. Carriers are confined in the plane of the strained QW layer formed by the In(Ga)As wetting layer (WL) and within the QDs. The QD level structure shown was analysed by combining the results of photoluminescence (PL), like shown in the inset of Fig. 4.3, and intra CB photocurrent (PC) spectroscopy presented in section 4.3. The data are not

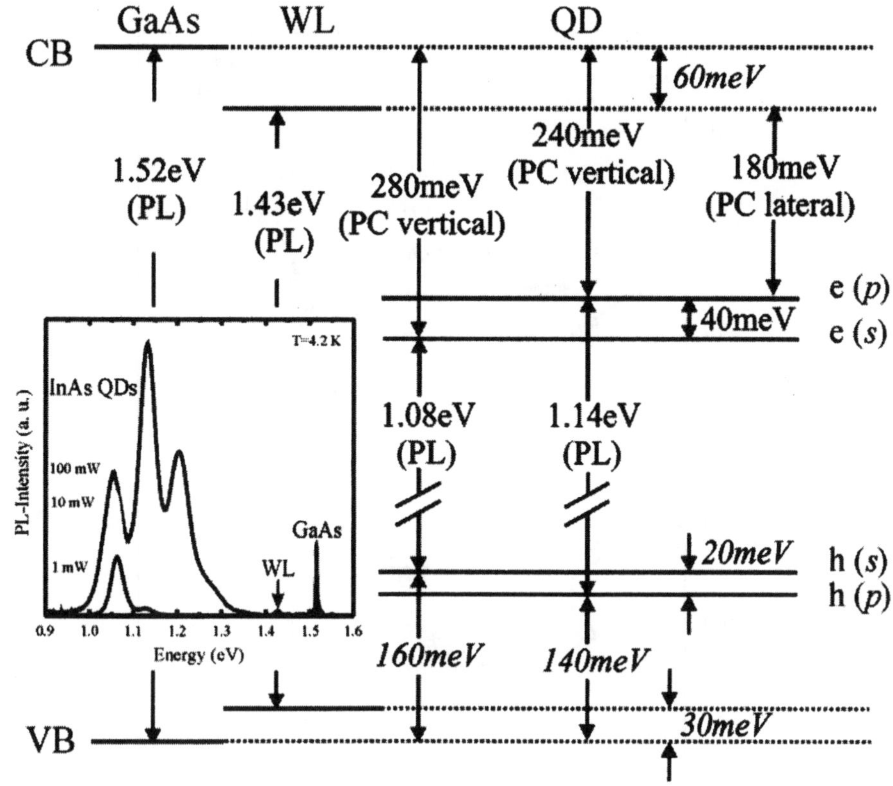

FIGURE 4.3. Level scheme of a typical GaAs/InAs QD structure deduced from optical spectroscopy. The inset shows PL spectra of an ensemble of InGaAs quantum dots embedded in GaAs observed at low temperature and varied optical pumping.

corrected for excitonic and many-particle Coulomb interactions. Electrons and holes in the QD ground state (*s*-shell) are bound by 280 meV and 160 meV, respectively. Their *s*-*p* level splitting due to lateral confinement is about 40 meV and 20 meV, respectively. This difference in quantum confinement energy is easily understood regarding the different effective masses and confinement potentials in the CB and VB. Such experimental results are also the basis for much more detailed calculations of the excitonic levels and the optical properties of quantum dots that reveal well-understood multi-exciton and charged exciton complexes, depending on the experimental conditions.[39] The results will be described in section 4.5.2.

The band gaps in Si and Ge are indirect in *k*-space with hole states at the Γ-point and electrons in Δ- and *L*-valleys with energy E_G(Si)=1.17 eV and E_G(Ge)=0.74 eV at low temperature, respectively. The band alignment in $Si/Si_{1-x}Ge_x$ quantum well structures is of type II with the CB minimum in Si and the VB edge in SiGe. The band edge energies are shifted and split by biaxial mismatch strain.[40] The resulting offset for heavy hole states forming the VB edge is about $x \cdot 0.75$ eV.[41] A strained SiGe layer on (001)Si represents a small (large) potential barrier for electrons in Si-like Δ(4)-valleys (Δ(2)-valleys) with in-plane (vertical) orientation.

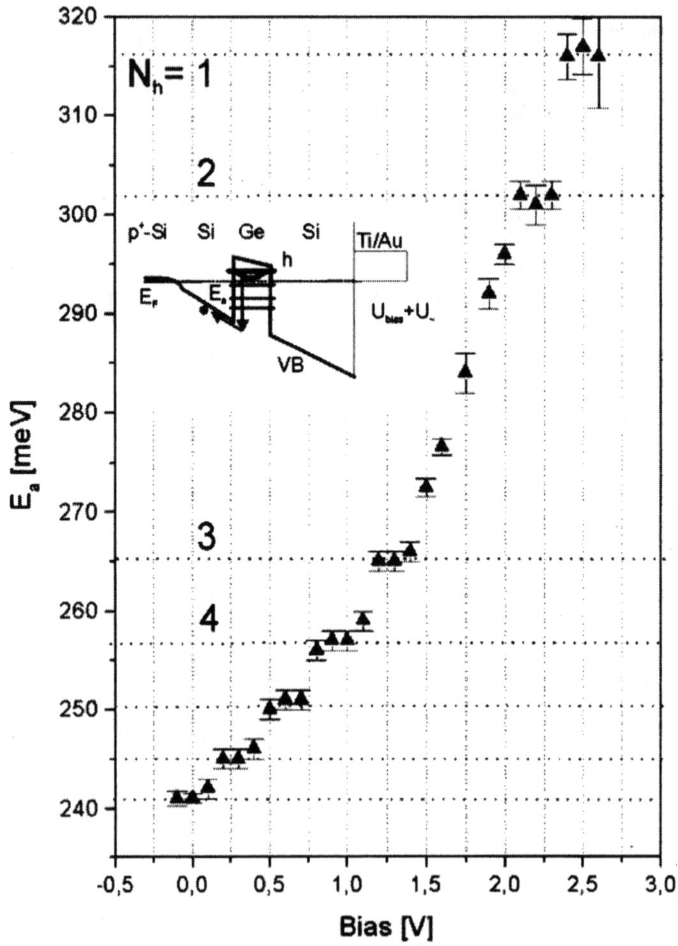

FIGURE 4.4. Quantized thermal activation energy of holes within discrete levels of Ge quantum dots depending on bias voltage of a *p*-Si Schottky diode with a single Ge dot layer embedded. The data result from admittance measurements.

Si/Ge QDs consequently localize mainly holes while electrons are just weakly bound by Coulomb attraction of localized holes or by shallow and narrow wells for Δ(2)-states due to tensile strain in the Si host surrounding relaxed Ge dots.[23] So far, the discrete levels within Si/Ge quantum dot structures could not be characterized by resolving spectral peaks in PL, PC, or PLE spectroscopy. Inhomogeneous broadening in QD ensembles and the lack of single QD spectroscopy data from just weak interband transitions, that are indirect in real and *k*-space, prevent detailed optical studies of Si/Ge QD levels. C-V and admittance spectroscopy of Si/Ge QDs embedded in the depletion region of a p-Si Schottky diode were demonstrated to be a very powerful technique for studying the VB structure.[42,43] Figure 4.4 shows the thermal activation energy, i.e. the hole localization energy of Si/Ge QDs depending on the bias voltage applied to the diode. With increasing bias, the QDs are more and more depleted and the observed localisation energy increases step-wise up to about 317 meV.[44] This discrete level corresponds to the QD ground state level filled by one hole. The next level at 302 meV corresponds to two holes within each dot. Thus, the Coulomb charging energy of a second hole in the *s*-state is about 15 meV. The energy plateau at 265 meV for lower bias that corresponds to an occupancy of three holes reveals a *s-p* state splitting of about 37 meV. These values are in very good agreement with simple model calculations assuming a lens-shaped dot of about 2 nm height and 20 nm diameter with a parabolic lateral confinement potential.

The effective excitonic levels within stacked QD structures show an even more complex behaviour due to the coupling of dot levels.[45] The experimental results, however, appear to be severely affected by the inhomogeneous island size and strain-induced material intermixing in stacked QDs fabricated by the modified S-K mode.[46,47]

4.3. INTRABAND PHOTODETECTORS

4.3.1. Vertical Photocurrent Structures

A fundamental type of semiconductor photodetectors is represented by unipolar structures in which the energy of detected photons in not given by the band gap energy of materials but by the engineered band offsets and subband energies in heterostructures. The optical intraband transition energies accessible in such quantum well infrared photodetectors (QWIP) made from SiGe, AlGaAs and InGaAs heterostructures enables detection in the mid infrared (MIR) spectral range of about 4 – 20 μm. The optical intersubband oscillator strength is given in general electric dipole approximation by:

$$f_{ij} = \left|\left\langle \Theta_j \middle| \hat{\varepsilon} \cdot p \middle| \Theta_i \right\rangle\right|^2 = \frac{2m_0 E_{ij}}{\hbar^2} \left|\mu_{ij}\right|^2; \quad with \quad \mu_{ij} = \left\langle \varphi_j(z) \middle| z \middle| \varphi_i(z) \right\rangle \hat{\varepsilon} \cdot \hat{z}, \quad (4.1)$$

where Θ_i, Θ_j are electron envelope wave functions, $\varphi_i(z)$, $\varphi_j(z)$ their separated part in a quantum well with confinement direction $\hat{z}$ (growth direction), m_0 the free electron mass and $\hat{\varepsilon}$ is the light polarisation unit vector. The calculated value of f_{12} for a quantum well with infinite barriers and isotropic, parabolic subband dispersion can be expressed by $f_{12}=0.96m_0/m_{eff}$ for light polarisation in confinement direction $\hat{z}$ and it is zero for in-plane polarisation. Thus, the absorption of normally incident MIR light by electron subbands in homogeneous QW layers is forbidden. III-V semiconductor QWIPs with electron subbands usually have to be processed with a grating coupler for inducing a vertical field component.

Novel quantum dot structures for infrared photodetectors (QDIP) were demonstrated to have the crucial benefit that carrier localisation in all dimensions lifts the optical selection rules given above and efficient absorption of normal incident light by bound-quasibound or bound-continuum transitions is observed.[48,49,50] An example of the polarisation dependent photocurrent spectra is given in Figure 4.5 for a Si/Ge island p-i-p structure excited in prism waveguide geometry at different bias voltages. Ten layers of p-doped Ge islands separated by 50 nm Si layers represent the isolating region between highly doped p-contact Si layers. The photoresponse increases with bias voltage and the

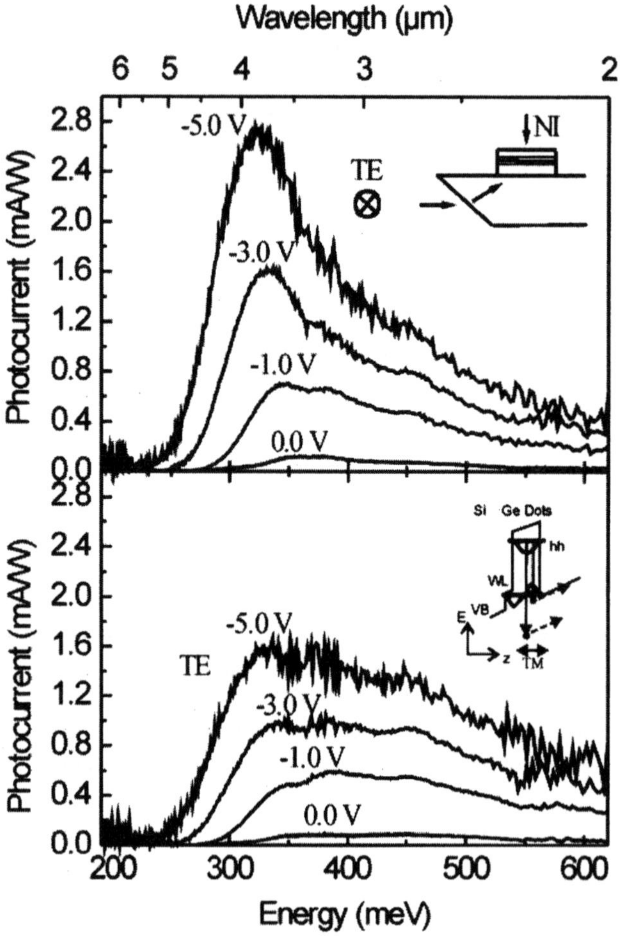

FIGURE 4.5. Vertical intra VB photocurrent spectra of a *pip* Si/Ge island structure (10x(8 ML Ge/ 50 nm Si) excited by MIR light in multipass waveguide geometry with TM or TE polarization at different bias voltages and T= 20K. The dots are p-modulation doped with Boron in the center of the 50 nm Si layers.

PC maximum observed especially for TM polarisation at about 340 meV shifts to lower energy. This peak is attributed to bound-quasibound transitions of hole states from the Ge dot ground state to levels close to the Si VB edge. The photoresponse for in-plane TE polarisation is well comparable to TM polarisation especially at high energies corresponding to bound-continuum states. Recent absorption studies and model calculations indicate that the initial states are hh-like dot levels and the continuum final states are of lh and soh symmetry enabling intra valence band absorption of TE polarised light, similar to the case of SiGe QWs.[51,52] The observed PC peak energy depending on the effective Si/Ge dot valence band offset, bias voltage, and doping level, is well understood considering the quantum confined Stark effect of dot and Si VB edge states in an asymmetric dot potential and the filling of dot levels.[50] A quite large spectral shift of the low-energy onset of PC (up to 80 meV/5V) with bias may be applied for spectrally tuneable QD detectors. The tunability may be enhanced in structures with self-ordered QD stacks or asymmetric coupled QD/QW structures.

Qualitatively similar results were observed from GaAs/InAs QD n-i-n structures.[49] The bound-quasibound transitions of electrons from dot levels to the GaAs CB edge in these structures, however, are much stronger than the bound-continuum transitions, in contrast to the results from Si/Ge dot structures with hole transitions. This may be related to the smaller effective masses enabling tunnelling of electrons from quasi-bound states close to the GaAs CB edge into the continuum. Further, the optical selection rules are more stringent for electrons than for holes with a mixing of states of different symmetries. At a doping level corresponding to about 4 electrons per dot, two PC peaks could be resolved. They correspond to excitation of electrons from the ground and from the first excited dot states to the GaAs band edge. Both levels are filled in average by 2 electrons due to the Pauli exclusion principle applied to the discrete electron levels within a GaAs/InAs quantum dot. The observed PC peak separation of 40 meV combined with PL results allows characterisation of the detailed dot level scheme like shown in Fig. 4.3.

All detector structures presented so far are based on drift of photoexcited carriers in a vertical electric field applied between biased bottom and front contact layers. The observed strong dependence of PC strength on bias voltage is similar for QDIP and QWIP structures and may be attributed to relaxation of excited carriers before leaving the dot or well and to recapture of drifting carriers in neighbouring layers at low field. An improved detector performance with very small dark current was proposed for such QDIP structures with detection of vertical photocurrent compared to QWIPs.[53] Detectivities up to about $D^*=2\times10^{10}$ $\mathrm{cmHz}^{1/2}/\mathrm{W}$ reached so far with not optimised Si/Ge dot structures at $T=77\mathrm{K}$ are indeed very encouraging.[50] Self-assembled quantum dot structures allow larger In or Ge content in the lattice mismatched heterosystems without defect generation. The resulting larger effective band offsets and transition energies, and smaller effective electron masses m_{eff} may enhance the optical oscillator strengths. A further optimisation of the epitaxial layer parameters and advanced device processing may offer high performance MIR detectors with quantum dots. A limitation of responsivity of such structures with vertical photocurrent perpendicular to the layers is given by a large recapture probability of excited carriers when passing neighbouring layers. This can be overcome by novel intraband photodetector structures with detection of lateral photocurrent described below.

4.3.2. Lateral Photoconductivity Structures

A novel type of detector structures with lateral photocurrent detected is based on the 3-dimensional localisation of carriers within quantum dots. In that sense, the novel detector structures are real QD devices that cannot be realised by QW structures. Carriers localise in the dots and result in high resistance and low dark current of Si/Ge and GaAs/InAs quantum dot multilayer structures at low temperature. Optical intraband excitation of carriers into extended states enables conductance as long as carriers keep mobile and a photocurrent between biased lateral electrodes can be measured. The theoretical photoconductive gain of detector structures is given by $G_{PC}=\mu\tau E/L$ for a carrier mobility μ, a lifetime of excited conducting carriers τ, an electric field E, and an electrode spacing L. Alloyed lateral electrodes of test structures are typically designed in an interdigital finger geometry with a finger spacing of 60 μm. The photoresponse observed from Si/Ge dot structures without doped contact layers is in lateral detection about twice as large as in vertical PC.[54] The spectral shape of both is very similar at high energy, but lateral photocurrent extends towards lower photon energy and its maximum is about 40 meV lower in energy than that for vertical hole transport. This redshift is attributed to the enhanced contributions of quantum well-like final states within the Ge wetting layer or the Si:B modulation doping layers. Such 2D subbands may contribute to in-plane conductance at low temperature (T= 20 K), but the holes are localised in vertical direction and do not contribute to vertical PC. Part of the increase in lateral PC may be tentatively attributed to large oscillator strengths of such dot-wetting layer bound-bound transitions. The main part of enhancement, however, is attributed to a prolonged lifetime of photoexcited carriers contributing to lateral photoconductivity.

The novel detection principle based on lateral photocurrent offers an engineering of the effective carrier lifetime by a more sophisticated band structure design. Figure 4.6(a) shows the lateral photocurrent spectra at T= 20K of Si/Ge QD structures without and with $Si_{0.75}Ge_{0.25}$ quantum well layers embedded within the 50 nm Si layers separating the dot layers. The photoresponse is enhanced by about a factor of 50 with incorporation of a conducting QW layer that is spatially separated from the dots. The fundamental operation principle of this structure is sketched in the inset of Fig. 4.6(a). Holes are optically excited from the dot ground (and maybe excited) states to the Si VB edge and are spatially transferred by diffusion, drift or tunnelling into the QW layers. There they relax to the subband bottom and contribute to lateral PC for a prolonged time τ until they may be excited and relax back into the energetically favourable dots. A long lifetime τ and a higher hole mobility μ within the SiGe QWs cause a drastic increase in PC compared to the reference structure without QWs. The PC maximum of the structure with QWs is about 40 meV higher in energy than in the reference structure without QWs. This energy correponds to excitation from the dot ground state to states at the Si VB edge. It indicates that holes in continuum states but not in bound states are efficiently transferred from the dots into the QWs and contribute to the large PC response. Responsivity values up to about 90 mA/W were reached in a study of structures with varied spatial separation of Ge QD and SiGe QW layers.[55] The largest response observed for small QD-QW separations may be due to a more efficient transfer of excited holes by diffusion or tunnelling. The dark current measured between interdigital electrodes is well comparable in the structures with and

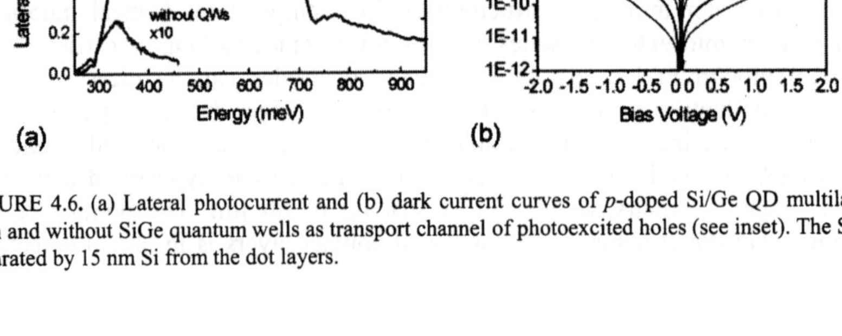

FIGURE 4.6. (a) Lateral photocurrent and (b) dark current curves of *p*-doped Si/Ge QD multilayer structures with and without SiGe quantum wells as transport channel of photoexcited holes (see inset). The SiGe layers are separated by 15 nm Si from the dot layers.

without QWs. The *I-V* curves in Figure 4.6(b) show just a slight increase of the quite low dark current at low temperatures by incorporation of QWs. Nearly identical curves with still low dark current are observed at *T*= 77K. The preliminary studies indicate detectivities in excess of 10^{11} cmHz$^{1/2}$/W, but a careful device characterisation concerning optimum response, dark current and noise is still required.

The fundamental operation principle of lateral PC structures for MIR detection is identical in GaAs/InAs and Si/Ge QD structures. The contributions of optical MIR absorption, spatial transfer of carriers, effective carrier lifetime, and carrier mobility to the photocurrent observed, however, is very complex and may differ significantly. Figure 4.7 shows the photoresponse at different temperatures from a structure with three absorbing InAs QD layers and a single InGaAs QW for lateral transport. The dots are about 25 nm in diameter, 9 nm in height, and have an areal density of about 1×10^{10} cm^{-2}. Details of the structure are described elsewhere.[56] The observed photoresponse is giant. Peak values up to about 10 A/W were reached at *T*= 30K. The observed moderate decrease of PC with temperature may indicate a decreasing lifetime or mobility of electrons. A lifetime in the range of a millisecond at *T*= 30K was analysed by time-dependent PC measurements, as shown in the inset of Figure 4.7. This slow temporal response is mainly attributed to the back transfer of electrons from the QW into the dots. The potential barriers formed by the GaAs layers efficiently keep electrons localised within the conducting InGaAs channel at low temperatures. The excitation, emission from the QDs and relaxation of electrons into the QW are expected to be much faster. A remarkable difference of the PC spectra observed from GaAs/InAs structures compared to Si/Ge is the distinct peak with a very narrow spectral width of about 20 meV. The relative strength of bound-continuum transitions at higher energy is negligible. The peak is tentatively attributed to transitions into weakly bound states in the dots or wetting layers that combine a large optical oscillator strength and a high electron transfer probability into the photoconductive QW channel. The high

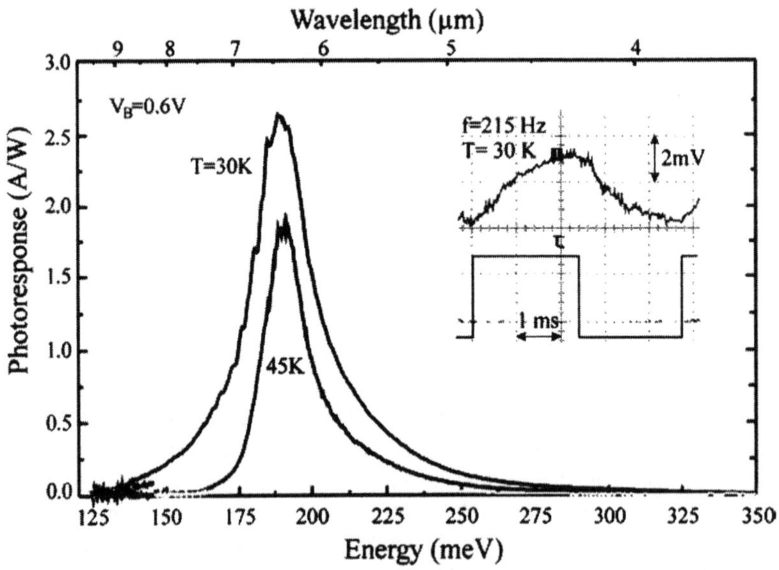

FIGURE 4.7. Lateral PC of a GaAs/InGaAs QD structure with a single InGaAs QW as conducting electron channel at different temperature. The inset shows the temporal response at T= 30K.

spectral resolution of such photoconductive QD-QW structures may enable detector structures with high spectral selectivity and a tunability by the bias voltage. This may be reached by utilising the Quantum Confined Stark effect, resonant tunnelling between QD and QW layers, and a further enhanced performance of vertically stacked QD structures. The complex mechanisms in QDIP structures with lateral photocurrent studied require further detailed studies of the optical and electronic processes involved. The structures offer a set of design parameters for even more advanced structures and for a further optimisation of performance regarding photoresponse, dark current, and novel functionality.

4.4. OPTICAL MEMORY DEVICE STRUCTURES BASED ON QUANTUM DOTS

4.4.1. Overview

Photoluminescence investigations on large ensembles of self-assembled QDs typically reveal a strongly inhomogeneously broadened PL linewidth in the range of 50 meV, which originates from size and composition fluctuations.[57] This contrasts strongly with the much narrower (down to about 10 μeV)[58], homogeneously broadened emission observed from single QDs, which reflects the δ function like density of states expected for zero-dimensional (0D) systems. Therefore, in the energy range of the QD groundstates, monochromatic light will interact only with a small fraction of a QD ensemble.

In 1995 Muto has suggested that an inhomogeneously broadened ensemble of absorbing centers may have potential applications as wavelength domain optical data storage media.[59] In order to develop such a technology, device concepts have to be established, which allow for efficient electrical or optical detection of optically written information. While optical writing has the clear advantage of opening the wavelength domain in addition to the dimensions available in realspace, the resulting excitations (excitons) are relatively short lived (typically 1 ns in a direct gap III/V semiconductor) and electrically neutral.

For memory applications on the basis of semiconductor structures those limitations can easily be overcome by implementing mechanisms for charge separation, which are typically based on the electric field induced ionization of excitons on timescales shorter than the exciton lifetime. The concept of field ionization and charge separation has been widely used in semiconductor devices over the last decades. Among the most important devices in this category are photodiodes and CCD detectors. More recently, related strategies have been proposed for 2-dimensional structures, such as charge separation in coupled quantum wells[60] and lateral charge separation in single quantum wells by electric fields or surface acoustic waves.[61] From a more general point of view however, important properties and quantum numbers of the incident photons are irreversibly lost by applying such schemes. Lost are not only the coherent properties of the photons, but also the information about the incident energy due to energy relaxation processes, the information about spin orientations, and spatial information due to carrier diffusion effects.

After the advent of high quality semiconductor QD systems some of those basic drawbacks can now be overcome. Whereas the associated improvements of the decoherence times are in particular important in the context of quantum computing,[62] some other basic properties of QDs have a more direct impact on memory devices. The most basic property of a QD is certainly its discrete density of states, which is often expressed in

the brief term "artificial atom". QDs, fully quantized semiconductor objects, exhibit therefore discrete absorption and emission spectra, which allow basically for energy conservation during an absorption-emission cycle. The temporal transfer of one of the charge carriers out of the QD, required for the realization of long storage times, does not destroy the energetic information deposited during the absorption process, as long as the second carrier resides in the QD. Subsequent back-transfer in turn retains an exciton in the groundstate of the same QD, ready to emit from the same quantum mechanical environment at the energy of absorption. The localization of a carrier in a defined nano-environment is therefore an important concept, which applies naturally also to spatial information, and – under suitable conditions concerning the spin lifetime - also to spin and hence polarization.

From those basic considerations it seems therefore most promising to implement optical write and read in order to get maximum benefit out from the underlying QD hardware. First exploratory investigations in this field were concentrated however on optical write and electric readout, an approach which seems to make sense in view of the challenges invoked in this new subject. Yusa and Sakaki have demonstrated an optically gated FET, which incorporates a layer of InAs QDs.[63] The operation of this device was attributed to the preferential trapping of nonresonantly photogenerated holes in a QD layer. Finley et al. have extended this approach to conditions of resonant excitation in the region of the ground state of the QD ensemble.[64] Shields et al. have demonstrated a scheme of electric readout in considerably smaller FET structures, which allows to monitor discrete charging events of single QDs in the FET transport.[65] A first demonstration of optical readout was obtained by Lundstrom et al., employing electric field induced charge transfer, both ways between a QD and a strain-induced potential minimum in an adjacent QW.[66]

In the remaining part of this section we will concentrate first on a more detailed description of a FET memory structure, which is designed for optical write und electric readout. As a second point we will discuss a more simple two-terminal device, which is designed of optical write and energy conserving optical readout.

4.4.2. *Optical Quantum Dot Memory with Electric Read out*

The concept of electric read out is based on electrostatic interaction between a chargeable QD and a mobile electron system. The introduction of net charge into the QD will have two different effects on the electron system, a change in carrier concentration and a change in mobility by introducing additional Coulomb scattering. In the following we will describe a specific QD memory device for electric read out and report the results of spectrally resolved photoresistance studies. The band diagram of the device is shown in Figure 4.8 for different operating conditions. As important ingredients the device contains a two-dimensional electron gas (2DEG), a layer of self-organized InAs quantum dots, and a back contact. The device is designed to store electrons in the QD layer, with the charging condition being monitored via the resistance of the 2DEG.

More specifically, the device was grown on a [100]-oriented semi-insulating GaAs substrate and consisted of the following layers: A 500 nm p-GaAs back contact ($p=2 \times 10^{18}$ cm^{-3}) followed by a 240 nm undoped GaAs spacer. This was followed by 2.25 ML of InAs grown at 530 °C (QD layer) and a composite tunnel barrier consisting of 30 nm undoped $Al_{0.3}Ga_{0.7}As$ and five periods of a 2 nm(AlAs)/2 nm(GaAs) superlattice. After this a 10 nm-thick $In_{0.1}Ga_{0.9}As$ electron channel was grown followed by a 40 nm wide modulation doped $Al_{0.3}Ga_{0.7}As$ barrier. The modulation doping was incorporated into two 1.5 nm δ-doped (2 x 10^{12} cm^{-2}) GaAs quantum wells. Finally, the structure was capped with 10 nm of n-doped (4 x 10^{18} cm^{-3}) GaAs. The wafers were processed into a Hall bar geometry using standard optical lithographic techniques and separate ohmic contacts were established to both the 2D electron channel and the p-type back contact. The p-i-n junction containing the QDs had a diode like characteristics with reverse bias breakdown voltages > 10 V at 300 K. The 2D electron channel was characterized using magneto transport techniques with an electron densitiy of N_s= 3.4 x 10^{11} cm^{-2}.

The top part of Figure 4.8(a) ("write" condition) shows a schematic representation of the device band diagram in the growth (z) direction, for a situation where the p contact is maintained at the same electric potential as the 2DEG channel by the external circuit. By illuminating the sample with monochromatic light, the QD ensemble is selectively excited as

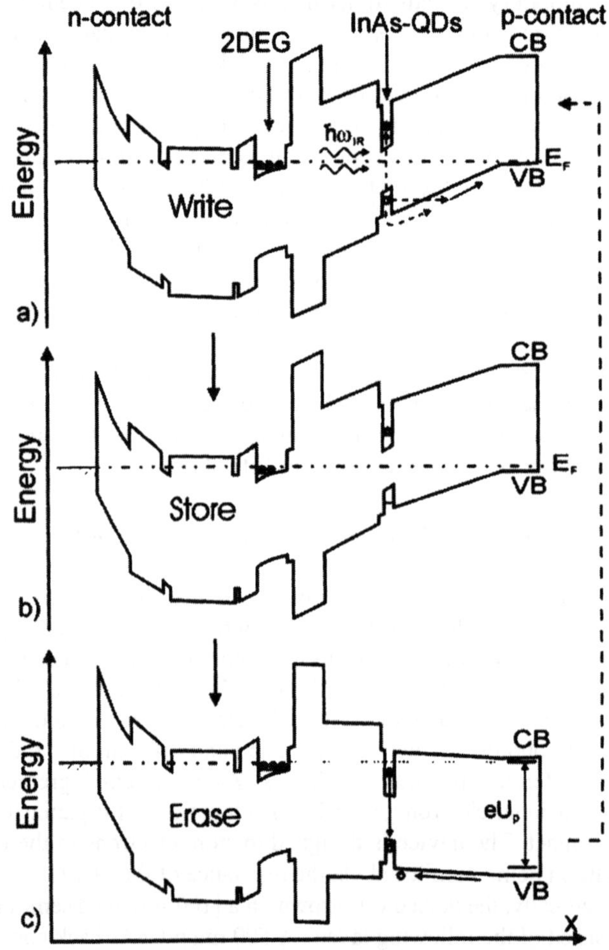

FIGURE 4.8. Schematic band diagram and operating principle of the optical memory structure. (a) Write cycle: Optical excitation of e-h pairs in the QD-layer, followed by hole escape by tunneling. (b) Store condition: Electrons remain trapped in the QD-layer. (c) Erase cycle: V_{pn} is switched to forward bias condition. Holes drift to the QD-layer followed by radiative recombination with the stored electrons.

described above. As a consequence of the internal electric field in this asymmetric hetero-type p-n junction, tunneling will be much more likely for holes as compared to electrons. This asymmetry is a natural consequence of the sample design, which provides a GaAs barrier for holes towards the p-type back contact and an $Al_{0.3}Ga_{0.7}As$ barrier with exponentially reduced transmission for electrons towards the 2DEG. Electrons remain therefore stored within the QD layer and will change the lateral resistivity of the 2DEG channel. The corresponding situation is depicted in the central part of Figure 4.8 (b) ("store"). The photo-generated charge in the QDs can be erased by application of a positive reset pulse $V_{pn} \geq 1.5$ V on the back contact with respect to the 2DEG (see Figure 4.8 (c) "erase").

A corresponding device structure view is shown in Figure 4.9(a) to (c). Near infrared illumination causes electron-hole pair generation in a subset of the InAs QDs (a).

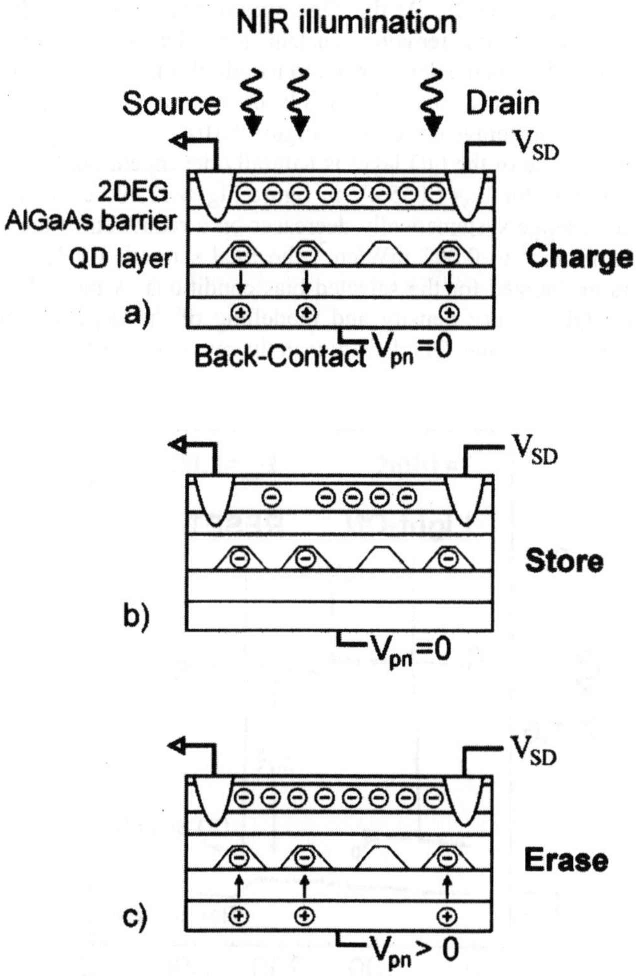

FIGURE 4.9. Schematic device structure and operating principle of the optical memory structure. (a) Write cycle: Optically excited e-h pairs are separated in real space, electrons remain stored in the QDs. (b) The stored negative charge acts electrostatically on the 2DEG and causes depletion, which can be monitored in the source-drain resistance. (c) Application of a reset pulse causes hole drift to the QD-layer followed by radiative recombination with the stored electrons.

After hole tunneling, the remaining electrons in the QDs are stored and cause a carrier depletion in the 2DEG channel (b). The stored electron density in the QDs will determine the amount of depletion in the channel and hence the resulting resistance change in a monotonic way, as expected for FET structures. The application of a reset pulse V_{pn} will cause hole drift towards the QD layer. This in turn causes an erase of the stored electrons by radiative recombination (c). At the end of the reset pulse, surplus holes will again tunnel out of the QDs, leaving the structure in an erased condition.

To demonstrate the charge storage effects in the presented structure we show time dependent resistance measurements consisting of a 100 s duration illumination phase, followed by a 100 s storage period, and a reset pulse in Figure 4.10. Illumination at E_{ex}=1.05 eV results in an increase of the channel resistivity $\Delta R = R_1 - R_0$ of about 25% with respect to the initial value R_0. At the beginning of the storage condition (light off) ΔR is slightly reduced first, but remains constant thereafter on this 100 s timescale at T=145 K. A reset pulse brings the device back to its initial state. It is important to note, that a reference sample grown with an identical layer sequence, but without the QD layer shows effectively no such storage effects (see Figure 4.10).

The charging rate of the QD layer is naturally dependent on the incident optical power density, as shown for E_{ex}=1.055 eV in figure Fig. 4.11. Whereas the slope of the temporal resistance change systematically decreases on a reduction of the optical power density from 0.5 mW/cm^2 to 0.003 mW/cm^2, the final saturation value of the channel resistance appears unchanged for the selected bias conditions. A more detailed analysis (AFM data of the QD surface density and modelling of the depletion effects on the 2DEG channel) shows, that the QD density is in the range of 1.5x10^{10} cm^{-2}, whereas the

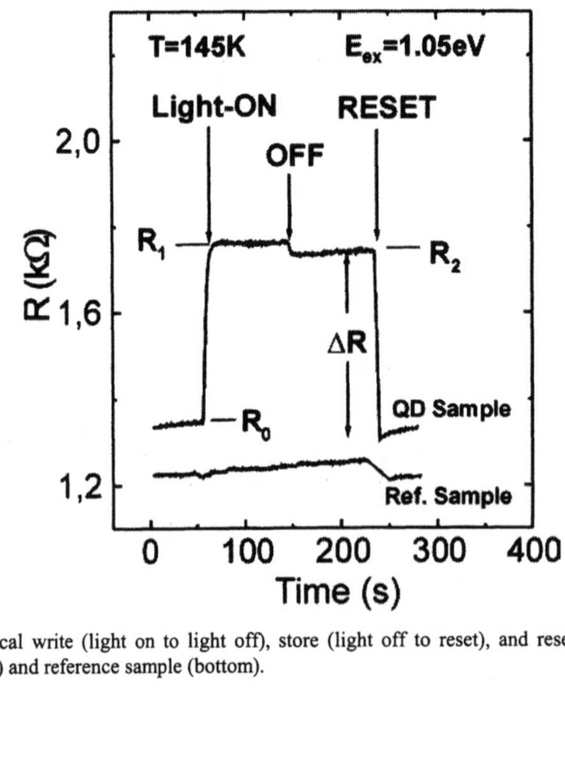

FIGURE 4.10. Typical write (light on to light off), store (light off to reset), and reset cycle for a structure containing QDs (top) and reference sample (bottom).

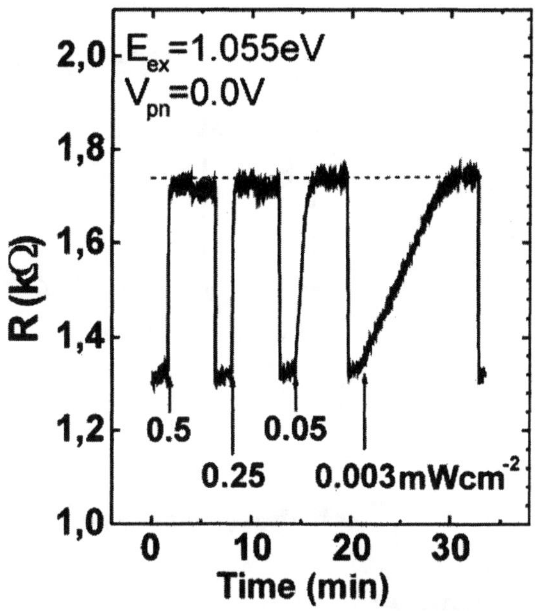

FIGURE 4.11. Write–store-eset cycles for different excitation power densities.

saturated surface density of the stored charge is in the range of 1.1×10^{10} cm^{-2}. This means, that on the average almost each QD carries one stored electron. However it remains unclear at this point, whether a smaller fraction of QDs is multiply charged, or excitation mechanisms exist, which eventually charge QDs at energies away from E_{ex} and thus distribute charge more or less over the whole ensemble.

An important point for the solution of this problem and for the understanding of the optically induced storage effect in general is its spectral dependence on the incident laser energy E_{ex}. For a discussion of this issue we show spectrally resolved data of the resistance change ΔR and, for comparison, of the photoluminescence (PL) response in Figure 4.12. The PL response (full line) arises from the recombination of excitons localized in the QDs, the two peaks correspond to ground (E_0) and first excited state (E_1) emission, respectively. The onset of the photoresistance effect observed for E_{ex}=1 eV clearly coincides with the direct excitation into the ground state. In the following, we demonstrate that phonon assisted processes are involved in the QD charging process. As the vertical p-i-n junction is more strongly reverse biased, the 2DEG channel becomes progressively depleted and consequently the magnitude of the observed photoresponse becomes strongly enhanced ($\Delta R/R_0$> 100% at V_{pn}= -2.5 V). Moreover, pronounced peaks labelled A_0 and A_1 become more clearly resolved from the background. The energy separation between A_0 and A_1 is about 75 meV, larger than the splitting between the ground and excited state PL emission (60 meV). Furthermore, these resonances are shifted from E_0 by 33 and 115 meV for A_0 and A_1 respectively. This indicates that they do not simply reflect the direct excitation of ground and excited QD states. The results shown in Figure 4.12 reflect the saturation value of the channel resistance. Thus, all

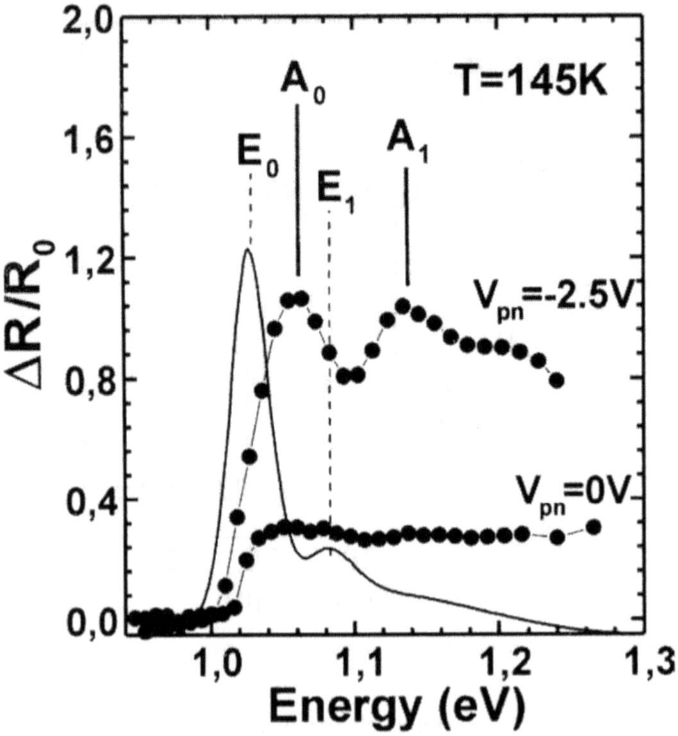

FIGURE 4.12. Spectral dependence of the photoeffect for different V_{pn}. PL data is shown for comparison (full line).

possible excitation mechanisms are included irrespective of the relative absorption probabilities for individual processes.

The energy shift between E_0 and A_0 is 33 meV, which is close to the LO phonon energy expected for InAs QDs in GaAs.[67] The participation of inelastic processes in optical QD excitation has been observed using PL excitation (PLE) spectroscopy on both large ensembles [68] and single dots,[69] with a number of vibrational modes involved. Thus, a high density of QD states is accessible via inelastic excitation. This may account for the relatively large charge storage densities observed, following selective optical excitation. A more direct manifestation of this interpretation towards phonon-assisted absorption will follow in the next paragraph, when device structures with optical readout are discussed.

On the basis of the presently discussed InAs/GaAs structures the above described storage effects are comfortably observable for hours at base temperatures below 150 K. At elevated temperatures thermionic emission will reduce the storage times. At room temperature storage times up to 0.5 ms have been observed[70], which are not off scale in comparison to storage times in modern DRAM memory devices. Significant improvements can be expected in the near future from slightly different III/V semiconductor material systems, which offer considerably higher band offsets.[71]

4.4.3. *Optical Quantum Dot Memory with Optical Read out*

The memory device discussed in the previous section was designed for optical write and electric readout. Now we introduce an alternative concept, which is based on a two terminal diode structure designed for optical write and read. As we will see in the following, this allows for the implementation of a wavelength domain memory on the basis of an inhomogeneously broadened ensemble of QDs. The basic working principle of the device can be inferred from the band diagrams shown in Figure 4.13. A layer of self assembled InGaAs QDs is embeded in the intrinsic region of a heterotype n-i-Schottky diode. The QDs are again sandwitched between a GaAs layer and an $Al_{0.45}Ga_{0.55}As$ layer with much higher bandgap. In the "write" condition, the device is unbiased (V_{reset}=0). The corresponding built in potential leads to high electric fields in the intrinsic region of the device. Optical excitation within the QD layer leads therefore to the generation of electron hole pairs with subsequent electron tunneling through the GaAs barrier. The remaining holes are stored because of the exponentially smaller tunneling probability through the $Al_{0.45}Ga_{0.55}As$ barrier. The memory structure is on hold under this condition.

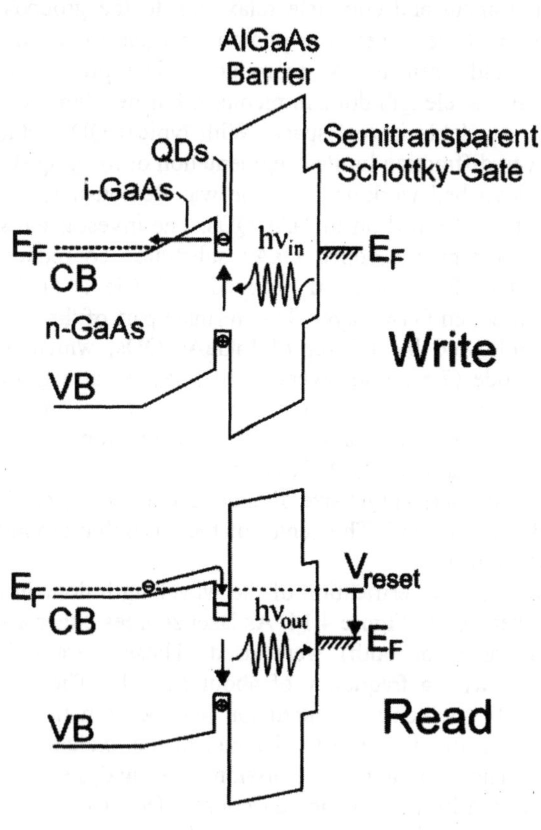

FIGURE 4.13. Schematic band diagram and operating principle of an all-optical, wavelength domain memory structure. Optical excitation of e-h pairs in the QD-layer followed by electron tunneling allows wavelength-selective write operations in an inhomogeneously broadened ensemble of dots (top). Read cycle: V_{reset} is switched to forward bias condition. Electrons drift to the QD-layer followed by radiative recombination with the stored hole. The outgoing photons appear basically at the wavelength of the preceding write process.

For resonant optical excitation in the region of the groundstates of the QDs (within the inhomogeneously broadened ensemble), only a small subset of QDs is expected to be charged. For a given laser energy $E_L=h\nu_{in}$ ideally only QDs with matching ground state energies will absorb and hence end up charged. Since excitonic ground state energies depend on the size and composition a QD, subsets of different morphologic structure will be selectively addressed by writing with different laser energies. Such manipulations can be safely performed, either parallel in time or time multiplexed, as long as the device remains biased in the hold condition ($V_{reset}=0$).

In order to readout the previously written information in the wavelength domain, an energy-maintaining readout scheme has to be employed. This is naturally obtained by spectrally resolved optical detection. The present memory device can be easily switched to the "read" condition by applying a reset pulse ($V_{reset}>0$), which brings the Schottky diode to a forward bias condition. At the rising edge of this pulse electrons from the n-GaAs backcontact start to drift towards the QD layer and lead, in QDs where stored holes are present, to radiative recombination ($h\nu_{out}$, see Figure 4.13). Light emission occurs therefore only from previously written QDs. Since each QD represents a unique quantum mechanical nano-enviroment and complete relaxation to the groundstate can be safely assumed for the present time scales, the photon energies of the originally employed storage procedure should basically be reproduced. The present device is therefore expected to perform as wavelength domain memory. Further dimensions for information storage are of course available in real space. With typical QD surface densities in the range of 300 μm^{-2}, even diffraction limited segmentation or imaging should be possible.

The above described working principle was demonstrated in low temperature (T=4.2 K) experiments performed on InGaAs QDs. The investigated sample was grown by MBE on a (100) oriented n-doped GaAs substrate. To reduce lattice defects a superlattice consisting of 25 periods of n-doped AlAs/GaAs (15/25 Å) was grown, followed by 2000 Å n-doped GaAs layer. The intrinsic part of the structure starts with a 500 Å GaAs layer followed by a layer of InGaAs QDs, which were grown in the Stranski-Krastanow mode (7.5 mono layers of $In_{0.4}Ga_{0.6}As$ at a growth temperature of 530°C). The QDs were covered by 10 Å of GaAs followed by a 500 Å thick $Al_{0.4}Ga_{0.6}As$ barrier and a final 100 Å GaAs cap layer. A Schottky junction was prepared by thermal evaporation of a semitransparent 100 Å Ti layer. Studies on control samples for atomic force microscopy exhibit a lateral QD size of about 20 nm, a height of 5.5 nm, and a QD surface density of about 300 μm^{-2}. The center of the ensemble groundstate PL emission occurs at 1.264 eV (980 nm).

For the experimental verification of the wavelength domain memory effect we have used the setup sketched in Figure 4.14. As laser sources either a 980 nm diode laser or a cw Ti:Saphire laser (or both) were used. Those were either electrically or mechanically chopped, with a frequency of about 50 kHz. The resulting on-off ratio exceeded a factor of 10^{10}, which is required for "monochromatic" electroluminescence (EL) experiments as performed here. The diameter of the excitation spot on the sample was about 100 μm. The resulting EL emission was analysed in a 0.64 m grating spectrometer and detected by a gated photon counter (Si avalanche photodiode). Light emission from the QDs was only recorded while the laser beam was off. A corresponding pulse sequence is shown in the lower left of Figure 4.14. The gating of the detector is activated about 4 μs after laser turn-off, to allow for complete PL decay. The electric reset pulse is applied in the center of the gate pulse ($V_{reset}>0$). EL photon emission is expected at the leading edge of reset pulse. With such a timing, only recombination events of stored holes in QDs with electrons supplied during reset are recorded.

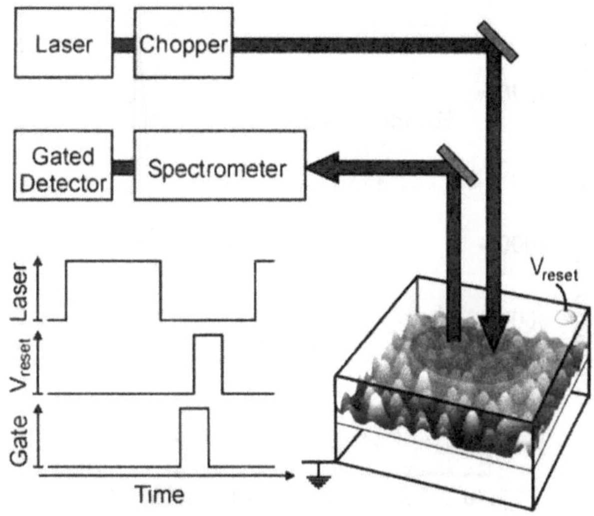

FIGURE 4.14. Experimental setup for the demonstration of a wavelength selective memory based on an inhomogeneously broadened ensemble of dots. After charging a subset of dots by resonant optical excitation in the groundstate the laser is switched off. Photon emission on readout occurs on the rising edge of V_{reset}. Prior to this the gating of the detector is activated.

Corresponding data are shown in Figure 4.15 for diode laser excitation at 980 nm. The resulting EL spectrum can be only detected in the presence of a readout pulse (as expected) and consists of two major contributions. The most intense peak appears at about E_{EL}=1.27 eV, slightly above the laser excitation energy E_L=1.264 eV (980 nm). The origin of this peak is the above described size selective absorption in a subset of QDs, followed by storage and recombination during reset. During this sequence the energy of the incident light remains in fact mostly conserved. A more detailed look reveals however a systematic blue shift by about ΔE_{Stark}=7meV, which is a consequence of the (periodic) time dependence of V_{reset}. The reason for this shift is easily explained in terms of the quantum confined Stark effect. During the “write” condition (V_{reset}=0), where optical absorption takes place, the QDs experience a huge perpendicular internal electric field (≈ 6×10^4 V/cm), which leads to a red-shift of the QD groundstate. During the “read” condition (V_{reset}=0.6V), where EL emission takes place, the QDs are in a flatband condition with almost vanishing electric field. The emission energy appears therefore blue-shifted with respect to the absorption energy.

In addition to the narrow main peak, the spectrum shown in Figure 4.15 contains also a broad emission feature, which appears red-shifted by 30 to 60 meV. This emission feature contains some substructure with a main peak E_{LO}, which is about 32 meV below the resonant main peak. Related phenomena have been observed previously in PL excitation spectroscopy experiments on single QDs. Phonon assisted absorption processes under the participation of InAs (32 meV) and GaAs (36 meV) LO phonons as

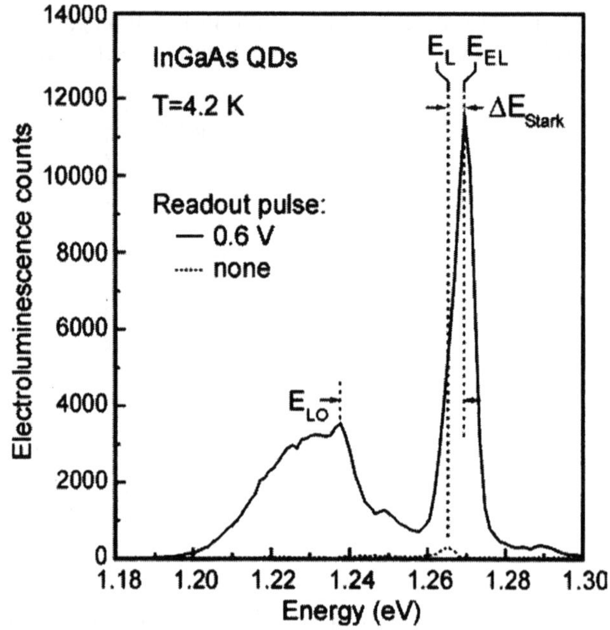

FIGURE 4.15. Spectrum of the electroluminescence output signal, which occurs on the rising edge of the readout pulse. The signal consists of a resonant part (E_{EL}), which is Stark-shifted with respect to the laser excitation energy E_L. The broad shoulder on the left corresponds to contributions from phonon assisted absorption and excited states. The sharp feature labeled E_{LO} is separated by 32 meV from the main line, which corresponds to the InAs LO-phonon energy.

well as excited state absorption have been shown to contribute about equally strong to the overall absorption of a comparable QD[72]. Resonant absorption in QDs with groundstate energies 30 meV and more below E_L will therefore result in charging as well, mediated by phonon assisted and excited state absorption.

For the demonstration of wavelength domain storage a two color experiment has been performed, as shown in Figure 4.16. Here two laser sources, a 980 nm diode laser and a tunable Ti:Saphire laser are used simultaneously for optical excitation. The resulting spectra exhibit the same general features as the single source spectrum discussed before. As expected, tuning of the Ti:Saphire laser results in an EL emission peak with similar tuning behavior. The EL emissions corresponding to both excitation sources can be clearly distinguished throughout the tuning range of the Ti:Saphire laser. The net emission can be described in first order as linear combination of independent contributions from two different QD sub-ensembles.

This remarkable result of multiple energy storage in a QD ensemble has been observed in low temperature experiment so far. Elevated temperatures will lead to thermionic emission and spectral out-diffusion of the EL response in the current GaAs/InGaAs system, as outlined before. Nevertheless the current results seem very encouraging and provide a basis for wavelength domain optical memory devices based on semiconductor QDs.

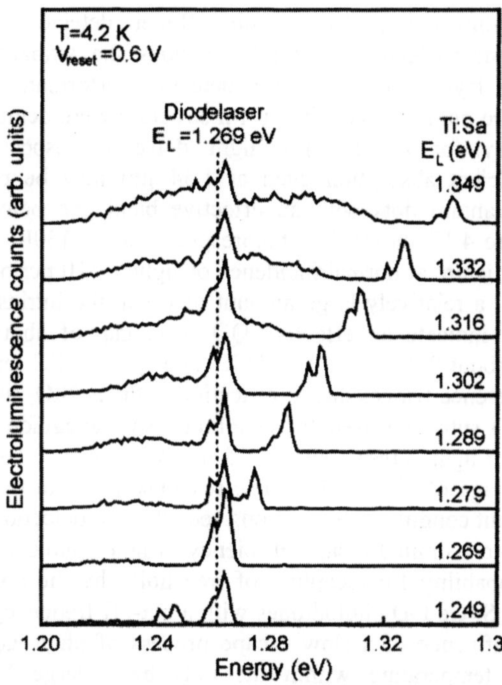

FIGURE 4.16. Demonstration of wavelength domain memory operation. Spectrally resolved readout signal for different two-color write pulses (tunable Ti:Saphire laser and fixed diode laser).

4.5. INTERBAND PHOTODETECTORS

An obvious advantage of self-assembled Stranski-Krastanov islands compared to quantum well structures is the enlarged amount of low band gap material incorporated in a host without generation of defects. Especially the larger height of islands compared to the thickness of a QW layer made from the same of material results in a reduced quantum confinement energy of localised carriers and by this in a lower effective absorption edge energy in heterostructures with QDs compared to QWs. This benefit of QD photodiode structures is exploited in Si/Ge QD structures summarised in the next section 4.5.1. A more advanced benefit of QD photodiodes is the specific band structure with discrete excitonic interband transitions in a quantum dot. Novel functionality of such QD photodetectors based on an extremely sharp spectral photoresponse, its tunability, and its saturation behaviour is studied with single quantum dot photodiodes (section 4.5.2) that eliminate the random scatter in transition energies caused by inhomogeneous size fluctuations in QD ensembles. Using a single QD excitonic level as an absorber or emitter further opens the path towards coherent quantum devices that may be applied for future quantum information processing and teleportation.

4.5.1. Si/Ge Quantum Dot Photodiodes

Very sensitive photodiodes from Si and Ge bulk material are widely applied for detection of light at wavelengths down to about 1.0 and 1.7 μm, respectively. Optical interconnects operating at λ=1.54 μm require efficient detectors on Si substrate for their integration in Si microelectronic circuits. Besides the straight-forward approach to deposit thick Ge layers on Si whose detector performance suffers from defect formation,[73] Si/SiGe multiple QWs in a p-i-n structure were demonstrated to show good quantum efficiency for λ= 1.3 μm light detection, especially in a waveguide geometry.[74,75] Efficient absorption even at 1.54 μm may be reached by embedding epitaxial Si/Ge quantum dots with an effective band gap below 0.8 eV in p-i-n Si photodiodes. Figure 4.17 shows the photoresponse at T= 150K observed from a Si/Ge QD p-i-n diode structure at normal incidence of light.[24] 110 periods of (10 nm Si/ 8 ML Ge QDs) represent a relatively large amount of Ge in the intrinsic region of a Si p-i-n diode. PL studies indicate an effective QD band gap of about 0.8 eV at T= 10K, corresponding to about 0.73 eV at T= 300K. The spectral PC response is undulated by Fabry-Perot interference fringes due to reflection at the Si/SiO_2 interfaces of the Silicon-On-Insulator (SOI) substrate used for enabling also waveguide detector operation. At normal incidence of light, a responsivity of 5 mA/W is observed at about λ= 1.54 μm for a reverse bias voltage of V = -5.0 V. The response is about two orders of magnitude lower at short circuit conditions. This strong reduction of detection efficiency at low bias may be attributed to a limited escape efficiency of holes photogenerated within the dots and to a high probability for recapture of free holes by the dots. Also the frequency behaviour of such Si/Ge QD photodiodes with a cut-off frequency f< 1 kHz at T= 150K indicates strong influence of a slow escape process of photo-generated holes that are localised at low temperature within the dots by a large VB offset. Admittance measurements of Si/Ge QD structures support this and suggest thermal escape times of about 10 ms at T= 150 K and of about 0.1 μs at T= 300K for holes to be thermally excited out of the dots. The time response and the efficiency depending on temperature for such Si/Ge dot NIR detectors with large hole localisation energies have to be further investigated. The spectral response of PC below about 1 eV indicates an indirect band

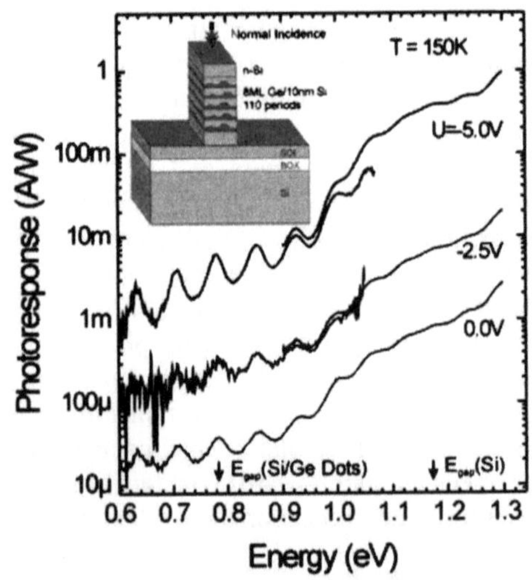

FIGURE 4.17. Photocurrent spectra for different bias voltages from a *pin* diode structure on SOI substrate with a 110 period (10 nm Si/8 ML Ge QD) multiple layer at T= 150K. The exciting light was at normal incidence resulting in interference fringes.

gap and the deduced absorption strength at about 0.8 eV is comparable to that across the gap in bulk Ge regarding the total amount of Ge within the islands. A significant increase of PC is observed at about 1 eV. It might be attributed to absorption at the direct band gap E_0 within Ge dots. An increase of E_0 from the bulk Ge value 0.8 eV is expected due to quantum confinement, a lowering of Ge content by Si/Ge intermixing and the deformation potential shift of E_0 with compressive strain. A tailoring of the direct band gap E_0 in Si/Ge dot structures, that was not investigated so far, and exploiting the abrupt onset of strong absorption at E_0 may promise very efficient Si/Ge QD NIR detector structures of high quality.

4.5.2. InGaAs Single Quantum Dot Photodiodes

Most of the QD devices discussed today make use of self-organized QD ensembles. Up to now the inhomogeneous broadening effects, which result from size and composition fluctuations, could not be reduced below about 15 meV.[76] Interband optoelectronic devices based on semiconductor quantum dots, such as lasers or modulators, could be improved tremendously, if homogeneous ensembles would be

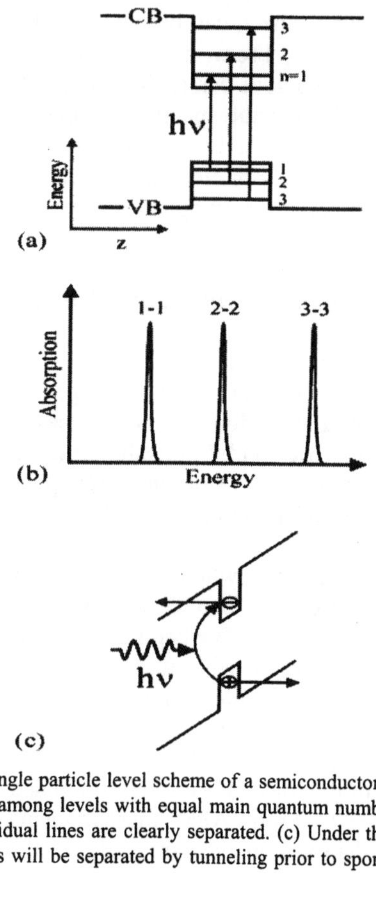

FIGURE 4.18. (a) Schematic single particle level scheme of a semiconductor QD. Optical interband absorption results mainly from transitions among levels with equal main quantum number n. (b) The absorption spectrum is delta-function like, the individual lines are clearly separated. (c) Under the influence of high electric fields photoexcited electron-hole pairs will be separated by tunneling prior to spontaneous recombination and hence contribute to photocurrent.

available. The same argument applies for interband photodiodes, but here it seems easier to engineer useful devices, which are based on a single QD – so called single quantum dot photodiodes.

With a single QD photodiode it is possible to benefit from the sharp spectral absorption features of semiconductor "artificial atoms" as known from the field of single QD spectroscopy.[77] The basic idea behind this device is sketched in Figure 4.18(a) to (c). The major allowed interband transitions in a QD appear between shells with equal principal quantum number n (a). The resulting absorption lines are sharp spectral features with basically lifetime limited linewidth (b). At high internal electric fields (typically ≥40 kV/cm in the GaAs/InGaAs material system), the tunneling time τ_t out of the QD is shorter as the radiative lifetime τ_r (c). According to $\tau^{-1}=\tau_r^{-1}+\tau_t^{-1}$, the total decay rate τ^{-1} will be dominated by tunneling under this condition. At sufficiently small tunneling times, effectively each exciton generated by an elementary absorption process will be photoionized by tunneling. In the electric circuit this process is detected as separation of a single electron-hole pair and hence as a PC, which is equivalent to the flow of a single elementary charge e through the diode. The maximum photocurrent, which can be obtained from a resonantly excited single QD photodiode will depend on the rate, at which successive cycles of absorption and photoionization can be performed. As we will see in the following, the tunneling time τ_t can be tuned by the external bias voltage V_B from infinity down to the ps range. From this we can estimate, that the amount of photocurrent from a single QD photodiode can easily exceed 10 nA.

In the following we will describe specific features of single QD photodiodes. Further on we will generalize the underlying concept und describe the single QD photodiode as a single 2-level system with electric contacts. Thereby we will show, that such systems can provide an essentially new optoelectronic functionality in the coherent regime.

4.5.3. Basic Properties of Single Quantum Dot Photodiodes

From a technological point of view, a single QD photodiode consists of a QD embedded typically in the intrinsic region of a p-i-n or n(p)-i-Schottky diode with optical access. Such an arrangement allows for different types of applications, like PL experiments as a function of electric field,[78] optically detected single electron charging ,[79,80,81,82] single QD light emitting devices,[83,84] or single QD photocurrent spectroscopy,[78,85] which was first demonstrated by Beham et al.[86] using local Schottky contacts induced by metallic scanning probes.

Here we describe now devices, which are defined by lithographic techniques. For controlled tuning of the electric field, the QDs are embedded in the intrinsic region of a n-i-Schottky diode (see Figure 4.19(b)). The band diagram under reverse bias condition (negative V_B on the Schottky contact with respect to the back contact) is shown schematically in Figure 4.19(a). The sample was grown by molecular beam epitaxy on a n^+-GaAs substrate. Nominally 8 monolayers of $In_{0.5}Ga_{0.5}As$ have been deposited at 530 °C on an intrinsic layer (here GaAs) above the n-doped GaAs layer (5×10^{18} cm^{-3}). The width and exact composition of the i-layers was varied to match the specific requirements for different experiments.[78,80,81,85] From AFM measurements on uncapped reference samples we determined dot densities in the range of 1×10^{10} cm^{-2} and an average dot height of 4 nm.

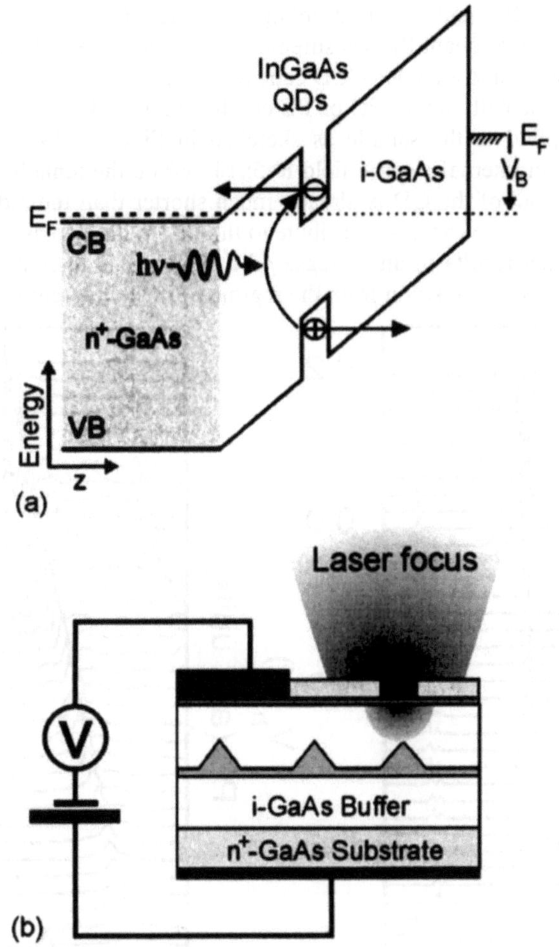

FIGURE 4.19. (a) Schematic band diagram of a n-i-Schottky diode. For detection of QD PC a reverse bias voltage is applied. (b) Device structure of a single QD photodiode. The QDs are embedded in a n-i-Schottky diode with a near field shadow mask on top.

The samples are processed as photodiodes combined with near field shadow masks as shown schematically in Figure 4.19 (b). Electron beam lithography and a dry chemical etch process ($SiCl_4$) are used to define an aluminium mask with apertures ranging from 150 to 500 nm. As a top contact we use a 5 nm thick semitransparent Ti layer (optical transmission ≈ 50%) below the electrically isolated shadow mask to achieve homogeneous electric field in the region of the investigated QD. As a back contact a Ge/Au/Ni/Au film was deposited and alloyed. For this device the whole mesa structure (300×400 μm) is contacted electrically, but only the QD under the selected aperture is probed optically in PC and PL experiments.

The devices were mounted in a confocal low-temperature microscope for analysis (see reference 87 for details) operated at temperatures down to T=2.3 K. In the PC experiments we used a power stabilized cw Ti:Sapphire laser for resonant optical excitation of the QDs and a current-voltage converter for detection of the photocurrent. For additional non-resonant PL measurements we used a HeNe laser (632.8 nm), a 0.7 m Dilor spectrometer, and a cooled CCD camera for detection.

Experimentally either PC or PL can be studied, depending on the external bias voltage V_B applied to the sample as sketched in Figure 4.19(a). For V_B<-1V, which corresponds to an internal electric field F>5x10^4 V/cm, the tunneling time τ_t of the photo excited carriers out of the QD is already much shorter than the radiative live time τ_r. In this regime most of the carriers contribute to the PC. With increasing V_B the electric field is reduced, which results in an increase of τ_t. Since τ_r is in first order independent on electric field we get a transition from the regime of PC to the regime of PL.

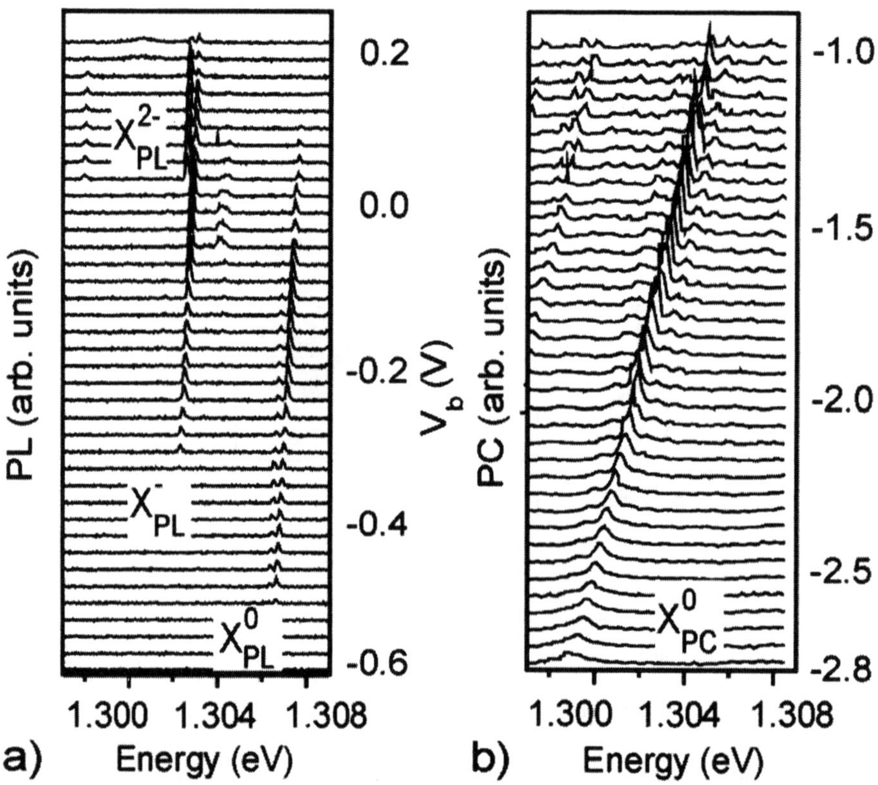

FIGURE 4.20. Photoluminescence (a) and photocurrent (b) spectra measured on a single QD photodiode as a function of the bias voltage V_B. In PC only the neutral QD is probed whereas in PL the X_0, X_-, and X_{2-} states are occupied according to the external voltage. For comparisons between PL and PC data Stark shift and bias controlled charging of the QD have to be taken into account.

In Figure 4.20 we present PL and PC spectra as a function of V_B taken at the same aperture of the shadow mask. In both cases we observe well separated sharp peaks attributed to single QD absorption (PC) and emission lines (PL). As a function of V_B a Stark shift is observed, which is stronger pronounced in the regime of PC. In PL (Figure 4.20(a)) two lines at 1302.8 meV and 1307.4 meV (V_B=0 V) marked as X^-_{PL} and X^0_{PL} dominate the spectra. In the PC spectra (Figure 4.20(b)) a single line, marked as

X^0_{PC}, appears at 1303.6 meV (V_B=-1.5 V). These lines arise from one and the same QD as discussed further on.

In PL the two main lines X^-_{PL} and X^0_{PL} are separated by 4.6 meV. Both lines are observed even for very low excitation power of only 5 Wcm^{-2}. In this limit we can exclude occupation of the QD with two or more excitons. Hence, biexcitons or higher multi-exciton states can not be the origin of the two lines. Moreover, typical biexciton binding energies for these QDs are about 3 meV.[88] In the regime of the PL, at low electric field, charging effects have been shown to play an important role on the characteristic features of the spectra. With increasing V_b the band flattens and the electron levels of the QD are shifted below the Fermi-energy of the n-GaAs back contact. This results in successive single electron charging of the QD (discussed more in detail in ref.[80]). The data in Figure 4.20(a) show the transition from the neutral to the single charged and the twofold charged exciton with increasing V_B. The charging of the QD is accompanied by discrete jumps in the emission energy due to few particle interaction of the exciton with an increasing number of electrons. Hence we can assign the two lines in the PL spectra for V_B<0 V to the charged (X^-_{PL}) and neutral (X^0_{PL}) exciton state of the QD separated by the X^- binding energy. In this sample, for V_B>-0.35 V the QD is already charged with one electron. Due to the statistical nature of the carrier capture process for non resonant excitation the X^-_{PL} and X^0_{PL} lines coexist in the PL spectra of this sample for a certain range of V_B. To compare PL and PC data, without disturbing energy shifts due to many body effects, the uncharged QD ground state has to be considered. This is represented by the lines X^0_{PL} and X^0_{PC}. The X^0_{PL} line can be traced in bias voltage ranges of –0.5 V<V_B<0 V. For V_B<-0.5 V τ_r exceeds τ_t and the regime of PC is entered.

In the PC regime the electron states of the QD are always lying far above the Fermi level of the n-GaAs back contact and the QD remains uncharged. To further study the influence of electric field on the uncharged exciton ground state the X^0_{PC} line as a function of V_B is considered in Figure 4.20(b). The PC signal can be observed in the voltage range from –1 V to -2.8 V. The magnitude of the PC saturates for V_b<-1.3 V (at a constant laser power of about 100 $kWcm^{-2}$). Under these conditions the experimentally observed PC peak magnitude is 140 pA (X^0_{PC}). This current corresponds to a generation rate of 10^9 e-h-pairs per second. Without electric field the exciton lifetime τ in a QD is about 1 ns.[89] For a generation rate of 10^9 e-h-pairs per second and τ=1 ns a QD would be occupied statistically with one exciton at a time. However, for a QD in electric field τ can be considerably smaller due to a reduced τ_t: $\tau^{-1}=\tau_r^{-1}+\tau_t^{-1}$. In this situation the statistical occupancy of the QD is well below one.

For a single QD, the reduction of τ with increasing electric field can be deduced from the PC line width. For increasing negative V_b the PC line width Γ becomes broader (Figure 4.20(b)). In the limit of high electric fields reduction of τ due to shorter τ_t is the main mechanism. τ can be estimated to 0.3 ps from the PC line width of about 1 meV at V_b=-2.5V ($\Gamma\cdot\tau=h/2\pi$). Assuming τ=0.3 ps, a statistical occupancy of the QD below 10^{-3} can be estimated from the PC magnitude. At the low field side, around $V_B\approx$-1V, the PC amplitude quenches abruptly. Here the line width of 250 μeV is certainly not lifetime limited, but restricted by the experimental conditions. In the transition region between PC and PL the condition $\tau_t\approx\tau_r\approx$1ns is valid. The observed strong variation of the exciton lifetime from typical 1ns at zero electric field to below 1ps at 83 kV/cm can be used to tune the exciton lifetime in a QD.

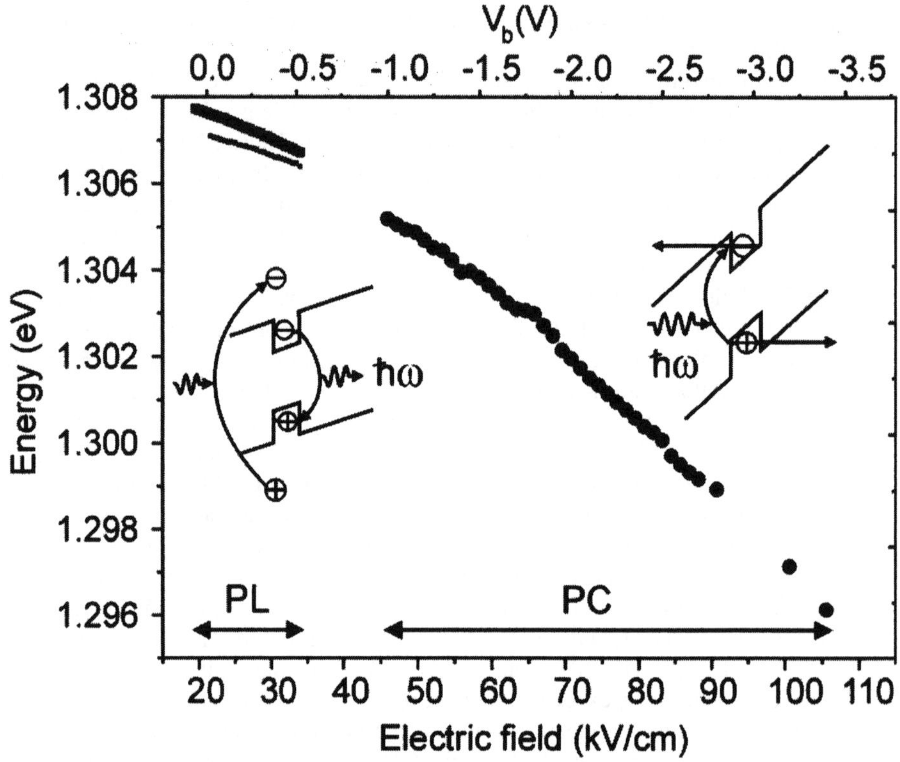

FIGURE 4.21. Line position of the neutral QD ground state as a function of the electric field measured in PL (main peak and satellite) and PC spectroscopy. The spectral shift due to the QCSE nicely continues in both regimes. The insets illustrate the different processes involved in PC and PL.

To further compare the X^0 line position in PL and PC, the quantum-confined Stark effect (QCSE) has to be taken into account. The QCSE causes a red shift of the QD ground state with increasing electric field F. The Stark shift continues monotonically over both regimes (see Figure 4.21). This nicely confirms, that in fact one and the same QD is observed. Hence, such a monotonic behaviour in absorption and emission spectra is expected only for an empty QD where renormalization effects are negligible.

The field dependence of the X^0 transition energy E_{X0} can be described by the expression $E_{X0} = E_{X0,F=0} + p \times F + \beta \times F^2$.[90] Here, $E_{X0,F=0}$ is the transition energy at zero field, p describes the dipole moment and β the polarization of the QD due to the QCSE. Taking PL and PC data into account the QCSE for this QD can be studied over a range of about 100 kV/cm. Using the PC data and the PL main peak we determined a nonzero dipole moment of $p=8\times10^{-29}$ Cm. These values are in good agreement with the work of Fry et. al who find $p=7\pm2\ \times10^{-29}$ Cm for an ensemble of InAs QDs suggesting a localization of the electrons (holes) in the base (apex) of the QD.[90]

We have seen now, that detailed insights into the charge distribution within a QD can be inferred form electric field tunable QD structures. For device applications, single QD photodiodes will inherently provide sharp resonant spectral sensitivity, a feature which is new in comparison to existing photo-detectors based on semiconductor

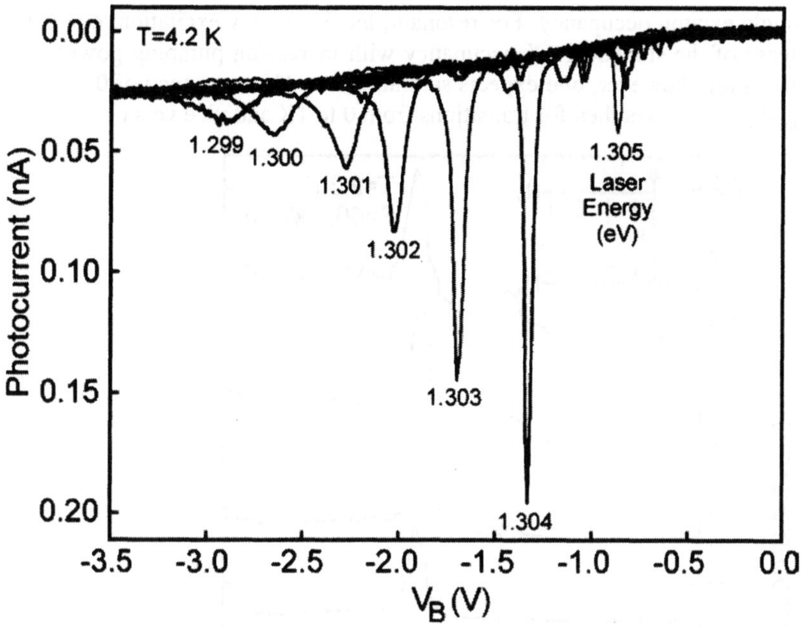

FIGURE 4.22. Photocurrent response (versus V_B) of a single quantum dot photodiode for different excitation energies.

structures with higher dimensionality. Even more important is the fact, that this spectral response can be tuned by the QCSE. Within a limited energy range, a single QD photodiode provides therefore the function of a mesoscopic, electric field tunable spectrometer without any additional dispersive elements. Corresponding data are shown in Figure 4.22 for excitation with various energies from a Ti:Saphire laser. Even with the current devices it is possible to tune the detection energy over a range, which corresponds to more than 10 linewidth. For applications in the nano-world it is further important to note, that the absorption volume of those devices is ultimately small for a semiconductor, allowing for detectors with highest spatial resolutions.

4.5.4. The Single Quantum Dot Photodiode as a 2-Level System with Electric Contacts

So far we have discussed the single QD photodiode in the context of conventional semiconductor devices. Taking a more general look from basic physics, a QD can be described within the general concept of a 2-level system.[91] In this framework a single QD photodiode is essentially a single 2-level system with electric contacts, a setup which has not been obtained so far on the basis of other 2-level systems, such as atoms or spins. In the following we analyse the response of a single QD within the model of 2-level systems. This approach is valid within a limited energy range around the single exciton ground state of a QD, where only one particular dipole allowed transition is able to modify the occupancies of the two considered levels. In the present case the energetically lower level "0" corresponds to the empty QD ground state, the higher level

"1X" to a single exciton occupancy. For resonant, incoherent cw excitation, one expects first an increase of the statistical 1X occupancy with increasing pumping power. In the high excitation limit, however, one expects saturation at a 1X occupancy of 0.5, because there are equal probabilities then for transitions from 0 to 1X and vice versa.

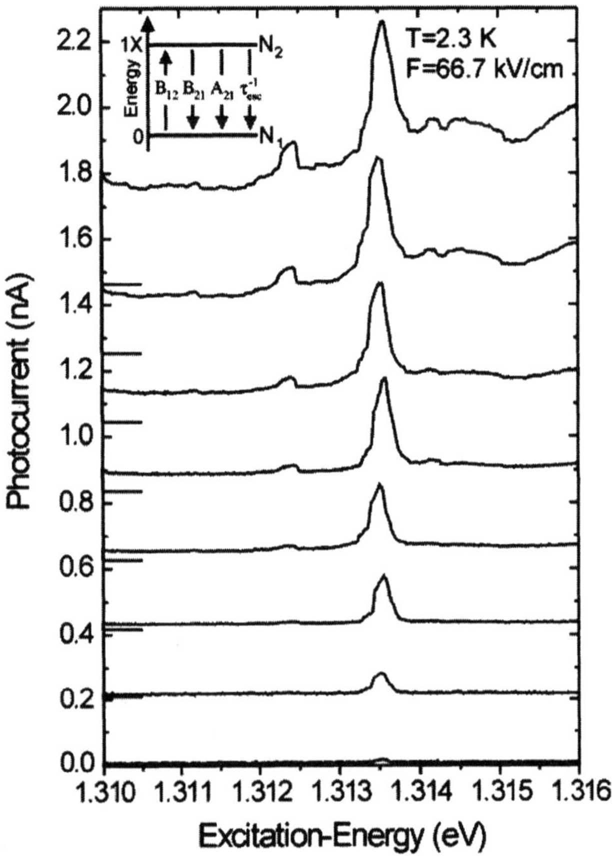

FIGURE 4.23. PC spectra of a single QD ground state under resonant excitation conditions. With increasing excitation power a saturation of the PC from the QD is observed. For each spectrum the zero PC line is indicated on the left. The inset shows a sketch of an excitonic 2-level system (see text).

In Figure 4.23 we show a series of PC spectra of a single QD photodiode for increasing excitation power. The dominating PC line results from the ground state absorption of the investigated QD. For further discussion we use the peak amplitude of the PC line as a measure of the QD ground state absorption at fixed electric field. Starting at a relatively low excitation level, the absorption line grows up significantly, but with increasing laser power we observe a saturation of the PC peak amplitude. Further absorption appears to be suppressed in the limit of high excitation. Additional background PC at high excitation intensities is probably due to additional excitation of other QDs by stray light, since all QDs are connected electrically in parallel. When a QD is occupied, its energy levels are subjected to a spectral shift (typically a few meV) due to

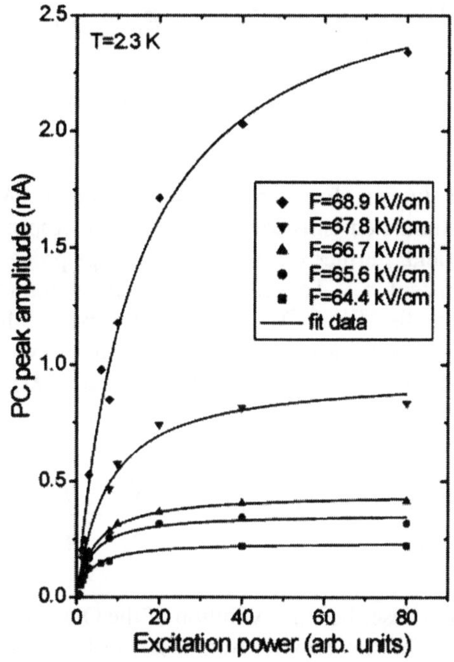

FIGURE 4.24. PC saturation as a function of excitation power for different bias conditions. For stronger electric fields a higher PC saturation value is observed. The experimental data are fitted by a specifically calculated function, based on the rate equation model described in the text.

renormalization effects.[91,92,93,94,95] Thus no second absorption process can take place at the same excitation energy as long as the QD is occupied. Experimentally, we are able to monitor the occupancy of the QD quantitatively by measuring the PC as described above. Similar series of power dependent PC spectra have been taken for different bias conditions. They all exhibit a saturation of the PC amplitude, measured from the peak of the dominating PC line (at ~1.3135 eV) with respect to the background PC. The experimental data are summarized in Figure 4.24. The saturation level of the PC amplitude shows a pronounced dependence on the electric field. At higher electric fields F the tunneling time is reduced and the QD returns faster in its initial empty state. Therefore a higher absorption rate is possible, leading to a higher saturated PC.

The saturation of the QD PC with increasing excitation power can be described by a fundamental rate equation model. In this model we concentrate on a specific QD ground state with an exciton occupation number N. The status of the QD is modeled in terms of a 2-level system. The QD is declared to be in level 1, if it is unoccupied, and in level 2 if it is occupied with one exciton. For these two levels we define two time-averaged occupation numbers N_1 and N_2, with values between 0 and 1, complementing each other to one: $N_1+N_2=1$. The underlying 2-level system is sketched in the inset of Figure 4.23. According to the theory of 2-level systems[96] we consider the following rate equations:

$$\frac{dN_1}{dt} = A_{21}N_2 + B_{21}N_2\rho - B_{12}N_1\rho + \frac{N_2}{\tau_{esc}} \tag{4.2}$$

$$\frac{dN_2}{dt} = -A_{21}N_2 - B_{21}N_2\rho + B_{12}N_1\rho - \frac{N_2}{\tau_{esc}} \tag{4.3}$$

Here the well known Einstein coefficients enter: A_{21}, B_{21} and B_{12} for spontaneous emission, stimulated emission and absorption respectively. ρ is the energy density of the radiation field. Further we include the escape process of the exciton out of the QD with a time constant τ_{esc}. To solve the rate equations (4.2) and (4.3) we use $B_{12}=B_{21}$ [96] and introduce the factor M reflecting the probability for the stimulated processes and we use $A_{21}=1/\tau_r$, where τ_r is the radiative recombination time of the exciton. Furthermore the strength of the radiation field ρ is expressed in terms of the excitation power P. Due to the relation $N_1+N_2=1$ we obtain from Eq. (4.3)

$$\frac{dN_2}{dt} = (1-2N_2)MP - \frac{N_2}{\tau_r} - \frac{N_2}{\tau_{esc}} \tag{4.4}$$

This rate equation describes the evolution of the QD occupation N_2. The two last terms in Eq. (4.3) stand for the two exciton loss mechanisms of the QD, radiative recombination and escape of the exciton. The first term on the right side of Eq. (4.4) stems from the power dependent resonant creation and annihilation of the exciton. It reflects the balance between absorption and stimulated emission. Since the resulting effective change of the occupation depends on $(1-2N_2)$, we recognize that the absorption behavior of the QD is affected by the occupation of the dot itself. This property leads to the nonlinear power dependence and saturation of the PC.

To return to the experimental data we calculate the steady state occupation N_0 of the QD and use the escape term in Eq. (4.4) to obtain the resulting PC

$$I = e \cdot \frac{N_0}{\tau_{esc}} = \frac{e}{2\tau_{esc}} \times \frac{P}{P + \frac{1}{2M} \times \left(\frac{1}{\tau_{esc}} + \frac{1}{\tau_r} \right)} \tag{4.5}$$

With this expression it is possible to fit the power dependence of the PC peak amplitudes in Figure 4.24 (see full lines) with good accuracy.

From Eq. (4.5) we obtain $I_{sat}=e/2\tau_{esc}$ as the high power saturation value of the PC. The tunnel escape time τ_t depends strongly on the chosen electric field. From a fit to the model we are able to obtain the tunneling escape time from the observed I_{sat}. Since there are both electrons and holes in the QD, it is of interest which escape time (electron or hole) is relevant here. Since the tunneling time is strongly affected by the energy barrier and the effective mass of the carrier, electron and hole exhibit different tunneling times. After tunneling of the faster of the two carriers, there is still one carrier left in the QD and the absorption is still bleached due to renormalization.[90] After the second carrier has also tunneled out, the QD is again in its initial state and ready for the next absorption process. This means, it is the lifetime of slower tunneling carrier that controls bleaching and determines the value of the saturated PC. Equation 3 is used to calculate the carrier lifetime from the saturated PC. For saturation values of 2.78 nA at F=68.9 kV/cm and 0.24 nA at F=64.4 kV/cm one obtains a tunneling lifetime for the slow carriers of 29 ps and 330 ps respectively.

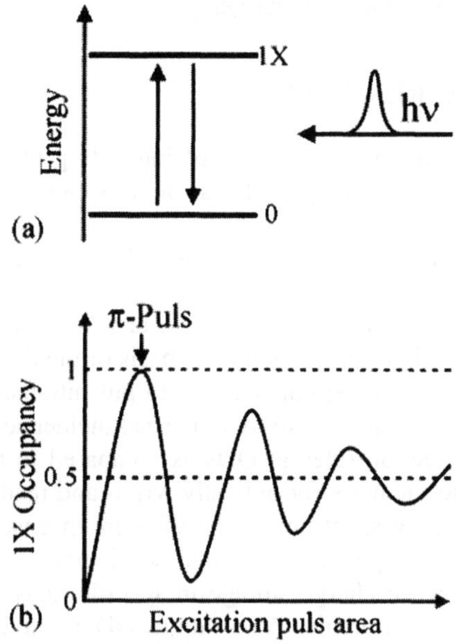

FIGURE 4.25. (a) 2-level system corresponding to the quantum dot ground state 1X in a photodiode with resonant, pulsed excitation. (b) Rabi-oscillations in the 2-level system versus the excitation pulse area in the presence of weak decoherence. Application of a π pulse results in a 1X occupancy close to 1.

An absolutely new optoelectronic functionality can be expected from a single QD photodiode, if it is driven by short laser pulses in the coherent regime. Recently coherent population oscillations, so called Rabi oscillations,[97] have been demonstrated in the exciton population of single QDs,[98,99,100] compare Figure 4.25. Low temperature dephasing times τ_p for excitons in self-assembled quantum dots have been shown to exceed several hundred ps, allowing for sufficiently high numbers of coherent manipulations with ps pulses.[101] Tunneling on a timescale $\tau_t < \tau_p$ will lead to enhanced dephasing and limits the available time range for coherent interactions (τ_t accounts here to the first carrier which escapes from the QD – after that the original excitation is dephased). Safe conditions for coherent excitation can be however obtained by using ps pulses from a mode locked laser. Coherent optical excitations of a single QD photodiode are then expected to result in Rabi oscillations, which can be directly and quantitatively evidenced in photocurrent (PC). Optical excitation with a π-pulse results in the coherent generation of a single exciton, which is subsequently dephased and photoionized by tunneling. In the electric circuit this process is detected as separation of a single electron-hole pair and hence as a deterministic PC, which is equivalent to the flow of a single elementary charge e through the diode. A repetitive optical excitation of a single quantum dot photodiode with π-pulses of frequency f has therefore been predicted to result in the quantitative PC of $I=f \times e$.[102] A coherently driven single quantum dot photodiode is therefore an optically triggered single electron source, which offers, just like the related single electron turnstile device ,[103] single electrons (or holes) on demand. Both devices have the potential to deliver frequency controlled currents according to $I=f \times e$. With circular polarized excitation, a single QD photodiode has further the potential to provide deterministic streams of spin polarized charges.

4.6. CONCLUDING REMARKS

In summary we have seen a variety of interesting QD devices, which emerge from self-assembled growth technologies. Those devices can be subdivided into different groups:

Into a more conventional group we like to put ensemble interband photodetectors. The introduction of self-assembled QDs results here in an extension of the available sensitivity range as compared to bulk or QW approaches. Detectors for the 1.5 μm regime can now easily be achieved on the basis of the Ge QDs on Si substrate.

Into a more QD specific group we like to put intraband photodetectors. Their performance, namely the inherent sensitivity to normal incidence radiation, is specifically based on the modified selection rules in QDs as compared to QW structures. Specific properties of QDs are also the basis for optically write- and readable memory structures. Here it is the discrete energy spectrum of the QDs in an ensemble, which allows for wavelength domain memory applications. Huge confinement energies are at the same time a necessary condition for charge storage in specific QDs and for isolation among individual QDs within the ensemble. Also single QD photodiodes are based on the discrete energy spectrum, which QDs offer. In addition to this however it became clear, that a single QD photodiode has to be described as a two-level system with electric contact in the limit of strong electromagnetic driving fields.

A coherently driven single quantum dot photodiode is an optically triggered single electron source, which offers, just like the related single electron turnstile device, single electrons (or holes) on demand. It has the potential to deliver frequency controlled currents according to $I=f \times e$. By means of single QD photodiodes it becomes possible to transfer optical excitations of single quantum systems into deterministic electric currents, which is essentially a new concept, relevant in general for optoelectronics with single quantum systems. This new and so far unattained opto-electronic functionality provides a link between the world of coherent optical excitations and electric currents on the single electron level. In the future this could allow for the electric readout of excitonic quantum gates.

Current activities in the field of quantum information technology[104] on the basis of excitonic phenomena in semiconductor QDs are mainly concentrated on coupled QD systems[105], single photon emitters[106,107,108], and high Q micro-resonators. Those new concepts for the implementation of future quantum information technologies require partly a coherent hardware, and they require input from the areas of quantum optics and cavity quantum electrodynamics. The use of coherent phenomena for the implementation of quantum information technology is expected to give plenty of room for advanced developments. This area of “applied quantum mechanics” however is very demanding and remains one of the biggest challenges for the future, also what the development of novel QD devices is concerned.

REFERENCES

1. Zhores I. Alferov Nobel Lecture: The double heterostructure concept and its applications in physics, electronics, and technology, Rev. Mod. Phys. **73**, 767 (2001).
2. Y. Arakawa and H. Sakaki, Appl. Phys. Lett. **40**, 939 (1982).
3. See for example: M. Bayer, A. Schmidt, A. Forchel, F. Faller, T. L. Reinecke, P. A. Knipp, A. A. Dremin, and V. D. Kulakovskii, Phys. Rev. Lett. **74**, 3439 (1995).
4. See for example: S. Tarucha, D. G. Austing, T. Honda, R. J. van der Hage, and L. P. Kouwenhoven Phys. Rev. Lett. **77**, 3613 (1996).
5. I. N. Stranski, and L. von Krastanow, Akad. Wiss. Lit. Mainz Math. Naturwiss. Kl. IIb **146**, 797 (1939).
6. For a review see: D. Bimberg, M. Grundmann, and N. N. Ledentsov, Quantum Dot Heterostructures, J. Wiley & Sons, Chichester (1999).
7. T. Demel, D. Heitmann, P. Grambow, and K. Ploog, Phys. Rev. Lett. **64**, 788 (1990).
8. K. Brunner, U. Bockelmann, G. Abstreiter, M. Walther, G. Böhm, G. Tränkle and G. Weimann, Phys. Rev. Lett. **69**, 3216 (1992).
9. K. Brunner, G. Abstreiter, G. Böhm, G. Tränkle, and G. Weimann, Phys. Rev. Lett. **73**, 1138 (1994).
10. A. Zrenner, L. V. Butov, M. Hagn, G. Abstreiter, G. Böhm, and G. Weimann, Phys. Rev. Lett. **72**, 3382 (1994).
11. B. J. van Wees, H. van Houten, C. W. J. Beenakker, J. G. Williamson, L. W. Kouwenhoven,, D. van der Marel, and C. T. Foxon, Phys. Rev. Lett. **60**, 848 (1988).
12. Y. W. Mo, D. E. Savage, B. S. Swartzentruber, and M. G. Lagally, Phys. Rev. Lett. **65**, 1020 (1990).
13. D. Leonard, M. Krishnamurthy, C. M. Reaves, S. P. Denbaars, and P. M. Petroff, Appl. Phys. Lett. **63**, 3203 (1993).
14. S. Fafard, Z. Wasilewski, J. McCaffrey, S. Raymond, and S. Charbonneau, Appl. Phys. Lett. **68**, 991 (1996).
15. B. R. Bennett, R. Magno, and B. V. Shanabrook, Appl. Phys. Lett. **68**, 505 (1996).
16. E. Alphandéry, R. J. Nicholas, N. J. Mason, and B. Zhang, Appl. Phys. Lett. **74**, 2041 (1999).
17. M. C. H. Liao, Y. H. Change, C. C. Tsai, M. H. Chieng, and Y.F. Chen, J. Appl. Phys. **86**, 4694 (1999).
18. O. G. Schmidt, C. Lange, K. Eberl, Appl. Phys. Lett. **75**, 1905 (1999).
19. T. I. Kamins, G. Medeiro-Ribeiro, D. A. A. Ohlberg, R. S. Williams, J. Appl. Phys. **85**, 1159 (1999).
20. V. A. Shchukin, N. N. Ledentsov, P. S. Kop'ev, and D. Bimberg, Phys. Rev. Lett. **75**, 2968 (1995).
21. S. Christiansen, M. Albrecht, H. P. Strunk, and H. J. Maier, Appl. Phys. Lett. **64**, 3617 (1994).
22. M. A. Cusack, P. R. Briddon, and M. Jaros, Phys. Rev. B **54**, R2300 (1996).
23. J. H. Seok, and J. Y. Kim, Appl. Phys. Lett. **78**, 3124 (2001).
24. M. Herbst, C. Schramm, K. Brunner, T. Asperger, H. Riedl, G. Abstreiter, A. Vörckel, H. Kurz, and E. Müller, Mat. Ci. Eng. B **89**, 54 (2002).
25. T. S. Kuan and S. S. Iyer, Appl. Phys. Lett. **59**, 2242 (1991).
26. Karl Brunner, *Si/Ge Nanostructures*, Rep. Prog. Phys. **65**, 27-72 (2002).
27. V. Le Thanh, V. Yam, P. Boucaud, F. Fortuna, C. Ulysse, D. Bouchier, L. Vervoort, and J.-M. Lourtioz, Phys. Rev. B **60**, 5851 (1999).
28. J. Tersoff, C. Teichert, and M. G. Lagally, Phys. Rev. Lett. **76**, 1675 (1996).
29. G. Springholz, V. Holý, M. Pinczolits, and G. Bauer, Science **282**, 734 (1998).
30. V. Holý, G. Springholz, M. Pinczolits, and G. Bauer, Phys. Rev. Lett. **83**, 356 (1999).
31. K. Brunner, J. Zhu, G. Abstreiter, O. Kienzle, and F. Ernst, phys. stat. sol. (b) **224**, 531 (2001).
32. G. Jin, J. L. Liu, and K. L. Wang, Appl. Phys. Lett. **76**, 3591 (2000)
33. J. Zhu, K. Brunner, and G. Abstreiter, Appl. Phys. Lett. **73**, 620 (1998).
34. Y. H. Phang, C. Teichert, M. G. Lagally, L. J. Peticolas, J. C. Bean, and E. Kasper, Phys. Rev. B **50**, 14435 (1994).
35. K. H. Goetz, D. Bimberg, H. Jürgensen, J. Selders, A. V. Solomonov, G. F. Glinskii, and M. Razeghi, J. Appl. Phys. **54**, 4543 (1983).
36. C. G. Van de Walle, Phys. Rev. B **39**, 1871 (1981).
37. B. A. Foreman, Phys. Rev. B **49**, 1757 (1994).

38. P. Bhattacharya, (Ed.), *Properties of Lattice-Matched and Strained Indium Gallium Arsenide*, 35-40, INSPEC, London (1993).
39. U. Hohenester and E. Molinari, Phys. Stat. Sol., **221**, 19 (2000).
40. M. M. Rieger and P. Vogl, Phys. Rev. B 48, 14276-14287 (1993).
41. G. Abstreiter, Semiconductors and Semimetals Vol. **49**, 37-76 (Academic Press, 1998).
42. S. K. Zhang, H. J. Zhu, F. Lu, Z. M. Jiang, and X. Wang, Phys. Rev. Lett. **80**, 3340 (1998).
43. C. Miesner, T. Asperger, K. Brunner, and G. Abstreiter, Appl. Phys. Lett. **77**, 2704 (2000).
44. T. Asperger, C. Miesner, K. Brunner, and G. Abstreiter, Thin Solid Films **380**, 227-229 (2000).
45. G. Schedelbeck, W. Wegscheider, M. Bichler, and G. Abstreiter, Science **278**, 1792-1795 (1997).
46. M. O. Lipinski, H. Schuler, O. G. Schmidt, K. Eberl, and N. Y. Jin-Phillipp, Appl. Phys. Lett. 77, 1789 (2000).
47. O. G. Schmidt and K. Eberl, Phys. Rev. Lett. 61 13721 (2000).
48. D. Pan, E. Towe, and S. Kennerly, Appl. Phys. Lett. **73**, 1937 (1998).
49. L. Chu, A. Zrenner, G. Böhm, and G. Abstreiter, Appl. Phys. Lett. **75**, 3599 (1999).
50. C. Miesner, O. Röthig, K. Brunner, and G. Abstreiter, Appl. Phys. Lett. **76**, 1027 (2000).
51. T. Fromherz, W. Mac, C. Miesner, K. Brunner, G. Bauer, and G. Abstreiter, Appl. Phys. Lett. **80**, 2093 (2002).
52. T. Fromherz, P. Kruck, M. Helm, G. Bauer, J. F. Nützel, and G. Abstreiter, Appl. Phys. Lett. **68**, 3611 (1996).
53. V. Ryzhii, Semicond. Sci. Technol. **11**, 759 (1996).
54. C. Miesner, K. Brunner, and G. Abstreiter, Infrared Physics & Technology **42**, 461-465 (2001).
55. D. Bougeard, K. Brunner, and G. Abstreiter, Proc. of ISCS 2001, IOP Conf. Series, (submitted).
56. L. Chu, A. Zrenner, M. Bichler, and G. Abstreiter, Appl. Phys. Lett. 79, 2249 (2001).
57. D. Leonard, M. Krishnamurthy, C. M. Reaves, S. P. Denbaars, and P. M. Petroff, Appl. Phys. Lett. **63**, 3203 (1993).
58. M. Bayer and A. Forchel, Phys. Rev. B **65**, 41308 (2002).
59. S. Muto, Jpn. J. Appl. Phys., Part 2 **34**, L210 (1995).
60. J. E. Golub, K. Kash, J. P. Harbison, and L. T. Florez, Phys. Rev. B **41**, 8564 (1990).
61. S. Zimmermann, A. Wixforth, J. P. Kotthaus, W. Wegscheider, M. Bichler, Science **283**, 1292 (1998).
62. P. Borri, W. Langbein, S. Schneider, U. Woggon, R. L. Sellin, D. Ouyang, and D. Bimberg, Phys. Rev. Lett. **87**, 157401 (2001).
63. G. Yusa and H. Sakaki, Appl. Phys. Lett. **70**, 345 (1997).
64. J. J. Finley, M. Skalitz, M. Arzberger, A. Zrenner, G. Böhm, and G. Abstreiter, Appl. Phys. Lett., **73** 2618 (1998).
65. A. J. Shields, M. P. O'Sullivan, I. Farrer, D. A. Ritchie, R. A. Hogg, M. L. Leadbeater, C. E. Norman, and M. Pepper, Appl. Phys. Lett. **76**, 3673 (2000).
66. T. Lundstrom, W. Schoenfeld, H. Lee, P. M. Petroff, Science **286**, 2312 (1999).
67. M. Grundmann, 0. Stier, and D. Bimberg, Phys. Rev. B **52**,11969 (1995).
68. M. J. Steer, D. J. Mowbray, W. R. Tribe, M. S. Skolnick, and M. D. Sturge, Phys. Rev. B **54**, 17738 (1996).
69. F. Findeis, A. Zrenner, G. Böhm, and G. Abstreiter, Phys. Rev. B **61**, R10579 (2000).
70. K. Imamura, Y. Sugiyama, Y. Nakata, S. Muto, and N. Yokoyama , Jpn. J. Appl. Phys. **34**, L1445 (1995).
71. H. Pettersson, L. Bååt, N. Carlsson, W. Seifert, and L. Samuelson, Appl. Phys. Lett. **79**, 78 (2001).
72. F. Findeis, A. Zrenner, G. Böhm, G. Abstreiter, Phys. Stat. Sol. B **224**, 337 (2001).
73. S. Luryi, A. Kastalsky, and J. C. Bean, IEEE Trans. Electron. Devices **31**, 1135 (1984).
74. K. Bernhard-Höfer, A. Zrenner, J. Brunner, G. Abstreiter, F. Wittmann, and I. Eisele, Appl. Phys. Lett. **66**, 2226 (1995).
75. B. Li, G. Li, E. Liu, Z. Jiang, J. Qin, and X. Wang, Appl. Phys. Lett. **73**, 3504 (1998).
76. H. Heidemeyer, S. Kiravittaya, C. Müller, N. Y. Jin-Phillipp, and O. G. Schmidt Appl. Phys. Lett. **80**, 1544 (2002).
77. A. Zrenner, J. Chem. Phys. **112**, 7790 (2000).
78. F. Findeis, M. Baier, E. Beham, A. Zrenner, and G. Abstreiter, Appl. Phys. Lett. **78**, 2958 (2001).
79. R. J. Warburton, C. Schäflein, D. Haft, F. Bickel, A. Lorke, K. Karrai, J. M. Garcia, W. Schoenfeld, and P. M. Petroff, Nature **405**, 926 (2000).
80. F. Findeis, M. Baier, A. Zrenner, M. Bichler, and G. Abstreiter, Phys. Rev. B **63**, 121309(R) (2001).
81. F. Findeis, M. Baier, A. Zrenner, M. Bichler, G. Abstreiter, U. Hohenester, and E. Molinari, Phys. Rev. B **63**, 121309(R) (2001).
82. J. J. Finley, P. W. Fry, A. D. Ashmore, A. Lemaître, A. I. Tartakovskii, R. Oulton, D. J. Mowbray, M. S. Skolnick, M. Hopkinson, P. D. Buckle, and P. A. Maksym, Phys. Rev. B **63**, 161305(R) (2001).

83. M. Markmann, A. Zrenner, G. Böhm, and G. Abstreiter, Phys. stat. sol. (a) **164**, 301 (1997).
84. Z. Yuan, B. E. Kardynal, R. M. Stevenson, A. J. Shields, C. J. Lobo, K. Cooper, N. S. Beattie, D. A. Ritchie, and M. Pepper, Science **295**, 102 (2001).
85. E. Beham, A. Zrenner, F. Findeis, M. Bichler, and G. Abstreiter, Appl. Phys. Lett. **79**, 2808 (2001).
86. E. Beham, A. Zrenner, and G. Böhm, Physica E **7**, 359 (2000).
87. A. Zrenner, A. Schaller, M. Markmann, M. Hagn, M. Arzberger, D. Henry, G. Abstreiter, G. Böhm, and G. Weimann, Appl. Surf. Sci. **123/124**, 356 (1998).
88. F. Findeis, A. Zrenner, G. Böhm, and G. Abstreiter, Solid State Commun. **114**, 227 2000).
89. R. Heitz, A. Kalburge, Q. Xie, M. Grundmann, P. Chen, A. Hoffmann, A. Madhukar, D. Bimberg, Phys. Rev. B, **57**, 9050 (1998).
90. P. W. Fry, I. E. Itskevich, D. J. Mowbray, M. S. Skolnik, J. J. Finley, J. A. Barker, E. P. O'Reilly, L. R. Wilson, I. A. Larkin, P. A. Maksym, M. Hopkinson, M. Al-Khafaji, J. P. R. David, A. G. Cullis, G. Hill, and J. C. Clark, Phys. Rev. Lett., **84**, 733 (2000).
91. L. Allen and J. H. Eberly, *Optical Resonance and Two-Level Atoms* (Wiley, New York, 1975).
92. E. Dekel, D. Gershoni, E. Ehrenfreund, D. Spektor, J.M. Garcia, and P.M. Petroff, Phys. Rev. Lett. **80**, 4991 (1998).
93. L. Landin, M.S. Miller, M.-E. Pistol, C.E. Pryor, L. Samuelson, Science **280**, 262 (1998).
94. F. Findeis, A. Zrenner, G. Böhm, and G. Abstreiter, Solid State Commun. **114**, 227 2000).
95. M. Bayer, O. Stern, P. Hawrylak, S. Fafard, A. Forchel, Nature **405**, 187 (2000).
96. A. Yariv, Optical Electronics, CBS College Publishing (1985).
97. I. I. Rabi, Phys. Rev. **51**, 652 (1937).
98. T. H. Stievater, Xiaoqin Li, D. G. Steel, D. Gammon, D. S. Katzer, D. Park, C. Piermarocchi, and L. J. Sham, Phys. Rev. Lett. **87**, 133603 (2001).
99. H. Kamada, H. Gotoh, J. Temmyo, T. Takagahara, and H. Ando, Phys. Rev. Lett. **87**, 247401 (2001).
100. H. Htoon, T. Takagahara, D. Kulik, O. Baklenov, A. L. Holmes, Jr., and C. K. Shih, Phys. Rev. Lett. **88**, 087401 (2002).
101. P. Borri, W. Langbein, S. Schneider, U. Woggon,R. L. Sellin, D. Ouyang, and D. Bimberg, Phys. Rev. Lett. **87**, 157401 (2001).
102. A. Zrenner, German Patent No. DE 100 06 909 C1, April 4, 2001.
103. H. Pothier, P. Lafarge, C. Urbina, D. Esteve, and M. H. Devoret, Europhys. Lett. **17**, 259 (1992).
104. See for example: "The Physics of Quantum Information", ed. By D. Bouwmeester, A. Ekert, and A. Zeilinger (Springer, Berlin, 2000).
105. M. Bayer, P. Hawrylak, K. Hinzer, S. Fafard, M. Korkusinski, Z. R. Wasilewski, O. Stern, and A. Forchel, Science **291**, 451 (2001).
106. J.-M. Gerard and B. Gayral, IEEE J. Lightwave Technol. **17**, 2089 (1999).
107. P. Michler, A. Kiraz, C. Becher, W. V. Schoenfeld, P. M. Petroff, Lidong Zhang, E. Hu, and A. Imamoglu, Science **290**, 2282 (2000)
108. E. Moreau, I. Robert, L. Manin, V. Thierry-Mieg, J. M. Gérard, and I. Abram Phys. Rev. Lett. 87, 183601 (2001).

5

Porous Silicon as an Open Dielectric Nanostructure: an Ensemble of Aspheric Silicon Nanocrystals

J. Diener, N. Künzner, E. Gross, G. Polisski, and D. Kovalev

Technische Universität München, Physik-Department E16, D-85747 Garching, Germany

5.1. INTRODUCTION

Reducing the dimensions of semiconductor structures is nowadays a well established field in solid state physics, not at least on demand by the preceding miniaturization in microelectronics. To confine carriers in two dimensions was one of the first steps towards the current nano-science and technology. At present, three dimensional confinement of carriers in quantum dots or nanocrystals (NCs) with typical dimensions of the order of nanometers is a standard technique to modify the physical properties known from the bulk material. One of the most prominent consequences arising from the reduced dimensionality is an altered density of states which is accompanied by a widening of the bandgap with decreasing size due to quantum confinement. Therefore the optical properties, especially the photoluminescence (PL), of such objects differ significantly from that of the bulk material.

One of the first observations of the size dependence of the optical properties of semiconductor nanocrystals was reported by Ekimov et al. [1] for semiconductor doped glasses and independently from Henglein [2] for colloidal solutions. It was shown for both systems that the color of the NCs strongly depend on their size. Since than a wide field of optical investigations on quantum dots evolved. Nowadays nanoscale modifications of a large variety of III-V and II-VI semiconductor compounds and group IV semiconductors like silicon (Si) and germanium are available and under intense investigation. Therefore the observation of visible PL at room temperature from porous silicon (PSi), a sponge like structure of partially interconnected Si NCs formed by electro chemical etching of bulk Si wafers, by Canham et al. [3] has also been attributed to quantum confinement effects. Actually at the same time Lehmann et al. [4] discovered an altered absorption behavior of PSi compared to bulk Si and also proposed quantum confinement modifications of the density of states. Since Si NCs, especially in the form of PSi, are compatible to Si based microelectronics a large amount of works where devoted to that topic [5] and up to now the vision of active light emitting Si-based electro optical applications is put forward [6,7,8]. However, their realization turned out to be quite difficult and working devices [9] are, so far, rare. Contrary, the first utilization of PSi was to realize passive optical elements, such as distributed bragg mirrors [10]. For these applications quantum confinement effects were of minor importance. The main reason to use PSi for such types of devices was that the refractive index (n) of PSi is reduced compared to that of bulk Si and technologically easy adjustable. n of PSi can be significantly smaller than that of Si because the porosity of PSi can be fairly large (up to more than 90% of Si is removed by the electro chemical etching). The porosity and consequently n is adjustable in a controlled manner by changing the etching parameters (for instance the current density). Further on, as the typical dimensions of the remaining Si and the voids are much smaller than the wavelength of visible light, PSi behaves as a macroscopic homogeneous dielectric medium in that spectral range - as long as the remaining Si is homogeneously distributed in space.

On microscopic scales the spatial modulation of the dielectric constant (ε) in PSi is quite large due to the big difference between the dielectric constants of the remaining Si and the surroundings. By controlling the spatial modulation of ε the opportunity arises to design artificial dielectric (nano-) structures.

While the concept of photonic bandgap materials [11, 12] is based on a regular, periodic modulation of the refractive index, we want to discuss an alternative way to produce artificial dielectric nanostructures by an overall spatial alignment of aspheric Si NCs. The structure of this chapter is as follows:

-The first part is a brief overview on the formation of PSi and the technological ways to influence the porosity and the morphology of the layers.

-The basic dielectric properties of PSi are introduced. It is shown how PSi acts as an effective dielectric medium (including theoretical considerations, experimental validations and devices based on the controlled variation of the dielectric constant of PSi).

- Consequences of the microscopic shape of the Si NCs on the PL properties of PSi - dielectric effects in the photoluminescence of PSi - are presented. Based on the fundamental PL properties of PSi the specifics of the polarization memory phenomenon (the PL from PSi is also preferentially linearly polarized when the luminescence is excited with linearly polarized light) are discussed. It is argued that the origin of this effect is the aspheric shape of the NCs in PSi and their overall alignment. How to gain control over the spatial distribution of luminescing NCs, their shape and their overall orientation is considered at the end of this section.

- Dielectric anisotropy of the refractive index of PSi (birefringence), arising from a uniaxial alignment of the aspheric Si NCs and devices based on birefringent PSi are debated in the last section of this chapter.

5.2. MATERIALS PROPERTIES OF POROUS SILICON

In recent years the number of works and review articles devoted to porous silicon was rising significantly. Therefore we will provide just to a short summary of the basic material properties of porous silicon (PSi), because they are already discussed in great details elsewhere [5,13,14], and focus on porous silicon as a dielectric nanostructure.

5.2.1. Formation of porous silicon

The formation of PSi has already been observed quite early [15] as a side-effect when studying the electropolishing of Si wafers. However, only after the reports of Canham [3] on visible room temperature photoluminescence (PL) and of Lehmann and Gösele [4] on an increased band gap compared to bulk Si due to quantum size effects did this material gain significant interest due to potential optoelectronic applications. In the same manner the procedures of preparing PSi samples were developed (details are for instance discussed in [5, 16, 17] Most commonly PSi is formed by the electrochemical etching of bulk Si wafers in an HF solution. The electrolyte is either an aqueous solution of HF or a mixture of ethanol and aqueous HF. The ethanoic HF solution is preferred because of its better wetability of the hydrophobic PSi and increased release of H_2 bubbles, formed during the etching on the sample surface. A Teflon beaker is pressed via an O-ring onto a Si wafer lying on a metal plate (the back contact). The front contact is established via a Pt mesh in the electrolyte. Details of the chemical reactions during etching and the very early stages of the pore formation from an initially flat Si wafer are still under debate. However, it is generally accepted that holes are required for the dissolution of Si. So far the model proposed by Lehmann and Gösele [4] is one of the most established as illustrated schematically in Fig. 5.1. It is assumed that the Si surface is H passivated until a hole reaches the surface which makes it possible for a F^- ion to replace the H of a Si-H bond. In the next step the second Si-H bond is attacked by a F^- ion, H_2 is released and an electron is injected back into the substrate. Finally the two Si backbonds are attacked by HF and the silicon atom is going in solution as a silicon tetrafluorid molecule. It reacts further with two HF molecules to give $SiF_4{}^{2-}$ and two protons.

For a p-type substrate, holes are majority carriers and therefore PSi is formed under anodic electrostatic conditions, with current densities below the electropolishing limits. The etching process is selflimited. With decreasing size of the remaining Si structures depletion occurs and the transport of holes is hindered. Consequently the dissolution of Si via the above mentioned process is not possible anymore. Therefore PSi layers are growing in depth while the already formed material is not dissolved (without additional illumination).

For an n-type substrate, holes are minority carriers and it is necessary to generate holes during etching near to the interface between the electrolyte and Si. This can be performed by additional illumination (photo generation of electron-hole pairs) during the etching process with visible light. We will restrict to PSi formed by electrochemical etching of (100) and (110) oriented p-type Si wafers in ethanoic HF

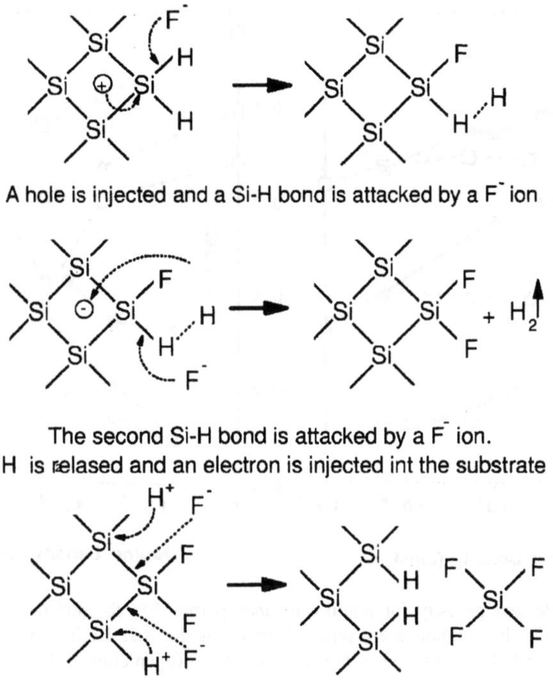

FIGURE 5.1. Dissolution of Si proposed by Lehmann and Gösele in [4].

solution. If not mentioned explicitly samples are produced from a (100) oriented substrate. PSi made from (110) oriented wafers is for clarity called (110) PSi.

5.2.2. *Tuning the porosity of porous silicon*

After the etching of Si wafers a sponge like network of Si is remaining. The porosity of the layer is defined as the fraction of empty space and can be determined by gravimetric methods. Porosities up to more than 90% can be achieved, but high porosity layers are mechanically fragile and it is necessary to apply supercritical drying [18, 19] to avoid stress during the removal of the electrolyte. For lower porosities the layers can be dried in ambient conditions, because they are mechanically more stable. With TEM (transmission electron microscope) and X-ray measurements it is possible to show that these particles are crystalline with their overall crystalline orientation like the Si substrate the layer was produced from ([5] and references in there). Due to the self limiting character of the electrochemical formation process caused by depletion of the etched Si the sizes of these nanocrystals (NCs) and the pores depend on the doping level of the substrate, the specifics of the electrolyte used (HF concentration, temperature) and the

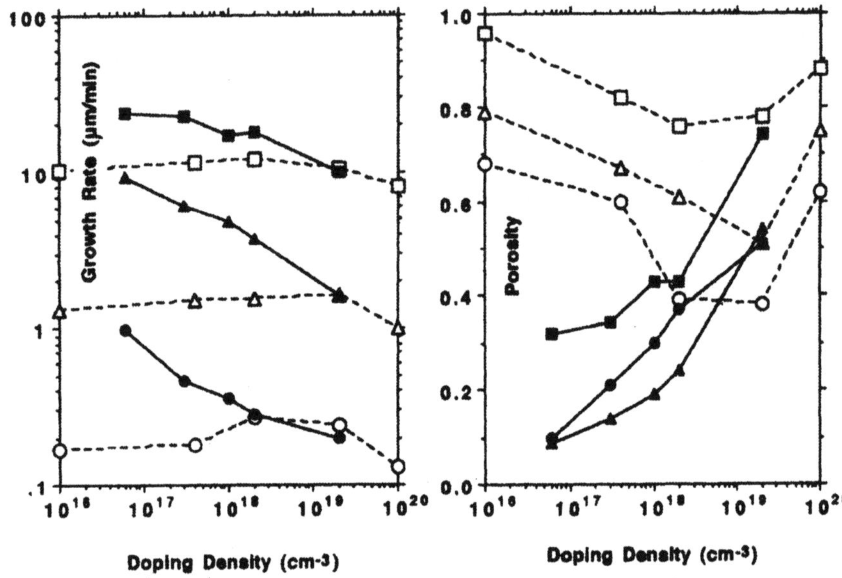

FIGURE 5.2. Growth rate and porosity for micro- and mesoporous PSi for different current densities and doping levels (from [20]). Solid symbols are n-type substrates anodized at 3 (circles), 30 (triangles) and 300 (squares) mA/cm^2. Open symbols are p-type substrates anodized at 3 (open circles), 30 (open triangles) and 300 (open squares) mA/cm^2.

applied current density. Porosity does not account for the typical dimensions of NCs and pores and therefore PSi is also sorted in three structural classes according to their mean pore size (see Table 5.1).

TABLE 5.1
Porosity does not account for the typical dimensions of NCs and pores and therefore PSi is also sorted in three structural classes according to their mean pore size.

main pore width (nm)	Type of PSi
≤ 2	Microporous Si
2-50	Mesoporous Si
≥ 50	Macroporous Si

The role of several parameters like for example doping level of the substrate, current density and HF concentration on the dimensions and the porosity of PSi has been studied in detail [13,20] (see for example Fig. 5.2 from [20]). There is no strict rule but the general trend is that the doping level of the substrate is mainly defining the structural sizes and the electrolyte (especially the HF concentration) and the corresponding current density determines the porosity of the layers. The upper limit of the current density is the electropolishing regime in which no PSi is formed any more. The thickness of the layer is determined by the etching time. Freestanding films are produced by detaching the PSi layer from the substrate by an electropolishing step after the layer is formed.

5.2.3. Morphology of porous silicon

The morphology of the remaining skeleton of Si nanoparticles is a result of the self limiting character of the etching process. Fig. 5.3 displays a set of scanning electron micrographs of PSi layers produced from (100) p-type substrates with different doping levels and current densities (from [20]). So far a detailed description of the arrangement of the remaining Si NCs is still missing. However, as evident from Fig. 5.3, the growth of the layer in depth is governed by a preferential pore propagation in the (100) growth direction.

This behavior is better defined for mesoporous Si (made from highly doped substrates, the right 2 columns of Fig. 5.3), but turns out to be also present for microporous silicon (made from low doped substrates). For (100) oriented substrates the growth direction and the direction of the current flow is parallel to the [100] crystalline

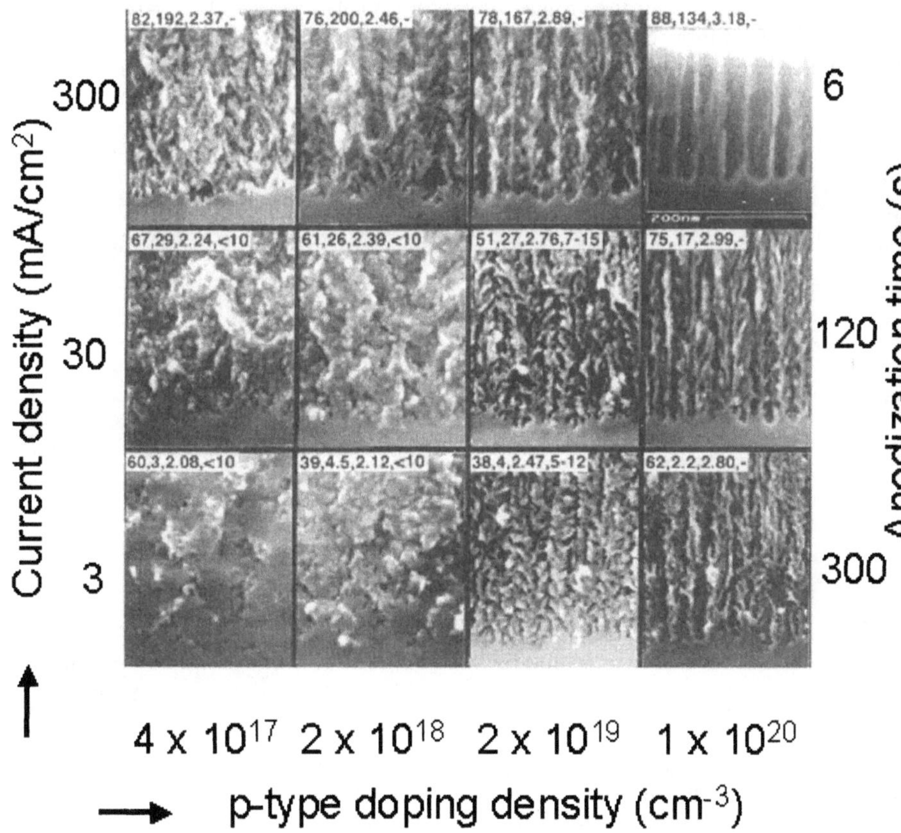

FIGURE 5.3. Scanning electron micrographs of the interface between bulk and porous silicon for PSi produced from p-type (100) oriented Si wafers with different current densities and doping density (from [20]).

direction perpendicular to the surface. Etching of a lower symmetry (110) oriented Si substrate shows that the layers grow in depth similar to (100) oriented Si, parallel to the direction of the current flow. However the pores preferentially grow in [100] crystallographic directions, not parallel to the direction of the current. The reason for the selectivity of the etching process, that pores tend to develop in equivalent [100] crystalline directions is so far not known but well established and documented[1] [5,21]. It is proposed that the termination of the Si interface with hydrogen is anisotropic which influences the current flow and the local dissolution of Si [21].

Related to the crystalline selectivity of the etching process is the shape of the formed Si NCs. They are more elliptical than round. Fig. 5.4 shows a TEM picture of a single NC. Also shown in Fig. 5.4 is the general arrangement of these Si ellipsoids for PSi produced from a (110) Si substrate in top view. Similar to the pore propagation, the long axes of these NCs are oriented in $[1\bar{1}0]$ direction, the projection of the [100] and [010] growth directions of the pores onto the (110) plane. However, visible light, having much larger wavelength than the typical structural dimensions in PSi, is averaging over the whole nanocrystalline ensemble. Therefore, arising from the large porosity of the layer, the dielectric response (for example the refractive index) of PSi differs from that of bulk Si.

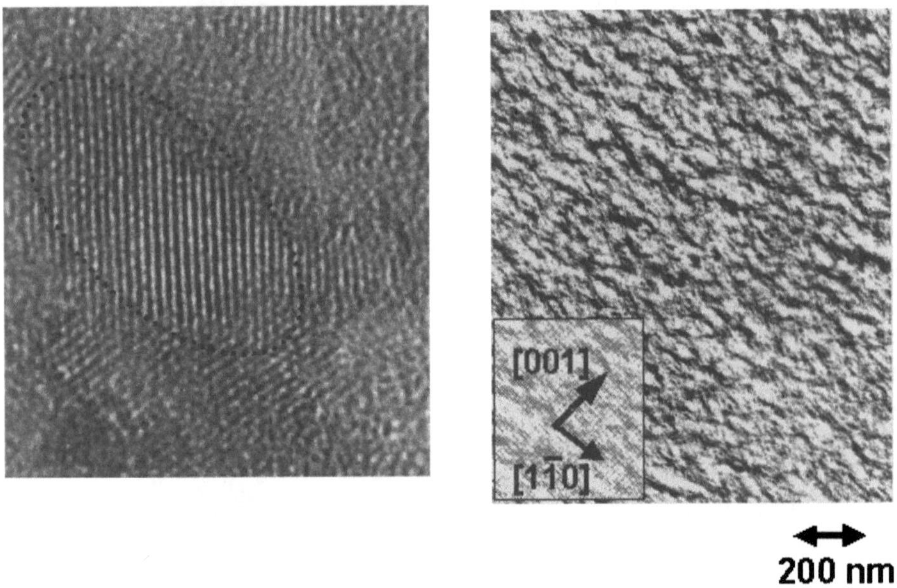

FIGURE.5.4. Left side: High resolution transmission electron micrograph of a single Si NC in PSi prepared from a (110) oriented Si substrate (p-type, 1 mΩ cm). The line is a guide to the eye to show the aspheric shape of the NC (the prolongation is parallel to the $[1\bar{1}0]$ direction as evident from the electron interference pattern).

Right side: transmission electron micrograph of the same sample with less magnification to show the overall alignment of the Si NCs.

---

[1] It is also possible to find extreme etching conditions where this behavior cannot be observed. It is mentioned for completeness but these exceptions are not discussed.

5.3. BASIC DIELECTRIC PROPERTIES OF POROUS SILICON

5.3.1. The refractive index of porous silicon

During the anodization process a large fraction of Si is removed. For mesoporous Si the remaining Si NCs have sizes in the range of tens of nanometers, significantly smaller than the wavelength of visible light. Their electronic states are not altered by quantum confinement and it is possible to treat this mixture of Si and voids as a homogeneous dielectric medium having an effective dielectric constant ε_{eff}.

Effective medium approximation

ε_{eff} can be estimated using different effective medium approximations. The simplest approach to treat such a problem is that of Maxwell-Garnett.

Maxwell-Garnett

In this consideration ε_{eff} is the sum of the dielectric constants multiplied with their relative volume fraction. For a mixture of i components with the ith fraction having the dielectric constant ε_i and the relative volume fraction f_i the effective dielectric constant is given by:

$$\varepsilon_{eff} = \sum_i f_i \varepsilon_i$$

This equation is only true if dielectric screening is neglected. It is valid for dilute gases or layered structures with the direction of the external electric field parallel to the layers. Normal to the layered structure dielectric screening is maximum and ε_{eff} is described by:

$$\frac{1}{\varepsilon_{eff}} = \sum_i f_i \frac{1}{\varepsilon_i}$$

The effective epsilon of PSi as a mixture of Si and voids is in between these two values, depending on the geometrical arrangement. Because the Maxwell-Garnett approximation is for PSi just a rough estimate a more detailed model to treat the dielectric properties of PSi is widely applied: the approach of Bruggeman [22].

Bruggeman

In this approach [22] the effective dielectric constant of a mixture of small spherical inclusions in a host medium (i components with the ith fraction having the dielectric constant ε_i and the relative volume fraction f_i) can be calculated according to:

$$\varepsilon_{eff} = \sum_i f_i \frac{\varepsilon_i - \varepsilon_{eff}}{\varepsilon_i + 2\varepsilon_{eff}}$$

This model successfully describes the reduction of the refractive index of PSi compared to that of bulk Si with increasing porosity on the basis of the refractive index of bulk Si. With the dispersion of the dielectric constant of bulk Si, the spectral behavior of the absorption coefficient of mesoporous PSi can be modeled. Fig. 5.5 shows the result of a calculation in comparison to a measurement of the absorption coefficient α of PSi and bulk Si. The spectral dependence of α for mesoporous PSi is similar to that of bulk Si and is well described by an effective medium calculation according to the Bruggeman model [23].

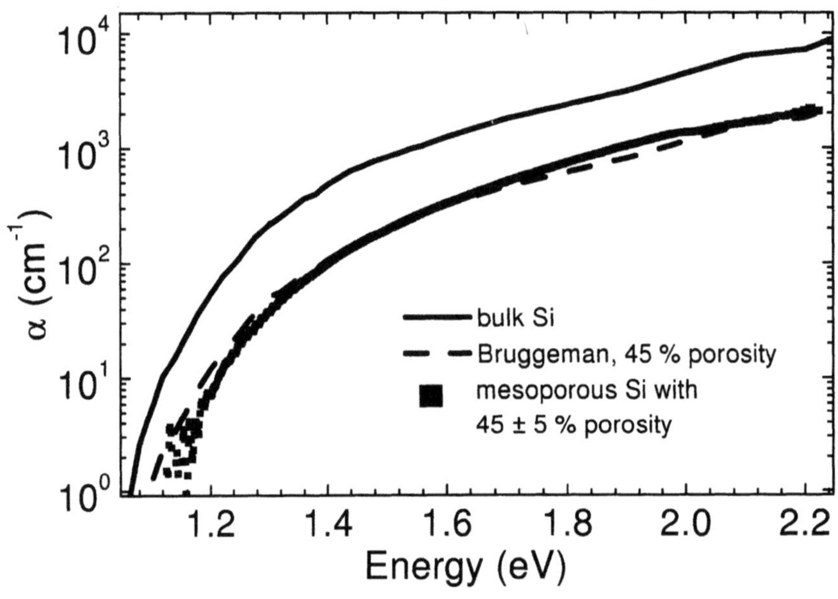

FIGURE 5.5. Absorption coefficient of mesoporous Si (squares) in comparison to that of bulk Si (solid line). The dashed line is an effective medium calculation (Bruggeman) explained in the text.

The essence of the Bruggeman formalism is that the electric field inside a small spherical volume in an "effective medium" is homogeneous. The polarization of the medium in this sphere can be described by the polarization of dipoles of an effective medium. Therefore it necessarily fails to describe the dielectric properties resulting from the aspheric shape of the NCs and consequences of their structural arrangement. As seen from Fig. 5.4 the long axis of the aspheric NCs in (110) PSi shows a uniaxial alignment in their projection onto the (110) surface of the PSi layer. The wavelength of visible light is significantly larger than the structural dimensions of such layers. Therefore visible light "averages" over the orientation distribution of the aspheric NCs. If even a partial uniaxial alignment exists in a large volume of the sample, the dielectric constant for linearly polarized light will dependent on the orientation of the polarization direction of the incident light with respect to this axis. To treat such an in-plane anisotropy of the dielectric constant it is necessary to extend the Bruggeman formalism to two dimensions.

Bruggeman formalism in two dimensions

As seen from the left side of Fig. 5.4 the shape of the NCs is complex but in a first approximation it is possible to assume that the Si NCs are ellipsoidal, for simplicity with two equal (x=y) and one larger axes (z, elongated ellipsoids of revolution). For a single NC an external electric field induces polarization charges at the surface of the NC. These charges create a depolarization field that decreases the strength of the electric field inside the NC. This effect is weaker if the electric field is applied along the long axis of the NC. The strength of this effect is determined by the shape (the ellipticity) of the NC and the difference between the dielectric constant of the nanocrystal and the surrounding medium. The electric field inside and outside each ellipsoid, E^i and E^o respectively, are related by

$$E^i{}_{x,y,z}(\omega) = \frac{E^0{}_{x,y,z}(\omega)}{1 + n^{(x),(y),(z)}\delta(\omega)}$$

where

$$\delta(\omega) = \frac{\varepsilon_i - \varepsilon_0}{\varepsilon_0}$$

with ε_i and ε_0 representing the dielectric constants of the ellipsoid and the surrounding medium, respectively, at frequency ω. $n^{(x),(y),(z)}$ are the depolarization factors along the three axes of the ellipsoid x,y and z. For ellipsoids of revolution:

$$n^{(x)} = n^{(y)} = [1 - n^{(z)}]/2$$

$$n^{(z)} \leq 1/3$$

$$n^{(x)} + n^{(y)} + n^{(z)} = 1$$

In a similar manner the macroscopic dielectric constant is related to the shape of the individual NCs, but for visible light the relative overall alignment of the ellipsoids is responsible for the in-plane variation of the dielectric constant. Extending the Bruggeman effective medium approximation [24] for a uniaxial composite medium containing ellipsoidal shaped inclusions leads to the following relation (equation 1):

$$fb(\varepsilon_i) + (1 - f)b(\varepsilon_n) = 0 \qquad (5.1)$$

with

$$b(\varepsilon_m) = \left[\begin{pmatrix} \varepsilon_1 & 0 \\ 0 & \varepsilon_2 \end{pmatrix} - \varepsilon_m \begin{pmatrix} 1 & 0 \\ 0 & 1 \end{pmatrix}\right] \cdot$$

$$\left[\begin{pmatrix} 1 & 0 \\ 0 & 1 \end{pmatrix} - \begin{pmatrix} L & 0 \\ 0 & \frac{1-L}{2} \end{pmatrix} \cdot \begin{pmatrix} \varepsilon_1 & 0 \\ 0 & \varepsilon_2 \end{pmatrix}^{-1} \cdot \left[\begin{pmatrix} \varepsilon_1 & 0 \\ 0 & \varepsilon_2 \end{pmatrix} - \varepsilon_m \begin{pmatrix} 1 & 0 \\ 0 & 1 \end{pmatrix}\right]\right]^{-1}$$

ε_m is either the dielectric constant of the inclusion medium (ε_i) or the host medium (ε_h). f is the porosity of the medium. L is the depolarization factor of the ellipsoids long axis and ε_1 (ε_2) are the dielectric constants for light linearly polarized parallel (perpendicular) to the long axis of the ellipsoids. In the limit of a macroscopic isotropic medium L=1/3 and ε_1=ε_2 (like for PSi produced from a (100) oriented Si substrate under normal incidence parallel to the growth direction) the two dimensional Bruggeman model gives the same results as the isotropic Bruggeman effective medium approximation.

5.3.2. Porous silicon multilayers

The first reliable optical devices made from PSi were based on the controlled variation in depth of the refractive index of a PSi specimen. The increase of the porosity results in a reduction of the refractive index of the PSi layers. This allows the modification of their dielectric properties in a controlled manner. Due to the reduced size of Si nanocrystals, in the range of a few nanometers, the sponge-like network of remaining Si and pores is a quasicontinuous optical medium. This concept is the basis of the variety of dielectric PSi structures like antireflection coatings for Si solar cells [25], waveguides [26], microcavities [27,28] and distributed Bragg reflectors [10,29] which have been realized so far. In all devices mentioned the refractive index is varied in depth via a time dependent etching current. This procedure results in a stack of alternating layers with different refractive indices and thicknesses. It has been demonstrated that these devices can be used as sensors for various types of chemical compounds [30,31] since a filling of the pores with dielectric substances modifies the average refractive index of each layer and thus the optical response of the stack.

5.4. DIELECTRIC EFFECTS IN THE PHOTOLUMINESCENCE OF POROUS SILICON

The microporous modification of PSi (prepared from low doped substrates) exhibits visible photoluminescence at room temperature. For Si as an indirect semiconductor with the indirect bandgap of the bulk material in the near IR (Γ_{25v}-Δ_{1c}= 1.1242eV at 300K) this observation was unexpected and initiated intense research on its origin. There was a controversial debate on the physical description of the photoluminescence in the early stages. However, nowadays it is the established well that the photoemission occurs from the recombination of spatially confined (photogenerated) excitons in Si NCs.

A detailed discussion of this point is beyond the scope of this article and can be found elsewhere [5]. This report will focus here on PL as a microscopic probe of the structural properties of the Si NCs since the Si NCs of standard p-type PSi are elliptical and therefore the specifics of their dielectrical anisotropy has to be considered. In that sense the PL of these objects serves as tool to gain deeper insight into their shape and geometrical arrangement [32, 33]. Besides the PL itself one of the most puzzling findings was that if the luminescence is excited with linearly polarized light the PL response is also preferentially linearly polarized [34, 35, 36]. Because of the unusual specifics of that behavior, this was called polarization memory effect.

5.4.1. Polarization memory phenomenon

To clarify the particularities of the polarization memory effect some basic PL properties of PSi are discussed.

Photoluminescence of PSi

The visible photoluminescence band of PSi is extremely broad and featureless as shown in Fig. 5.6. This inhomogeneous linewidth is to a large extend caused by the huge size distribution of the NCs. Due to quantum confinement the band edge of the individual NCs is shifted to larger energies according to their reduced size. Consequently the PL at increasing detection energies originates from smaller NCs. If the NCs are excited with photon energies well above the visible luminescence band, the whole ensemble of NCs having different shapes and sizes are photoexcited. They respond with visible PL in the vicinity of their individual band edge. If the excitation energy chosen to fall within the luminescence band, only NCs with an absorbing state at the excitation energy give rise to PL. Under such (resonant) excitation conditions PL of a subset of NCs is monitored.

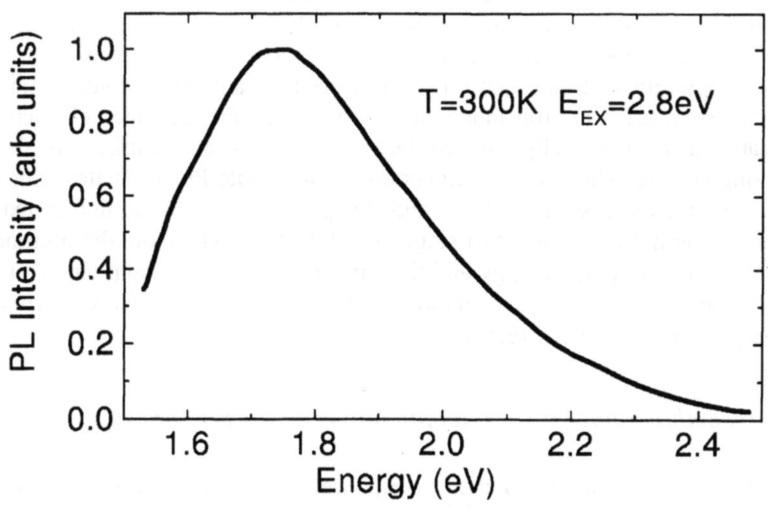

FIGURE 5.6. Room temperature PL spectrum of standard PSi.

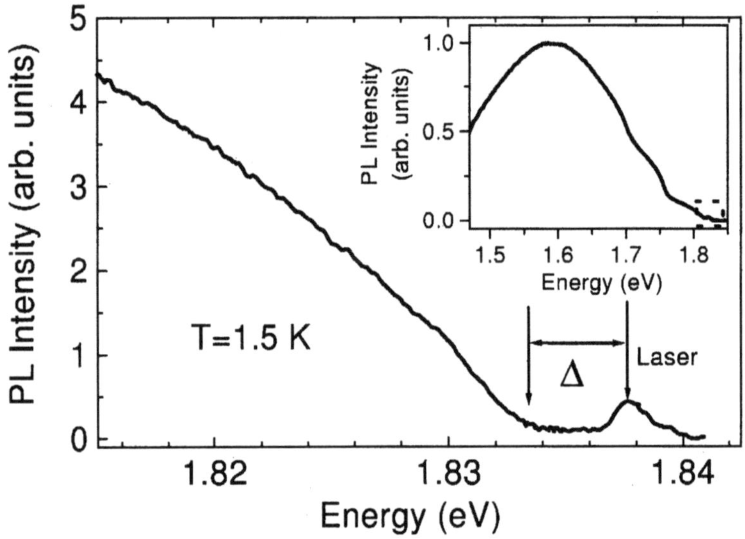

FIGURE 5.7. PL of PSi under resonant excitation conditions at low temperatures. The inset shows the overall resonant PL spectrum with the box indicating the spectral range displayed in the enlarged in the figure. Δ denotes the spectral gap between the resonant excitation (laser) and the PL onset.

Besides the consequences of quantum confinement on the energetical position of the bandgap, the fine structure of the exciton in a Si NC of reduced size is also changed. The exciton ground state is split into a spin singlet and a spin triplet state with the triplet state lowered in energy by electron-hole exchange splitting Δ. This electron-hole exchange splitting is enhanced by quantum confinement and can be readily seen as spectral gap between the resonant excitation and the PL onset [37] as shown in Fig. 5.7.

Δ is increasing with increasing confinement (reduced size) from 0.15 meV for a NC emitting near the band gap of Si, up to approximately 15 meV at a detection energy of 2.3 eV [14]. Luminescence occurs from these exciton states. Optical transitions via the singlet state are spin allowed and therefore fast while they are spin forbidden for the triplet state. However, the spin-orbit interaction partially lifts the forbidden character of the ground state making it partially allowed, but the transitions are significantly slower than via the singlet state. Thus, at low temperatures the visible PL originates from the slow triplet exciton ground state while at high temperatures the PL is dominated by recombination of thermally occupied fast singlet exciton states. Therefore the lifetime of the PL depends on the detection energy and the temperature [38]. At room temperatures typical PL lifetimes are in the range of hundreds of microseconds and slow down with decreasing temperatures. For 4.2 K they are on the order of milliseconds.

The recombination statistics of optically generated excitons in PSi

Besides the spectral and dynamic response, the recombination statistics of optically generated excitons is also modified [39, 40]. In the recombination statistic of bulk semiconductors the measured decay times τ_{mean} of the PL is defined by the relative rates of radiative τ_r and nonradiative τ_{nr} optical transitions:

$$\frac{1}{\tau_{mean}} = \frac{1}{\tau_r} + \frac{1}{\tau_{nr}}$$

where both times can be affected by temperature. The same holds for the value of the PL quantum yield η:

$$\eta = \frac{\tau_{nr}}{\tau_r + \tau_{nr}}$$

Therefore a correlation between the values of η and τ_{mean} is usually observed. However in porous silicon a correlation is only reported in the temperature range from 300K to 500K when thermally activated transport through the nanocrystalline network becomes important. Fig. 5.8 shows the PL intensity detected at 1.7 eV and the corresponding τ_{mean} in the temperature range from 300 K to 4.2 K.

In the temperature range from 150 to 5 K, despite the significant increase of the PL lifetime due to thermal depopulation of the fast exciton state, the external quantum yield remains nearly constant. Therefore at T<150 K τ_{mean} can be considered to a large extend as the radiative time. Taking into account that PSi is a granular material the NCs are either individual isolated luminescing crystallites (having no nonradiative defect) or dark crystallites (having at least one nonradiative defect). At low temperatures the

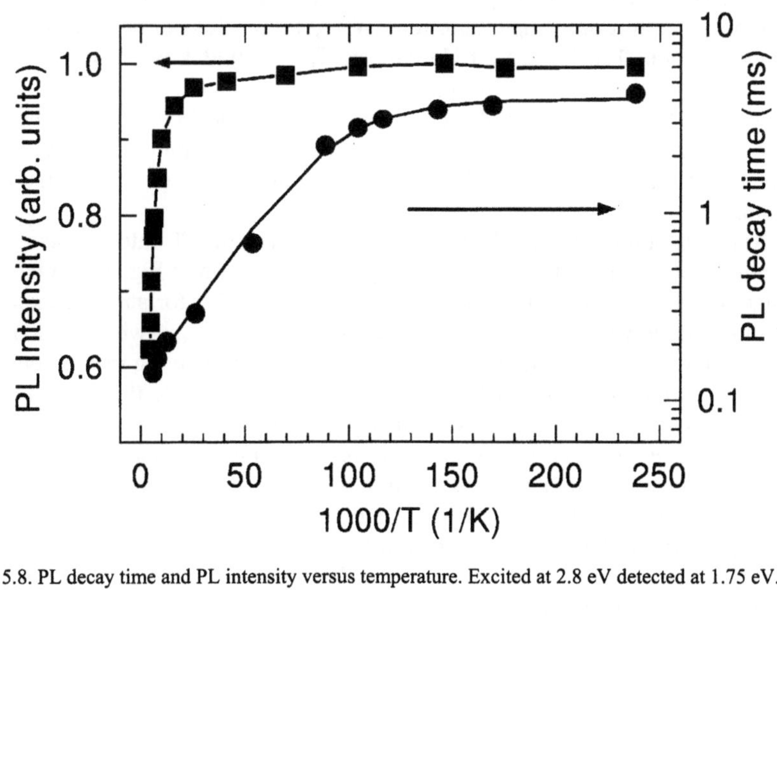

FIGURE 5.8. PL decay time and PL intensity versus temperature. Excited at 2.8 eV detected at 1.75 eV.

transfer of carriers between the crystallites is inefficient and the external quantum yield is only determined by the ratio of these two types of crystallites and is independent of the value of $\tau_{mean} = \tau_r$.

The geminate character of the light emission process in these Si NCs (photoexited carriers recombine in the same NC they were generated) combined with the large radiative lifetime at low temperatures are the main reasons for the pronounced observation of Auger saturation phenomena in PSi.

Auger saturation of the PL intensity of PSi

The essence of Auger saturation is that in a high density electron-hole pair system one electron-hole pair (exciton) recombines and transfers all of its energy to a third particle (electron or hole) as additional kinetic energy. This nonradiative Auger recombination limits the quantum efficiency under high power optical excitation. At low excitation levels, when only one electron-hole pair is present per NC, the Auger rate is zero. This is in contrast to bulk semiconductors were the Auger rate W_{Auger} is nonzero at any pumping power: $W_{\mathrm{Auger}}=A\ n^2\ p+B\ p^2$ n with A, B as material dependent parameters and n, p as the concentration of electrons n and holes p respectively. In that sense, at low temperatures and weak excitation power, the granular character of the material and the absence of carrier transport makes the Auger process less efficient in a nanocrystalline system than in a bulk semiconductor. On the other hand geometrical confinement of carriers became important at higher excitation levels: double occupation of nanocrystal by electron-hole pairs implies efficient carrier concentration in the range of 10^{18}-10^{20} cm^{-3}. Therefore nonradiative Auger recombination starts to take place when more than one electron hole pair exists in a NC. The long radiative lifetime τ_r makes it possible to excite a second electron-hole pair before the first one has recombined. A simple estimate shows that at room temperature for CW light intensities of the optical excitation in the range of about 100 W/cm^2 (at He temperatures around 1 W/cm^2) all NCs are on average already singly occupied. In addition the slow radiative recombination cannot compete with the fast nonradiative Auger process whose typical times are on the order of nanoseconds. Therefore an electron-hole pair created in a luminescent NC will only recombine radiatively when this NC did not already contain an electron hole pair or a free carrier before.

Auger saturation can easily be seen by measuring the PL intensity as a function of temperature for different optical excitation powers as shown in Fig. 5.10. Only at very low excitation density a constant η is observed while its value for stronger excitation is decreased towards lower temperatures. As mentioned earlier τ_{mean} (which is at T<150 K the radiative lifetime τ_r) increases with decreasing temperature giving rise to nonradiative Auger recombination and a reduction of the PL intensity, because the created electron-hole pairs exists longer in a NC making a double occupation more probable. For the description of the PL quantum yield η as the result of luminescing and dark NCs this implies that the quantum yield of a single luminescing NC at high excitation intensities is not constant any more and averaged over time is smaller than unity. In the saturation regime the quantity of *effectively* dark crystallites depends on the excitation power. Since the maximal emissivity of a NC is limited by one photon per exciton lifetime the decrease of τ_r towards higher temperatures leads to the increase of η with higher temperatures.

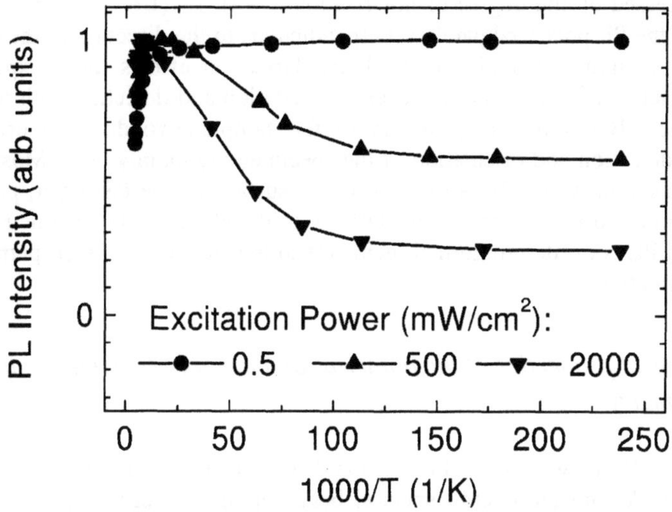

FIGURE 5.9. PL intensity for different excitation intensities versus temperature. Excited at 2.8 eV and detected at 1.75 eV.

Most of the optical studies of Si NCs have been concentrated on quantum confinement modifications of their absorbing and emitting states. Other important consequences of the reduced dimensionality arising from the shape of the Si NCs in porous Si (PSi) are optical effects due to dielectric nanostructuring of bulk Si.

Degree of linear polarization of the PL

For bulk semiconductors polarization resolved PL measurements are widely used to investigate the electronic symmetry of the luminescing state and energy relaxation processes. Polarized PL can be a result of the crystalline lattice structure or external perturbations. To observe polarized light emission from a multivalley cubic semiconductor like silicon (Si) it is necessary to lift the degeneracy of the hole band or valleys, respectively, by an external uniaxial perturbation [41]. Excitation of the PL by linear or circular polarized light selects an individual state among spin-, angular momentum-, or linear momentum-degenerate states. If relaxation is absent this results in polarized PL [42]. Low dimensional structures exhibit PL polarization as well since quantum confinement effects can be regarded as a strong perturbation that lowers the crystal symmetry. Besides the electronic contribution to the polarization of the PL, the modification of the electric field via dielectric screening effects has to be taken into account. For quantum well structures this effect is minor since the dielectric constants of the layers are almost the same. On the contrary, the lateral modulation of the dielectric constant in nanocrystalline assemblies can be fairly large due to a large difference in the dielectric constants of the nanocrystals and their surroundings [43]. Therefore, both the electric field and the dipole moment of the interband transitions contribute to the polarization of the light emission from nanocrystalline assemblies. PSi exhibits strongly linearly polarized PL when excited with linearly polarized light. The strength of this effect is expressed by the degree of polarization $\rho = (I_{\|}-I_{\perp})/(I_{\|}+I_{\perp})$, where $I_{\|}$ ($I_{\perp}$) are the intensities of the PL polarized parallel (perpendicular) to the direction of the electric field of the exciting light. To avoid artificial depolarization effects these experiments are usually conducted using normal incidence of excitation and detection with respect to the sample surface. Based on the geometry of the sample, two different orientations of illumination/detection and orientation of the specimen are mainly used. Most common is the top geometry in which PL is excited on the surface of the PSi layer, parallel to the growth direction, normal to the sample surface. In the cleaved edge geometry the sample is broken and PL is excited in normal incidence to the edge of the layer, perpendicular to the growth direction.

General properties of the polarization memory phenomenon in the top geometry

Fig. 5.10 shows the visible PL band of PSi produced from (100) oriented Si excited at 2.8 eV and the overall spectral behavior of the degree of polarization. ρ is positive over the whole visible spectral range and can reach values up to 16 %. An in-plane rotation of the polarization direction of the incident light does not alter ρ for standard samples[2].

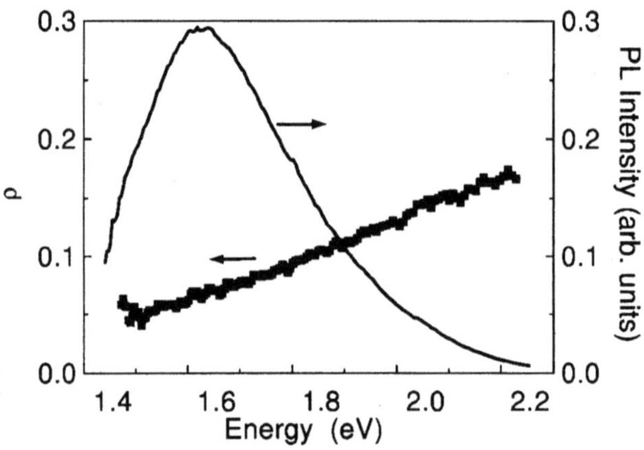

FIGURE 5.10 PL of PSi at room temperature excited at 2.8 eV and the corresponding degree of polarization memory ρ.

Fig. 5.11 displays the angular profile of the PL intensity if the detection polaroid is turned. 0^0 correspond to the detection polaroid parallel to the [100] direction in the sample surface (see the scheme of the experimental arrangement).

---

[2] Please note that there are some exceptions. For instance, strongly oxidized, samples exhibit an in plane anisotropy of ρ due to an anisotropical oxidation behavior of Si [44]. However, this is not the usual characteristic of the polarization memory phenomenon and is therefore only mentioned for completeness.

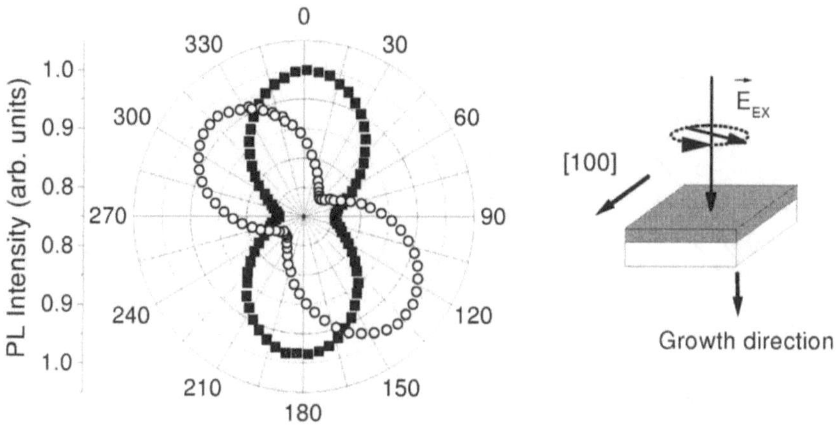

FIGURE 5.11 Polar plot of the light intensity when the detection polaroid is turned. 0^0 corresponds to the detection polarization parallel to the [100] direction indicated in the right part of the figure (a scheme of the geometrical arrangement). The direction of the incident light is parallel to the indicated [100] direction (squares) or -45^0 rotated with respect to this direction (circles).

The direction of the polarization of the incident light is parallel to the in-plane [100] direction (squares) and -45^0 rotated with respect to this direction (circles). In this configuration the value of ρ is independent of the orientation of the polarization direction of the exciting light. ρ is constant and $I_{||}$ is always a maximum: the angular polarization profile follows the polarization vector of the excitation. For a system of randomly in-plane oriented perfect linear dipoles the theoretical limit of ρ is 50%. Larger values imply uniaxial alignment which is not the case for this geometry. Therefore the observed values in the range of tens of percents can be considered as quite large. However, the PL polarization phenomenon cannot be understood in traditional terms like exciton alignment or selective optical excitation of those nanocrystals whose transition dipole moments are oriented parallel to the polarization vector of the exciting light. The nature of the absorbing and emitting dipoles has to be quite different because, as can be seen in Fig. 5.10, the photoexcited e-h pair loses an energy of the order of an electron volt in inelastic relaxation processes and therefore the observed PL polarization is not a coherent effect. Another confirmation of this incoherence can be deduced from polarized photoluminescence excitation (PLE) experiments.

Polarized photoluminescence excitation spectroscopy of the PL of PSi

For the PLE experiment the detection energy is fixed at 2 eV and the energy of the linearly polarized exiting light is varied. The PL response is normalized to the intensity of the incident light. The orientation of the polarization of the detected light is defined by a polaroid placed in the detection path. Fig. 5.12 depicts the intensities of the PL polarized parallel and perpendicular to the polarization of incident light as a function of its energy.

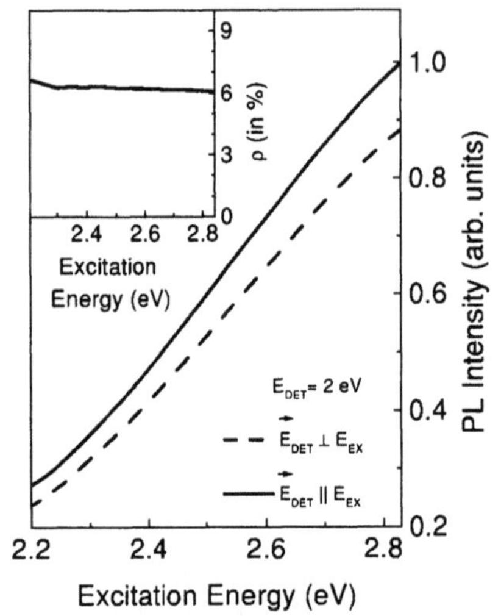

FIGURE 5.12 Polarization resolved PLE spectra. The inset shows the calculated ρ.

If PL is excited in parallel polarization conditions its intensity is always higher than that in the perpendicular polarization configuration. The resulting ρ is shown in the inset of Fig. 5.12. Over the whole energy range, from 2.2 to 2.8 eV, no distinct variation of the degree of linear polarization is observed within the accuracy of the measurement (1-2 %). For other nanocrystalline systems like CdSe NCs, polarized PL is only observed under resonant excitation conditions where the energy relaxation of the carriers is small and, therefore, the memory of the initial dipole orientation is conserved [45]. Contrary to that in PSi the level of the PL polarization is more or less energy independent within an energy interval of hundreds of millielectron volts adjacent to the detection point.

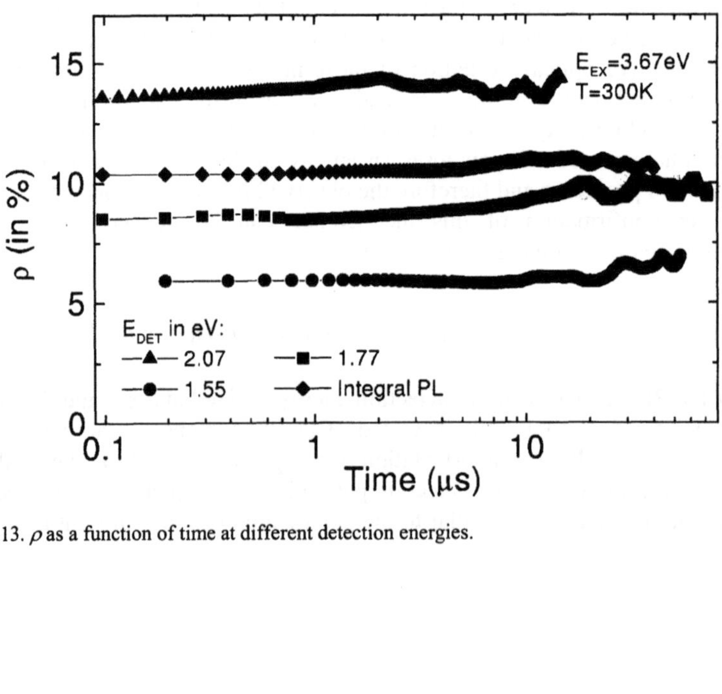

FIGURE 5.13. ρ as a function of time at different detection energies.

Time-resolved measurements of the degree of linear polarization in PSi

Even under the absence of energy relaxation the initial electronic dipole orientation can be scrambled if the lifetime of the optical transition is long enough. For instance, high positive values of ρ have been observed only at the very early times of the PL decay (nanosecond time domain) [46]. Fig. 5.13 depicts the degree of linear polarization versus time for different detection energies and for the integrated PL signal. PL is excited by a linear polarized pulsed laser (3.67 eV, 300 ps) and its decay is measured for parallel and perpendicular polarized PL components. The value of linear polarization of the PL from PSi does not depend on the time within a millisecond time scale.

Therefore this unusual behavior of polarized PL from PSi cannot be attributed to an alignment of the excited transition dipole. Taking into account that other Si based nanocrystalline materials like Si NCs prepared by Si implantation in SiO_2 layers or by chemical synthesis, while having the same general luminescence properties, do not show this effect, it is possible to conclude that the shape of the nanocrystals rather than quantum confinement effects seem to be the major reason for this phenomenon.

The degree of linear polarization in PSi under resonant excitation conditions

For PSi an enhancement of ρ under resonant excitation conditions is expected, because the energy relaxation of the carriers is small and, therefore, the memory of the initial dipole orientation is conserved. Fig. 5.14 compares the value of ρ under resonant and nonresonant excitation conditions and also displays the resonant PL spectrum.

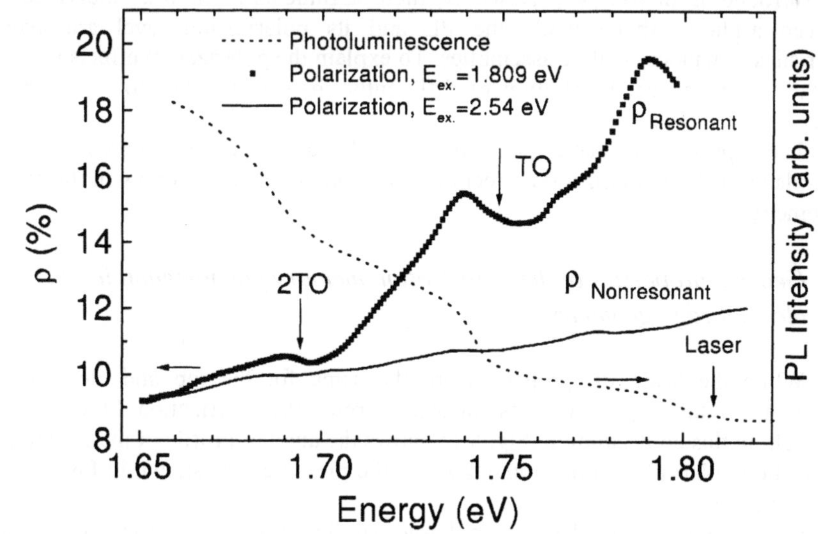

FIGURE 5.14. ρ under resonant (ρ $_{Resonant}$ - squares, excited at 1.809 eV) and nonresonant excitation conditions (ρ $_{Nonresonant}$ - solid line, excited at 2.54 eV). The dotted line shows the resonant PL spectrum excited at 1.809 eV. TO indicates the energy position of transverse optical Si phonons participating in the depolarization process (details are discussed in [47]).

Near to the exciting laser a relative enhancement compared to the nonresonant ρ is observed. However, this effect vanishes within an energy interval of 150 to 200 meV below the excitation point. This behavior and the fine structure of the spectral dependence of the resonant ρ is discussed in details elsewhere [47]. It is important that for PSi under nonresonant excitation conditions, a resonant contribution to the observed ρ can be neglected as long as the excitation energy is ≈ 200 meV larger than the detection energy. Consequently the dielectric properties of small aspheric NCs have to be examined further.

The polarization memory phenomenon - the result of a nanocrystalline assembly of aspheric NCs

In general nonspherical NCs will preferentially absorb and emit light that is polarized parallel to their long axis. Since all dimensions of the NC are smaller than the wavelength of light this problem can be solved in classical electrostatic terms. An external electric field induces polarization charges at the surface of the NC. These charges create a depolarization field that decreases the strength of the electric field inside the NC. This effect is weaker if the electric field is applied along the long axis of the NC. The probability of optical absorption is proportional to the square of the electric field inside the NC and therefore NCs with their long axes aligned in the direction of the light polarization will preferentially absorb light. The probability of spontaneous emission is proportional to the square of the electric field inside the NC. This field is again largest for light with a polarization vector along the large dimension of the NC. Therefore the emission is also preferentially linearly polarized. The strength of this effect is determined by the shape of the NC and by the difference of the dielectric constants of nanocrystals and the surrounding matrix [43]. However, these arguments refer to an individual NC with given aspheric shape while the PL and its polarization level are average characteristics of nanocrystalline assemblies. To explain the polarization effects in the PL of PSi the overall space distribution of differently shaped NCs has to be taken into account.

The importance of that can be demonstrated in an experiment where ρ is referred to the in-plane rotating polarization direction of the exciting light, but now in the cleaved edge geometry.

General properties of the polarization memory phenomenon in the cleaved edge geometry

While the basic PL properties are the same for the top and cleaved edge geometries, the value of ρ is not constant when turning the polarization direction of the incident light in this configuration. Fig. 5.15 shows the angular profile of the PL intensity if the detection polaroid is turned (analogous to the measurement shown in Fig. 5.11 for the top geometry).

0^0 correspond to the detection polaroid parallel to the [100] growth direction of the sample (see the scheme of the experimental arrangement). The direction of the polarization of the incident light is parallel to the [100] growth direction (squares), -45^0 and -90^0 rotated with respect to this direction. In the edge geometry the polarization

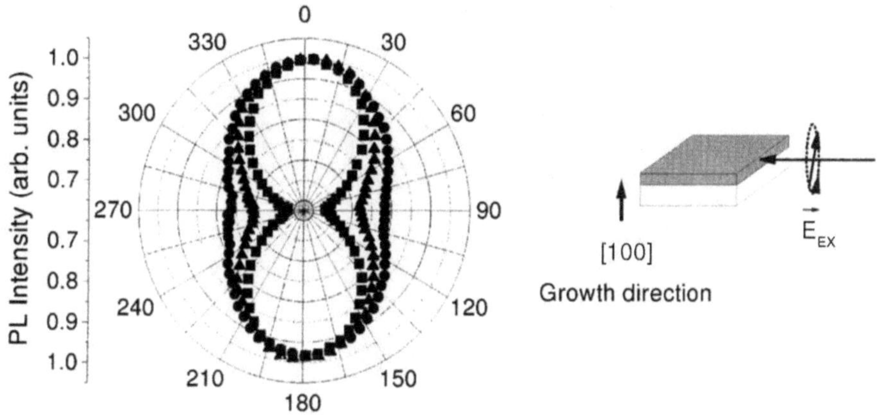

FIGURE 5.15. Polar plot of the light intensity when the detection polaroid is turned. 0^0 corresponds to the detection polarization parallel to the [100] direction indicated in the right part of the figure (a scheme of the geometrical arrangement). The direction of the incident light is parallel to the indicated [100] growth direction (squares) or -45^0 (triangles) and -90^0 (circles) rotated with respect to this direction.

distribution of the PL components does not follow the orientation of the polarization vector of the exciting light. On the contrary, the PL is always partially linearly polarized parallel to the [100] growth direction of the sample. The value of ρ can even be negative (-90^0 configuration, $I_{\parallel} < I_{\perp}$).

Within this dielectric model [48, 49] the two geometry's reflect two limiting cases of the average alignment of the NCs. In the edge geometry a large fraction of the NCs is oriented with their long dimension preferentially along the [100] growth direction. Therefore the efficiency of the excitation of NCs with their long axes parallel to the polarization direction of the exciting light will strongly depend on their orientation with respect to the [100] growth direction. As a result the degree of linear polarization is smaller if the polarization of the exciting light is turned perpendicularly to the growth direction and a maximum of ρ is achieved otherwise. However as depicted in Fig. 5.15 the main contribution to the PL comes from crystallites emitting light that is mainly polarized along their long dimensions aligned in the [100] growth direction. In the cleaved edge geometry this preferential orientation of the PL polarization direction is independent of the polarization direction of the incident light. On the contrary, in the top geometry the probability of exciting NCs with their long axis parallel to the orientation of the polarization direction of the incident light does not depend on the direction of this polarization. ρ is independent of the orientation of the polarization of the exciting light (Fig. 5.11) This is a consequence of a random distribution of the NC shapes and orientations in this projection.

In the framework of the dielectric model [48,49] PL polarization memory is a result of a selective excitation of a subset of crystallites with their long dimension parallel to the polarization direction of the exciting light. They will radiate light preferentially polarized in the same direction giving rise to polarized PL. Fig. 5.16 is a schematic drawing of the orientation of the ellipsoids in (100) PSi and in different projections. The

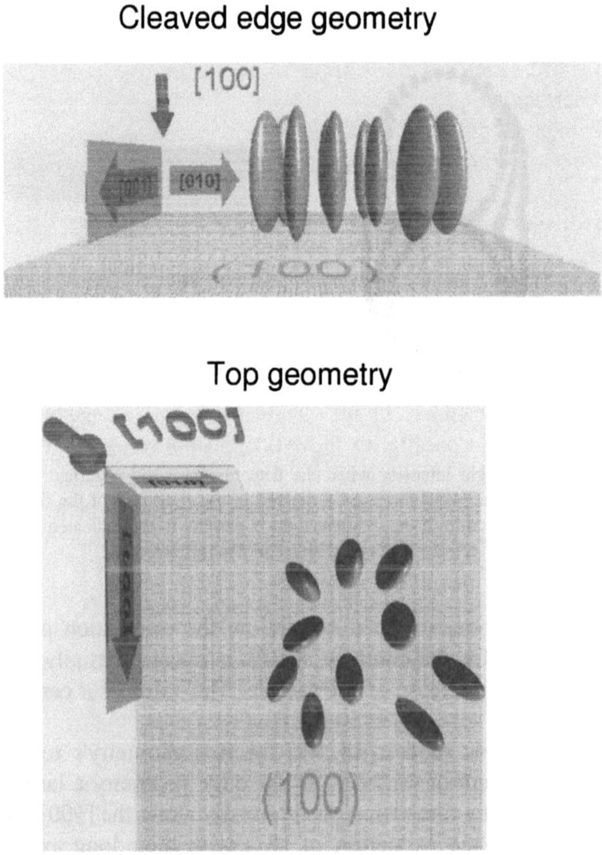

FIGURE 5.16. Schematic drawing of the orientation of the Si ellipsoids in (100) PSi in the top and cleaved edge geometry.

upper part and the lower part are the projections in the cleaved edge and top geometry of the same arrangement of the ellipsoids. While in their projection onto the sample surface no uniaxial alignment is present, they are well aligned in the growth direction.

The spectral dependence of the degree of ρ

So far only a subset of NCs, emitting at a given wavelength is considered. As shown earlier ρ is also spectral dependent, rising towards higher detection energies. Several effects can contribute to this behavior:

- The dielectric constant of Si (ε_{Si}). ε_{Si} is growing with increasing energy in that spectral range and therefore the difference in the dielectric constant of NCs and their surroundings is increasing giving a small increase to ρ for the same relative dimensions (long axis/short axis) of the NC.

- The shape (ellipticity) of the NCs. PL at higher detection energies is originating from smaller NCs. Their ellipticity can be different from the NCs emitting at lower energies - smaller ones are more elliptical than larger ones leading to an increase of ρ towards higher detection energies.
- The overall alignment of the NCs. Smaller crystallites can show less uniaxial alignment in the growth direction and therefore their *projection* onto the sample surface shows higher ellipticity.

Until now a detailed separation of these contributions to the spectral dependence of ρ is still missing. However, spectral resolved measurements of ρ in the top and cleaved edge geometry indicate that a loss of the uniaxial alignment for smaller NCs might contribute significantly to the spectral dependence of ρ observed in the top geometry. Fig. 5.17 shows the PL spectrum and ρ in the top geometry (part c).
Part (a) of the same figure illustrates the spectral dependence of ρ (calculated from the PL spectra shown there) observed in the cleaved edge geometry for the polarization direction of the exciting light parallel to the (100) growth direction. Contrary to the top geometry ρ is only slightly increasing towards higher detection energies. The PL component polarized parallel to the polarization direction of the incident light (Fig. 5.17 part a, dashed line) is always larger than the one polarized perpendicular to the direction of the excitation polarization (Fig. 5.17 part a, dotted line). This behavior is reversed if the direction of the excitation polarization is perpendicular to the [100] growth direction (Part b of Fig. 5.17). In this configuration the intensity of the PL polarized perpendicular to the direction of the electric field of the incident light (parallel to the [100] growth direction, see Fig. 5.17) is largest. Consequently ρ is negative over most of the detection energy range. Negative ρ indicates a uniaxial alignment of the long axis of the ellipsoidal Si NCs, which preferentially will emit light polarized parallel to their long axis, independent of the orientation of the polarization of the incident light. Thus the increase of ρ with larger detection energies (Fig. 5.17 part b) shows that this alignment in growth direction is lost for smaller NCs and the overall orientation is becoming more random with decreasing size. This is also reflected in the increase of ρ in the top geometry (Fig. 5.17 part c) with larger detection energies. With increasing randomization of the orientation of the long axis of the NCs the projection of the long ellipsoidal axis onto the (100) surface plane is contributing more and more to the polarization memory effect observed in top configuration. Therefore the loss of uniaxial alignment in the growth direction with decreasing size of the NCs is giving rise to the value of ρ in the top geometry, as depicted in Fig. 5.17 part c. Because the origin of ρ is caused by the dielectric properties of aspherical Si NCs it seems appropriate to introduce some theoretical aspects to treat such a problem.

Theoretical considerations

An external electric field induces polarization charges at the surface of the NC. These charges create a depolarization field that decreases the strength of the electric field inside of the NC. The electric field inside each ellipsoid and the external electric field, E^i and E^o respectively, are related by [48]

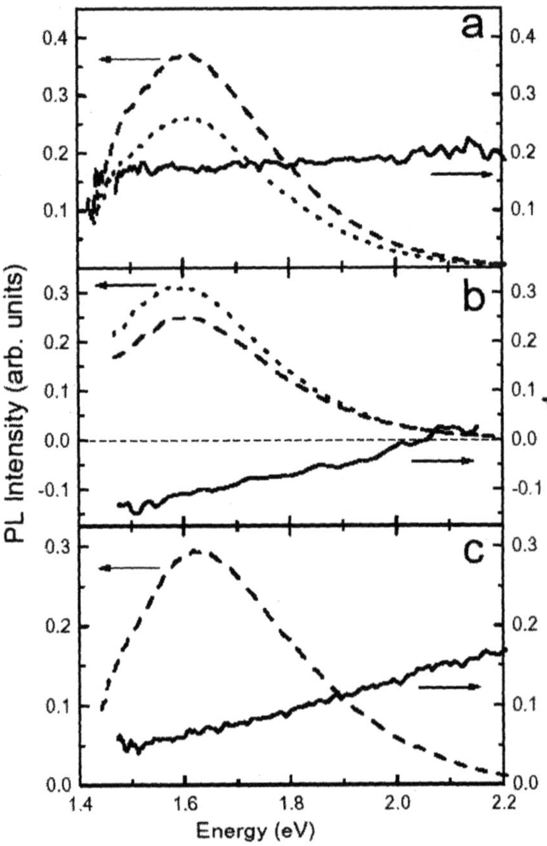

FIGURE 5.17. Part a,b, cleaved edge geometry - PL intensity detected with polarization parallel and perpendicular to that of the exciting light (dashed and dotted line, left axis) and the degree of polarization ρ (solid line, right axis). Part a, excitation polarization parallel to the [100] growth direction. b, excitation polarization perpendicular to the [100] growth direction. c, the top geometry - PL spectrum (dashed line, left axis) and the degree of polarization ρ (solid line, right axes).

$$E^{i}{}_{x,y,z}(\omega) = \frac{E^{0}{}_{x,y,z}(\omega)}{1 + n^{(x),(y),(z)}\delta(\omega)}$$

where

$$\delta(\omega) = \frac{\varepsilon_i - \varepsilon_0}{\varepsilon_0}$$

with $\varepsilon_i(\omega)$ and $\varepsilon_0(\omega)$ representing the dielectric constants of the ellipsoid and the surrounding medium at frequency ω. The depolarization factors $n^{(x),(y),(z)}$ along the three axis of the ellipsoid x, y and z depend on the shape of the ellipsoid. The electric field is a maximum along the longest axis and minimal along the shortest axis. This leads to preferential excitation of ellipsoids with their long axis parallel to the polarization of the incident light. The probability of an excitation of an ellipsoid at a fixed orientation is given by:

$$P_e(\omega_e) \propto \left|\vec{D}^T \vec{E}^i\right|^2 = \left|D^T\right|^2 \left|E^i\right|^2$$

with

$$\left|E^i\right|^2 = \left|E_x^i\right|^2 + \left|E_y^i\right|^2 + \left|E_z^i\right|^2$$

and $\vec{D}^T$ being the (isotropic) transition dipole matrix element. The emission probability $P_e(\omega_e)$ for PL can be calculated in a similar way because absorption and emission are mirror like processes. Therefore, as an ellipsoid preferentially absorbs light polarized along its long axis it also emits light polarized along its long axis. This leads to the preferential polarization of the PL parallel to the long axis of the NC. Here it is assumed that in decaying from the absorbing state to the luminescing state the carriers have lost all their polarization memory. As shown in Fig. 5.14 this is a good assumption from approximately 200 meV below the excitation energy. The overall PL intensity (I_{PL}) from a *collection* of ellipsoids is obtained by averaging over the shape and orientational distribution of the ellipsoids:

$$I_{PL} \propto < P_e(\omega_e)\ P_d(\omega_d) >$$

where the angular brackets denote averaging. An azimuthal isotropic distribution (the projection onto the top of the sample is therefore also isotropic) and a certain correlation with the shape and polar orientation of the ellipsoids (the projection onto the cleaved edge of the specimen is therefore anisotropic) is usually assumed. Solving the problem for a 3D orientational distribution of NCs with a given shape for two geometrical arrangements (top - edge geometry) is not trivial [48,50]. Most experiments are conducted in the top geometry because first of all the sample area is much larger (of the order of cm^2) and second it is experimentally more accessible than the micrometer thick stripe in the cleaved edge geometry. The computation of PL from a system of aspheric NCs only in top projection is significantly easier. On the other hand the results of this calculations are not as conclusive as a complete 3D mapping of ρ. For example, in the top geometry a larger tilt of the long axis of the NCs with respect to the (100) growth direction leads to an increase of ρ as well as an increased ellipticity while keeping the orientation fixed. Because it is not possible to separate these parameters the results can only be treated qualitatively.

The simplest shape assumed for the NCs are prolonged ellipsoids of revolution ($x=y<z$). For this shape the depolarization factors can be calculated [51]:

$$n^{(z)} = \frac{1-\alpha^2}{2\alpha^3}\left(\ln\frac{1+\alpha}{1-\alpha} - 2\alpha\right) \leq \frac{1}{3}$$

$$n^{(x)} = n^{(y)} = \frac{1-n^{(z)}}{2}$$

with

$$a = \sqrt{\left|1 - x^2 / y^2\right|}$$

and

$$n^{(x)} + n^{(y)} + n^{(z)} = 1$$

The depolarization factor for the longer axis is smaller than those for the shorter axis; thus the decrease in the component of the electric field along a long axis is less than along the short axes. Clearly the assumption of ellipsoids of revolution does not account for the exact shape of the Si NCs in PSi but it allows a description of the most basic effects connected to the asphericity of the NCs. In that sense, if NCs in PSi are called afterwards elliptical this is just a rough approximation of the real shape and is done for convenience and for clarity.

Predictions and experimental validations

Geminate/nongeminate recombination processes

In the dielectric approach, ρ in the top geometry is only observed for geminate recombination processes because it is based on the assumption that NCs with their long axis parallel to polarization vector of the exciting light are preferentially excited and the recombining excitons reradiate in a similar manner in the same NCs, where they were generated, leading to preferentially linear polarized PL parallel to the long axis of the NCs. If the photogenerated carriers would recombine in *any* NC and not only in the one where they were generated (nongeminate recombination) the model predicts that ρ is absent in the top geometry. Linearly polarized light also selectively excites the NCs with their long axis parallel to the polarization direction of the exciting radiation, but radiation occurs from *all* NCs leading to unpolarized PL. Because the NCs are randomly distributed in this projection their overall PL is unpolarized.

Nonlinear excitation processes

The shape dependent depolarization factors occur in the calculation of the PL intensity in the excitation $P_e(\omega_e)$ and in the detection $P_d(\omega_d)$ probability. Under excitation by one-photon absorption $P_e(\omega_e)$ is proportional to E^2. In a similar manner $P_d(\omega_d)$ is calculated, which is also proportional to the internal electric field but now calculated at ω_d. Nonlinear excitation processes (e.g., two-photon absorption) are characterized by a higher order dependence on the electric field (e.g., E^4). The dielectric model predicts that for the same $P_d(\omega_d)$, ρ under nonlinear excitation should be significantly enhanced.

The dielectric constant of the medium surrounding the NC

The importance of the depolarization factors depends on the shape of the NC and on the difference between the dielectric constant of the NC and the surrounding medium. $\delta(\omega)$ takes the role of a weight function; the larger the dielectric constant of the surrounding medium the smaller $\delta(\omega)$ and the smaller the importance of the depolarization factors. Therefore ρ should decrease with increasing dielectric constant of the surrounding medium. In the dielectric model PSi is treated as an aggregate of Si NCs embedded in an effective medium (PSi as a mixture of Si-voids). According to effective medium calculations (Bruggeman effective medium theory) the effective dielectric constant of PSi (which is the dielectric medium surrounding the NC) is increasing after, even partial, filling of the pores with a liquid having higher dielectric constant than air. Therefore filling of the pores should lead to a reduction of ρ.

Experimental validations

Degree of linear polarization for nongeminate recombination processes

At room temperature, geminate excitonic recombination processes are the major part in the PL from PSi. For T < 150 K the lifetime τ_{mean} of the PL is the radiative PL lifetime τ_{rad} which is at low temperatures in the range of milliseconds. For low optical excitation intensities the PL is again a pure geminate excitonic recombination process. At higher optical excitation levels, nonradiative Auger processes become important. If a second electron-hole pair is created while the NC is still occupied, one exciton recombines nonradiatively transferring all of its energy to one of the carriers (electron or hole) of the other exciton as kinetic energy.

The excited carrier has several opportunities to loose energy:

(i) It relaxes into the ground state of the same NC and either recombines with the remaining carrier of the opposite type or is again excited via Auger recombination of another exciton. The resulting PL is therefore dominated by geminate recombination processes.

(ii) It has high enough energy to overcome the confinement potential and leaves the core of the NC either into its surrounding, (e.g., surface) or into a neighboring NC where it relaxes into its ground state. In the first case the carrier, let us say an electron, is usually trapped in a deep state (which has therefore long lifetimes) and the remaining NC stays singly charged. As long as this is the case this NC will not emit light anymore because if an electron - hole pair is created in this NC it will again recombine nonradiatively via Auger recombination. This behavior is called the photodarkening effect and was first observed in the PL from semiconductor doped glasses [52] and is also found in PSi [53-56]. If the ejected electron is transferred back into the NC, this NC will again be neutral and luminescing with the recombination of this exciton. But there is also a finite probability that the NC is discharged by the emission of the remaining carrier (in this case a hole) which has gained energy from the Auger recombination of a subsequently created electron-hole pair. After this event the NC will also start to be luminescing again, but now in the presence of an electric field of the two previously ejected carriers of opposite sign located presumably in the surroundings of the NC. For a single CdSe quantum dot this leads to a "blinking" [57, 58] of an individual NC ("PL intermittency") and to a shift of the PL emission lines [59, 60] depending on the electric field in which the QD is located ("spectral diffusion"). The subsequent ejection of an electron and hole from a NC can also lead to another scenario. If both carriers are ejected *into* another NC having a different size, the PL of a smaller or larger NC is mediated by successive Auger recombinations and carrier ejection from the initially excited NC. The PL from larger NCs occurs as additional PL intensity below the excitation point and is therefore in the same energy range as the PL from the photoexcited NC. However, under resonant excitation conditions the PL from the smaller NC happens to appear *above* the energy position of the exciting laser. This effect (upconverted PL) is observed in the PL from PSi [61] but only in absolutely freshly prepared PSi. The surface of these NCs is mainly covered by hydrogen and therefore the barrier between these NCs is not as large as in the case of partial coverage by oxygen (after natural oxidation of the specimen in air). This luminescence is purely nongeminate.

Fig. 5.18 shows the spectrum of the resonantly excited PL for different energies of the exciting laser within the PL band (shown in the inset). The upconverted PL (UPL) is predominantly observed when the PL is resonantly excited at the low energy side of the PL band and nearly vanishes if the PL is excited at the high energy side. This is a consequence of the different probabilities to find a NC of smaller size, emitting at higher energies. This probability is large at the low energy side of the size distribution.

Fig. 5.19 shows the degree of linear polarization of the PL in the vicinity of the exciting laser. ρ shows a sharp drop at the energy position of the exciting laser and is absent for the upconverted PL. Fig. 5.19 also displays the resonant PL spectrum to show the presence of UPL. The value of ρ detected below the excitation point is quite small but the PL in this energy range is an indistinguishable mixture of preferentially linearly polarized PL and presumably unpolarized downconverted PL lowering the level of ρ. As predicted, nongeminate PL processes in PSi lead to unpolarized PL.

The degree of linear polarization for two-photon excitation of the PL

Nonlinear spectroscopy is a standard technique to gain additional information about the optical properties of NCs [62,63]. However, it can also serve as a tool to verify the understanding of the origin of the polarization memory phenomenon and to get further insight into the structural properties of PSi [64]. While PL is most commonly

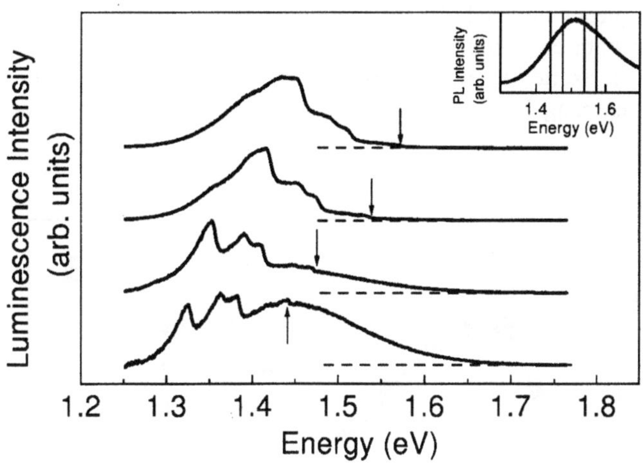

FIGURE 5.18. Development of the UPL versus the position of the exciting light (40 W/cm^2, T = 4.2 K) within the distribution of NCs. Resonant spectra are normalized to the same value. Inset: PL spectrum under nonresonant (2.8 eV) excitation. The lines indicate the energy of the exciting light.

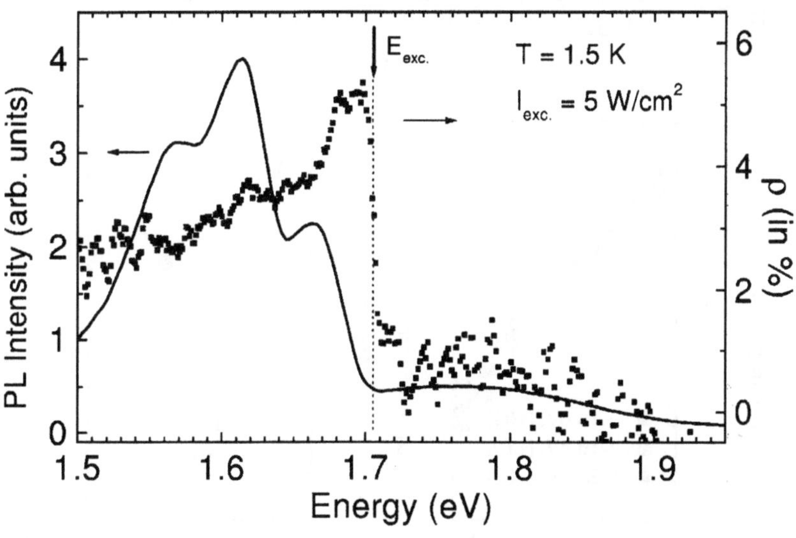

FIGURE 5.19 UPL excited at 1.705 eV (solid line) and the corresponding degree of linear polarization (squares).

excited via the absorption of a single photon (one-photon absorption), being proportional to the square of the electric field of the exciting light (E^i), an excitation by the simultaneous absorption of two-photons (two-photon absorption) is proportional to the fourth power of E^i. Fig 20 shows the PL of PSi excited via two-photon absorption. It compares the PL spectrum and the lifetimes of the PL excited with the fundamental

(1.165eV) or frequency doubled (2.33eV) radiation of a pulsed Nd:YAG laser. While the PL response for both excitation modes is basically the same (Fig. 5.20 a, and b), the intensity of the PL excited by two-photon absorption is increasing nonlinearly with the intensity of the exciting laser (Fig. 5.20 c), as expected for a two-photon absorption process.

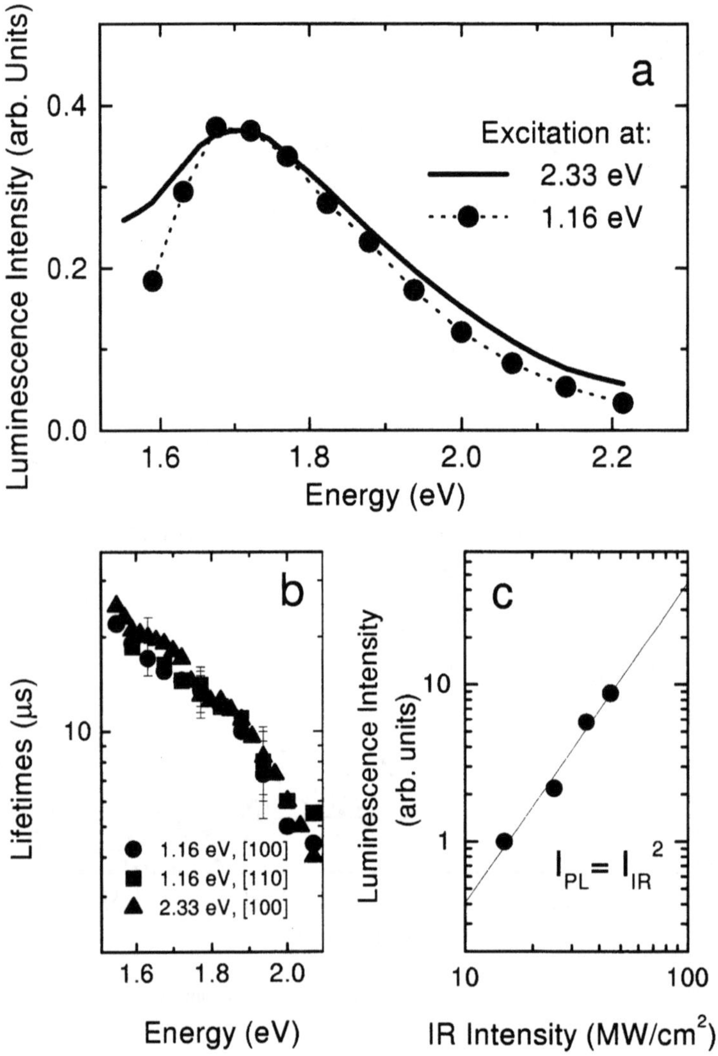

FIGURE 5.20. Part a: Photoluminescence spectrum of porous Si obtained by one-photon excitation excited at 2.33 eV and two-photon excitation with photon energy at 1.165 eV (at an intensity of 30 MW/cm^2). Part b: Lifetimes of PL from two-photon excitation at 1.165 eV with polarization along [100] and [110] and one-photon excitation at 2.33 eV. Part c: Magnitude of PL (detected at 1.77 eV) versus input pulse intensity (at 1.165 eV). The solid line describes a quadratic dependence of the relation.

For the two-photon excited PL from a Si ellipsoid, the emission probability $P_d(\omega_d)$ is the same as that under one-photon excitation, but the excitation probability is more complex because it depends not only on the orientation of the ellipsoid but also on the orientation of the crystalline structure in the ellipsoid. Treating two-photon absorption as a third-order nonlinear optical process, the two-photon excitation probability in a medium can be written as [65]:

$$P_e^{2\hbar\omega}(\omega_e) \propto \mathrm{Im}\left[\sum_{\alpha,\beta,\gamma,\delta} \chi^{(3)}_{\alpha,\beta,\gamma,\delta}(\omega = -\omega + \omega + \omega) E_\alpha^{i*}(\omega) E_\beta^{i*}(\omega) E_\gamma^{i}(\omega) E_\delta^{i}(\omega)\right]$$

where $\chi^{(3)}$ is the third order nonlinear susceptibility. The subindices $\alpha, \beta, \gamma, \delta$ refer to the crystalline axes of the medium, and Im denotes the imaginary part. In cubic crystals of point symmetry $\bar{4}m3$, like Si, $\chi^{(3)}$ has 3 independent elements [65] $\chi^{(3)}{}_{\alpha\alpha\alpha\alpha}$, $\chi^{(3)}{}_{\alpha\beta\beta}$ and $\chi^{(3)}{}_{\alpha\beta\alpha\beta} = \chi^{(3)}{}_{\alpha\beta\beta\alpha}$ with the crystalline axes along [100], [010] and [001]. The two-photon excitation probability then becomes:

$$P_e^{2\hbar\omega}(\omega_e) \propto \mathrm{Im}\left[\chi^3_{\alpha\alpha\alpha\alpha}\left(\left|E^i\right|^4 - \sigma \sum_{\substack{\alpha,\beta \\ \alpha\neq\beta}} \left|E^i_\alpha\right|^2 \left|E^i_\beta\right|^2\right)\right]$$

where

$$\sigma = \frac{\chi^{(3)}_{\alpha\alpha\alpha\alpha} - (\chi^{(3)}_{\alpha\alpha\beta\beta} + \chi^{(3)}_{\alpha\beta\alpha\beta} + \chi^{(3)}_{\alpha\beta\beta\alpha})}{\chi^{(3)}_{\alpha\alpha\alpha\alpha}}$$

is a parameter that describes the anisotropy in the two-photon absorption process. (The same description applies to isotropic media, which, by symmetry, have $\sigma = 0$). Because $\sigma \neq 0$ and E^i_α depends on the orientation of the ellipsoid with respect to the input polarization, $P_e^{2\hbar\omega}$ is a function of not only the orientation of the ellipsoids but also the orientation of the crystalline axes. As discussed earlier, the luminescence emission probability $P_d(\omega_d)$ is independent of excitation and is therefore the same as in the linear case. The two-photon excited luminescence intensity from a collection of ellipsoids is again obtained by averaging over the orientational distribution of the ellipsoids and is given by:

$$I_{PL}^{2\hbar\omega} \propto \left\langle P_e^{2\hbar\omega}(\omega_e) P_d^{2\hbar\omega}(\omega_d)\right\rangle$$

It is expected to show both a finite ρ larger than that observed under one-photon excitation because of the $|E^i|^4$ dependence of $P_e^{2\hbar\omega}(\omega_e)$,and an anisotropy with respect to the crystal orientation. Fig. 5.21 shows the degree of linear polarization under one- and two-photon excitation. Under nonlinear excitation ρ is always significantly larger than

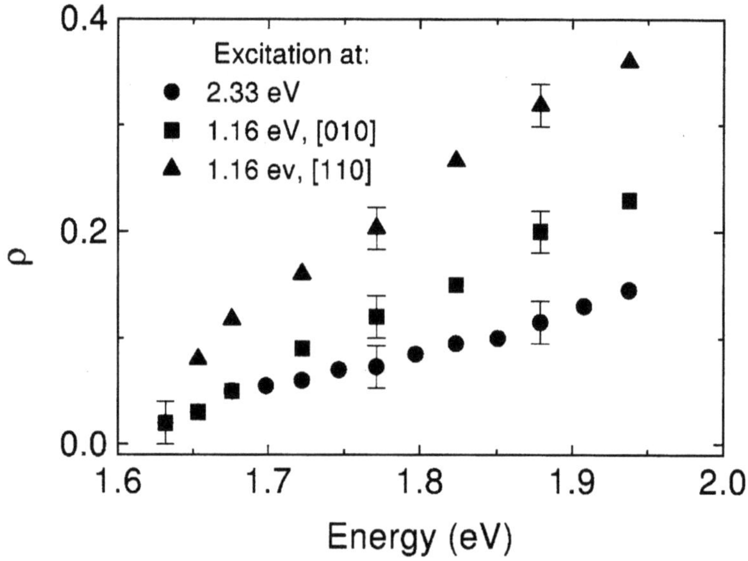

FIGURE 5.21. Degree of polarization ρ excited by one-photon (at 2.33 eV) and two-photon (at 1.165 eV) excitations with the polarization direction of the incident light parallel to the [110] and [010] direction.

under linear, one photon excitation, as predicted. However, while ρ is isotropic for a rotation of the polarization direction of the incident light in the top geometry for one photon excitation, it becomes anisotropic for an excitation by two photon absorption. ρ is largest for the polarization of the incident light parallel to the [110] direction (Fig. 5.21) because in this configuration the two photon absorption coefficient is also a maximum ($\sigma \neq 0$, more details can be found elsewhere [64]). Since ρ is enhanced under an excitation by two photon absorption this technique is a very sensitive measure of the shape of a light emitting particle.

Filling the pores with liquids having larger dielectric constant than air increases the dielectric constant of PSi [28]. This effect can easily be seen as a shift of the interference fringes observed in the reflection of thin PSi layers
in the range of low self absorption [66, 67].

The degree of linear polarization observed after filling the pores of porous silicon with dielectric liquids

The refractive index of PSi after filling the pores with a dielectric liquid

Fig. 5.22 shows the reflection spectrum of a 6.5 μm thick microporous sample before and after filling with benzene. For the known thickness d the refractive index n can be evaluated according to the simple relation [68]:

$$n = \frac{1}{2d}\left(\frac{1}{\lambda_r} - \frac{1}{\lambda_{r+1}}\right)^{-1}$$

where λ_r is the wavelength of the rth maximum in reflection. Already from the number of fringes in the same spectral interval it is clear that therefractive index is larger for the sample filled with benzene. Using the Bruggeman effective medium approximation and assuming a constant refractive index of Si (n_{Si}=3.564) in this spectral range (the dispersion of ε_{Si} is weak) allows one to estimate the porosity of the sample to be around 72 % based on the measured refractive index of PSi (n_{PSi}=1.522). This value is in good agreement with the porosity calculated from gravimetry. The measured value of n_{PSi} filled with benzene is 1.73, which is lower than the number predicted from effective medium calculations for a 100 % filled sample. However, for microporous Si, incomplete filling due to its small pores is well known [31]. Therefore the Bruggeman effective medium theory for three media (Si, benzene, air) is applied to deduce the filling factor to be 0.32. Filling the same sample with acetone leads to a measured refractive index of 1.68. This corresponds to the same filling factor and is to be expected if some regions of the sample are not accessible by liquids.

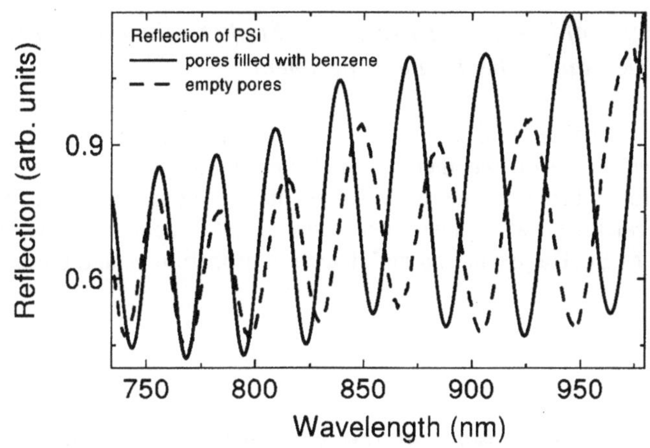

FIGURE 5.22. Reflection of PSi before (solid line) and after filling with benzene (dashed line).

The degree of linear polarization of PSi after filling the pores with a dielectric liquid

Fig. 5.23 shows the degree of linear polarization measured for the sample in filled (acetone, benzene) and empty conditions excited at 2.39 eV. The points are a result of a calculation explained later in the text. Filling the pores of PSi with liquids having larger dielectric constant than air reduces ρ. The larger the effective epsilon of PSi the smaller ρ. This is in good agreement with the predictions of the dielectric model.

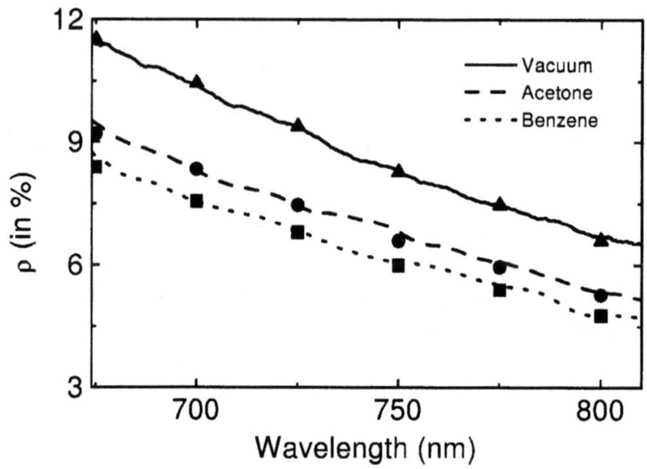

FIGURE 5.23. Degree of linear polarization of PSi before (solid line) and after filling with benzene and acetone. The points are results of a calculation explained in the text.

The degree of linear polarization of PSi after filling the pores with a dielectric liquid - theory and experiment

Besides a qualitative description of the observed behavior these measurements can be employed to check whether the dielectric model also succeeds in giving a quantitative estimate. From the measured values of the volume fraction of Si, liquid and air, ε_{eff} can be calculated at a given energy taking into account the dispersion of the refractive index of Si. The excitation and detection probabilities and the intensity of the PL:

$$I_{PL} \propto \langle P_e(\omega_e) P_d(\omega_d) \rangle$$

are calculated assuming ellipsoids of revolution with a spatial orientation function described before (gaussian distribution of the angle of the long ellipsoidal axis with respect to the [100] growth direction with the center parallel to the growth direction, and azimuthal isotropic distribution of the crystalline alignment). The ρ of the empty sample was used to adjust the calculated ρ by fitting the ellipticity of the ellipsoids, *i.e.,* the ratio of the long axis with respect to the dimension of the two short axes - the z/x value. As discussed in detail before, it also is possible to adjust the width of the gaussian distribution or the center of its angle with respect to the growth direction (tilt of the NCs). When fitting only the measured ρ in the top projection all these parameters are indistinguishable and the z/x value is chosen only for convenience. z/x is decreasing towards lower detection energies, which is consistent with the size and therefore spectral dependent uniaxial alignment of the NCs established before. After fitting z/x at a given detection wavelength the same value is used but the calculation is now done for different δ calculated with the changed ε_0 of the surrounding. The results are shown as points (circles and squares) in Fig. 5.23. As evident from the comparison of calculation and measurement the dielectric model succeeds in describing quantitatively the drop of ρ in this experiment.

Up to now the dielectric model, treating PSi prepared from (100) Si substrate as an anisotropic dielectric medium comprising uniaxial aligned Si ellipsoids (with their longer axis parallel to the growth direction) in an effective dielectric medium, has turned out to be very robust. It has proven in several cases to predict and even to evaluate the polarization properties of the PL of PSi. However, these dielectric properties discussed are intrinsic features of dielectricaly nanostructured (100) Si and mainly caused by the specifics of the etching process. Aiming for potential applications of PSi in optical devices makes it necessary to gain active control of these material specific parameters. The first step in this direction is to control the value of ρ and the angular in-plane distribution of light emitting NCs. In that sense Auger saturation and Auger autoinization of NCs (photo darkening effect) turnes out to be a strong tool to manipulate the polarization properties of the PL of PSi.

5.5. OPTICALLY INDUCED POLARIZATION ANISOTROPY IN POROUS SILICON

Due to the long radiative lifetime of the photogenerated excitons in PSi and the large carrier densities achieved due to geometrical restriction of the carriers in a nanometer sized object, nonradiative Auger processes limit the PL intensity at high optical excitation intensities. This effect can easily be seen in the dependence of the PL intensity at low temperatures on the optical excitation power shown in Fig. 5.9. The same holds for room temperature.

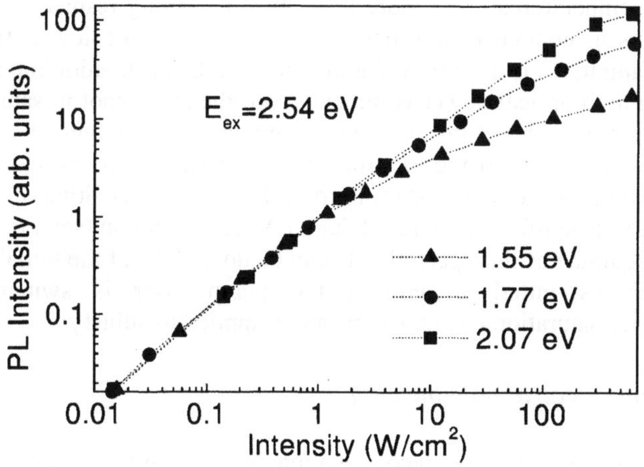

FIGURE 5.24. PL intensity, detected at different energies, as a function of the optical excitation power. T=300K

Fig. 5.24 shows the PL intensity detected at different detection energies as a function of the optical excitation power. At high enough optical excitation power the PL saturates, which leads in Fig. 5.24 to the deviation from a linear power dependence at around 1 W/cm^2 for PL detected at 1.55 eV. The optical pumping intensity at

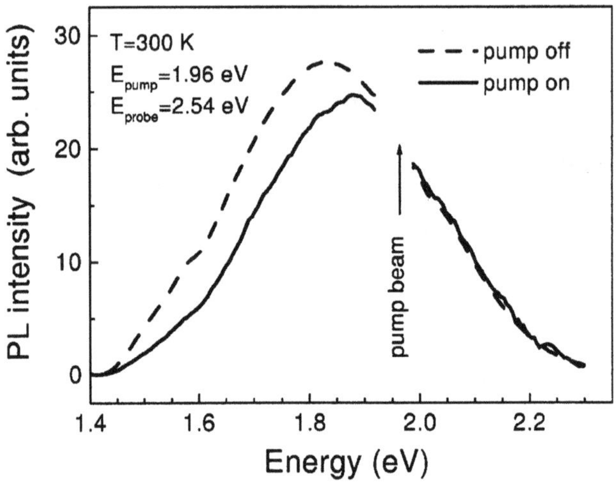

FIGURE 5.25. The influence of intense resonant illumination on the PL. The dashed (solid) line is the PL spectrum without (in the presence of) the pump beam.

which saturation is observed depends on the exciton lifetime and is therefore spectral dependent with the low energy range of the PL, having long radiative lifetimes, saturated at lower optical powers (see Fig. 5.24). Consequently the PL spectrum under optical excitation becomes pump power dependent for a high level of optical excitation. At room temperature, under resonant excitation conditions, the PL intensity can be reduced by up to 20% (at low temperatures up to more than 80%) by strong resonant illumination. In these experiments a modulated high energy probe laser having low optical power (in order to avoid saturation) is used to probe to whole PL band. In addition a CW resonant pump laser with high optical power is directed onto the same spot to saturate the PL of NCs having absorbing states at this particular pump energy. Applying a standard lock-in technique allows separating of the PL from the pump and the probe laser and only the response of the probe is measured. Starting from the energetic position of the pump laser the PL of the NCs is significantly reduced due to Auger saturation (see Fig. 5.25).

The magnitude of this spectral hole can be up to 20% of the initial signal but the PL recovers to its initial strength if the pump laser is switched off. This photoluminescence saturation phenomenon can be applied to adjust ρ.

Dynamic spectral hole burning

The probability of Auger recombination depends on the average carrier density in a NC. The larger the carrier density the larger the Auger rate. For elliptical NCs the excitation probability $P_e(\omega_e)$ and therefore the average carrier density is largest if the polarization direction of the exciting light is tuned to be parallel to the long axis of the NC. Therefore the PL of NCs with their long axis parallel to the polarization direction of the pumping laser are preferentially saturated. Consequently the PL component polarized parallel to the polarization direction of the pumping laser $I_{\|}$ is in stronger saturation (larger deviation from the linear power dependence) than the one polarized perpendicular to the pump polarization $I_\perp$. Therefore ρ ($\rho = (I_\parallel - I_\perp)/(I_\parallel + I_\perp)$) is successively reduced with increasing pump power. Fig. 5.26 displays this behavior at three different detection energies. ρ is decreasing with increasing pump power and thus adjustable. ρ can be reduced roughly 8% to 10%, which is around 50% of the absolute value of ρ. The temperature stability of the sample limits the maximum laser intensity used (at too high sample temperatures (T > 300 K) irreversible chemical oxidation and hydrogen effusion processes are mediated that can change the PL properties of the sample).

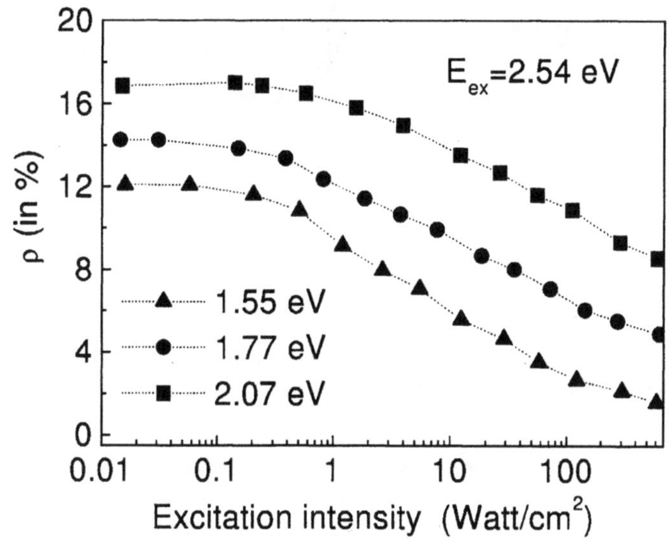

FIGURE 5.26. Degree of linear polarization at different detection energies as a function of the optical excitation power.

However, a relative *increase* of ρ is also desired. To achieve this situation an optical pump-probe technique has to be applied. As already shown a linearly polarized pump laser preferentially saturates the PL of the Si ellipsoids having their long axis parallel to the polarization direction of the laser light. If PL is excited using two laser beams - one linearly polarized unmodulated CW pump beam with high optical power and a second linearly polarized modulated probe beam with low intensity (to avoid saturation effects) – the standard lock-in technique allows separating of the optical response of both excitation beams. Only the PL signal of the modulated probe beam is detected. Without pump beam ρ in top configuration is isotropic for (100) PSi; ρ is independent on the orientation of the polarization direction of the probe laser within this plane. This is shown as squares in Fig. 5.27.

However, in the pump-probe configuration the polarization direction of the pump beam specifies a direction in the (100) surface plane of PSi prepared from (100) Si. In Fig. 5.27 this corresponds to 0^0. If the polarization of the probe beam is chosen to be parallel to the polarization direction of the pump beam a relative decrease of ρ is observed, similar to the behavior discussed before. But now the saturation is caused by the pump laser and the perturbation of the system by the probe laser can be neglected due to its low intensity. Consequently, if the polarization direction of the probe beam is tuned

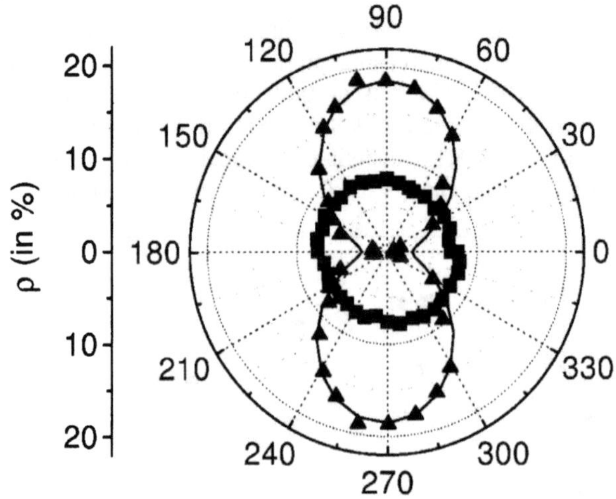

FIGURE 5.27. Degree of linear polarization in pump (2.54 eV) - probe (2.54 eV) conditions with (triangles) and without pump beam (squares) detected at 1.55 eV. 0^0 denote the polarization direction of the high intensity pump light. ρ is measured with a low intensity probe beam as a function of the direction of the polarization direction of the incident light within the surface plane (the top geometry).

to be perpendicular to that of the pump beam the PL of elliptical Si NCs with their long axis perpendicular to the polarization direction of the probe beam are saturated by the pump beam. Therefore, the PL component polarized perpendicular to the polarization direction of the probe beam is weakened by Auger saturation caused by the pump beam and thus the observed ρ is enhanced. Depending on the power of the pump beam ρ can be enlarged by a factor of two (see Fig. 5.27). Using optical techniques it is possible to adjust the level of ρ and even the angular in-plane distribution of ρ [55,56] . But still, at room temperatures, these effects are only present if both pump and probe laser are directed onto the sample. However, for optical memory devices based on the polarization properties of the PL, it is desired to prepare a subset of luminescing NCs by optical techniques (write-in cycle), which will be stable afterwards (during read-out cycles) but can also be erased (delete-cycle). This can be done by polarization selective Auger autoionization (photo darkening) of the PL of PSi at low temperatures.

Persistent spectral hole burning

At room temperature Auger saturation of the PL, which can be accompanied by Auger autoionization of the NC, is a dynamic effect mainly due to thermal reinjection of the trapped carrier back into the NC. However at low temperatures the ejected carrier can be trapped for a significantly long time, leaving behind a singly charged and therefore dark NC. The consequence of the photo darkening effect is for instance persistent spectral optical hole burning [69,70] observed at low temperatures [71]. The experimental arrangement is basically the same as for detecting upconverted photoluminescence (UPL). At low temperatures a strong laser beam is tuned to fall within the visible PL band of PSi (resonant excitation). While UPL can be detected *during* illumination, the

PL response of the whole ensemble is investigated by optical excitation at high energies with low intensity light before and after extended illumination under resonant conditions. Fig. 5.28 shows the PL spectrum excited at 2.54 eV before and after the spectral hole burning procedure with the strong resonant pump laser at 1.916 eV. After the hole burning procedure the PL spectrum is altered at energies lower than that of the degrading laser.

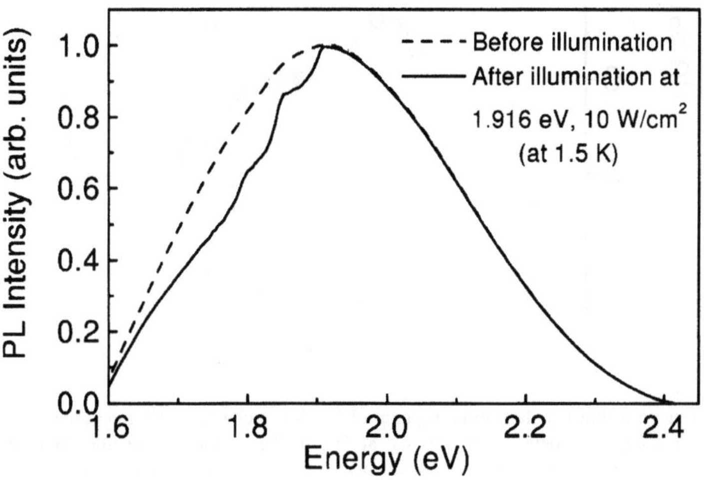

FIGURE 5.28. Low temperature (1.5 K) PL spectrum excited at 2.54 eV before and after intense illumination at 1.916 eV.

The PL in this energy range is weakened by Auger autoionization and shows even a finestructure similar to that observed under resonant excitation conditions. This effect is at low temperatures stable in time and can be monitored up to several hours. In some sense persistent hole burning and resonant PL are similar but reverse processes. Under resonant excitation conditions NCs with an absorbing state at that particular energy are photoexcited and luminesce but in the same way this also gives rise to Auger saturation because excitons are generated in those NCs. Consequently the luminescent particles are the same one that are also "darkened" by Auger autoionization with the strength of the both effects (resonant PL and resonant persistent hole) depending on their density of states at the degradation (resonant excitation) energy point [72]. This explains the similarity between the resonant PL spectrum and the spectral distribution of the persistent hole introduced at resonant excitation conditions. However, it should be mentioned that UPL also limits the strength of (resonant) spectral hole burning because it acts as discharging mechanism of NCs which where previously singly charged due to Auger autoionization. Persistent spectral hole burning at low temperatures is analog to hole burning at room temperatures selective to the orientation of the long axis of the NCs with respect to the polarization direction of the pump laser and the same arguments apply. But now the distribution of still luminescing NCs can be probed *after* the illumination with intense linearly polarized laser light. Typical results are shown in Fig. 5.29.

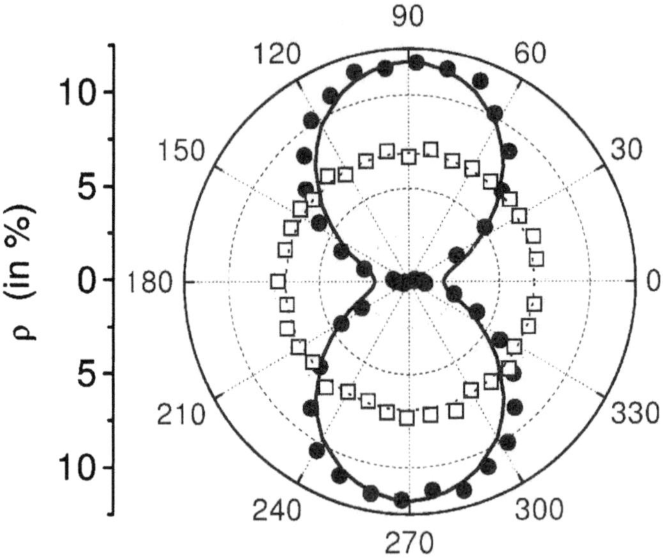

FIGURE 5.29. Degree of linear polarization in pump (2.54 eV) probe (2.8 eV) conditions with (circles) and without pump beam (squares) detected at 2.07 eV at 5K. 0^0 denote the polarization direction of the high intensity pump light. ρ is measured with a low intensity probe beam as a function of the direction of the polarization direction of the incident light within the surface plane (the top geometry).

Again, a relative enhancement or reduction of ρ depending on the orientation of the polarization direction of the probe laser with respect to that of the pump laser is observed. The illumination with the degrading laser serves as write-in cycle, probing the angular variation of ρ is the read out-cycle (which carries the information of the polarization direction of the degrading laser) and heating of the sample to elevate temperatures erases the whole effect due to thermal reinjection of the ejected carriers back into the autoionized NC (delete-cycle). But so far all the effects mentioned and all opportunities shown to adjust the parameters of interest are based on the orientation and shape selective quenching (switching off/on) of the PL of Si NCs. The material specific overall alignment and shape of the NCs are determined by the etching process and are not affected by these methods. Therefore the next step is to gain control of these material parameters by adjusting the etching conditions.

5.6. STRUCTURAL INDUCED POLARIZATION ANISOTROPY IN POROUS SILICON

Etching of a p-type Si substrate is usually done in the dark because the supply of holes, necessary for the etching process, come from the substrate. However, illumination with light having higher photon energies than the band gap of a NC can also photogenerate electron-hole pairs. Therefore combining the etching of p-type material with additional illumination during or after the etching is a promising way to influence the shape of the Si NCs in PSi [73].

Light assisted etching

If the etching process of (100) Si generates elliptical particles that are oriented isotropically in their projection onto the (100) surface plane (top-geometry), an illumination with linearly polarized light selects those having their longer axis parallel to the polarization direction of the incident light. Consequently these NCs are preferentially excited and more electron-hole pairs are photo generated, which leads finally to a larger etching rate of those NCs. Enhanced photo induced etching of elliptical particles will presumably lead to a rounding of them, because dissolution of Si by the electrolyte (F^- ions) is expected to take place preferentially at the sharp, point ends similar to the removal of surface roughness by electropolishing. Consequently ρ is in-plane anisotropic after etching. A minimum of ρ is observed when the polarization direction of the illuminating laser is parallel to the polarization direction of the linearly polarized light present during etching. Fig. 5.30 displays the in-plane (top-geometry) variation of ρ of PSi, prepared with the assistance of light.

The light source was a Xe lamp with a set of low pass filters to cut the high energy component above 2.2 eV. This assures that the absorption of the PL excitation source at 2.8 eV is homogeneous while rotating the polarization direction in the surface plane. If photogenerated electron-hole pairs are roundening the ellipsoids this also opens a way to adjust the overall value of ρ using unpolarized light during etching. ρ is isotropic in the (100) surface plane (top-geometry) because all Si NCs are excited, but the value of ρ is successiveıy reduced with rising light intensities as shown in Fig. 5.31.

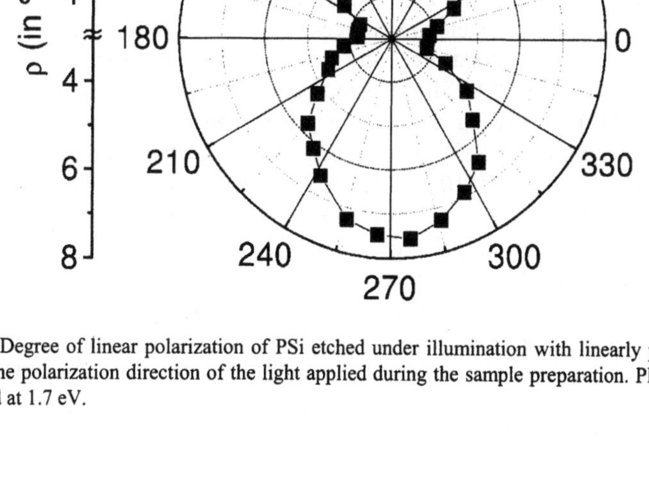

FIGURE 5.30. Degree of linear polarization of PSi etched under illumination with linearly polarized light. 0^0 correspond to the polarization direction of the light applied during the sample preparation. PL is excited at 2.8 eV and detected at 1.7 eV.

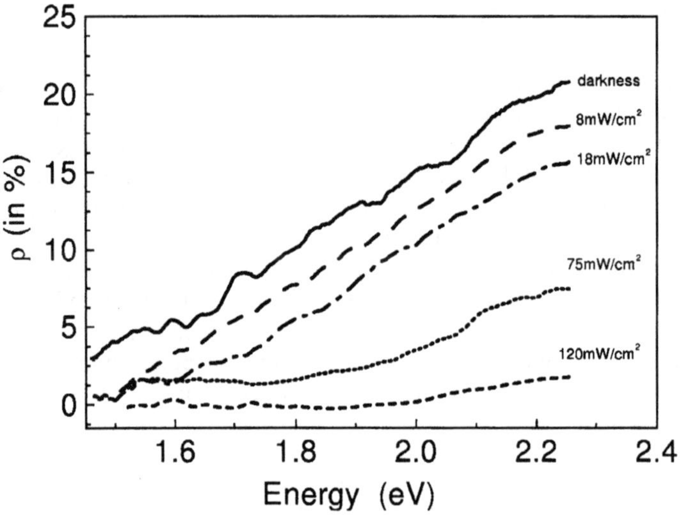

FIGURE 5.31. Illumination induced quenching of the degree of linear polarization by increasing the intensity during the anodization.

(110) PSi

This technique allows the adjustment of ρ and the in-plane variation of ρ by changing the structural composition of the sample. But it presumably addresses only the luminescing NCs having long radiative lifetimes. NCs in which the photo generated exciton is recombining fast due, for instance, to the presence of a nonradiative defect are first of all dark and second basically not influenced by the illumination, because the photogenerated hole has no time to take part in the etching process. Therefore the structural modifications are only partial. Even so, the projection of the long axis of the luminescing Si ellipsoids onto the surface plane (in the top geometry) have a uniaxial alignment. This is similar to the in-plane anisotropy of ρ observed for the cleaved edge geometry, which is also caused by a uniaxial alignment of the long axis of the NCs. In this case the selectivity of the etching process - the preferential pore propagation in equivalent [100] directions - determines the overall alignment of the NCs. The in-plane isotropy of ρ for standard (100) PSi in the top geometry is caused by the *equivalence* of the [010] and [001] directions in the surface plane. This is not the case for (110) PSi etched from lower symmetry (110) Si substrate. For clarity the geometrical arrangement of equivalent (100) axes in (110) oriented Si is illustrated in Fig. 5.32. The two orthogonal axes on a (110) PSi surface in the top geometry are the $[1\bar{1}0]$ direction, which is the projection of the [100] and [010] direction onto the surface of the specimen, and the [001] direction. Therefore the preferential growth of pores and Si NCs in equivalent [100] directions should lead to an in-plane anisotropy in the top geometry of ρ for (110) PSi.

While a part of the NCs are lying with their long axis parallel to the [001] direction, within the surface plane, the NCs pointing into the [100] and [010] are tilted with respect to the surface plane (Fig. 5.32 shows a schematic drawing of the arrangement of the elliptical NCs for (110) PSi for the top and edge geometries). In their

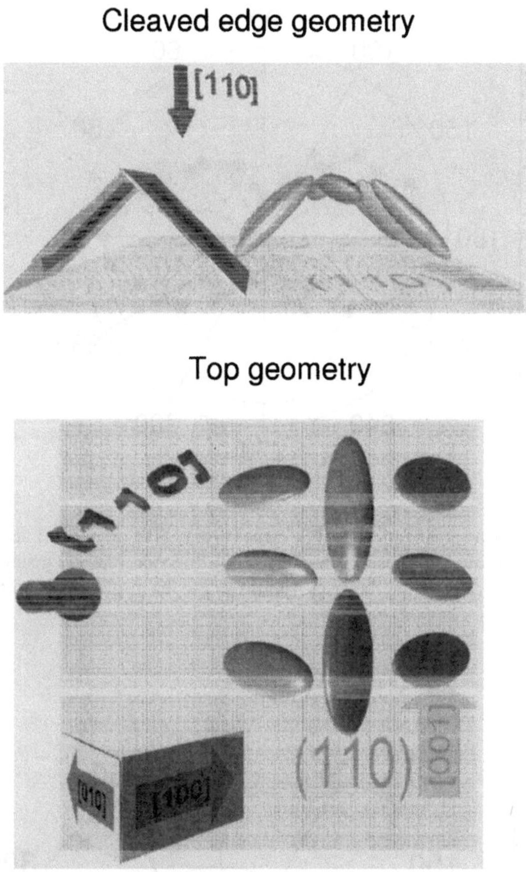

FIGURE 5.32. Schematic drawing of the arrangement of the Si ellipsoids in (110) PSi.

projection onto the sample surface (as for normal incidence) this leads to a different ellipticity of the NCs in $[1\bar{1}0]$ and [001] direction. Consequently ρ is in plane anisotropic when rotating the polarization direction of the PL exciting laser in the top geometry within the surface plane of (110) PSi as displayed in Fig. 5.33. However, also the absorption of linearly polarized visible light is anisotropic for (110) PSi as shown in Fig. 5.34.

This is a result of the anisotropic morphology, the pronounced uniaxial alignment of the long axis of the NCs, of (110) PSi. The absorption coefficient of linearly polarized light differs for the polarization vector parallel to the $[1\bar{1}0]$ or to the [001] axis in the surface plane of (110) PSi. To describe the above observations it is necessary to consider the anisotropic morphology (the absolute gravimetrical measured porosity is not a suitable parameter for such layers). The simplest assumption is that the porous layers consist of dielectric ellipsoids with a uniaxial alignment (two dimensional Bruggeman model). To correlate the porosities of PSi layers and their in the top geometry directional

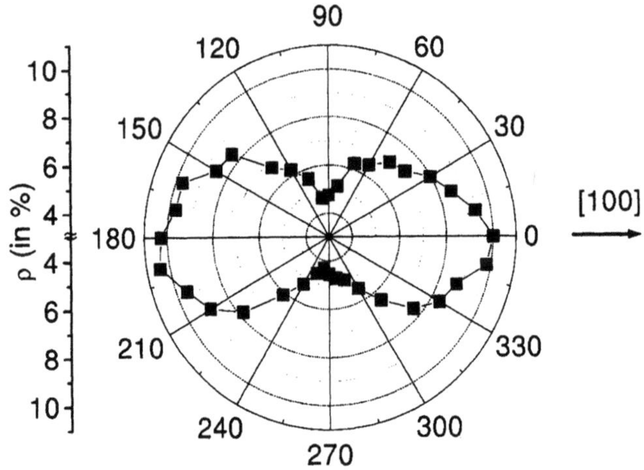

FIGURE 5.33. Degree of linear polarization for PSi etched from (110) oriented Si substrate excited at 2.8 eV, detected at 1.77 eV. The arrow indicates the in-plane [100] direction (the top geometry). Angles denote the orientation of the polarization direction of the laser with respect to the in-plane [100] direction.

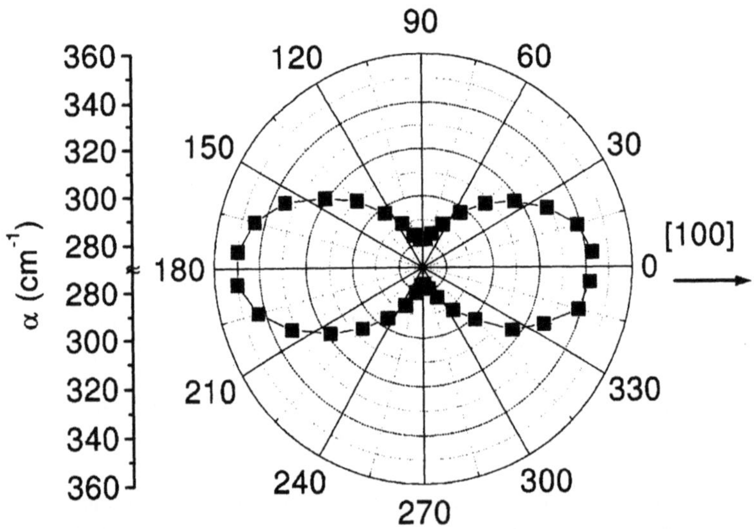

FIGURE 5.34. Absorption coefficient (α) for linearly polarized light for PSi etched from a (110) oriented Si substrate at 1.96eV. The arrow indicates the in-plane [100] direction (the top geometry). Angles denote the orientation of the polarization direction of the laser with respect to the in-plane [100] direction.

refractive index values the two dimensional Bruggeman effective medium approximation is successfully used. $\alpha_{[1\bar{1}0]}$ and $\alpha_{[001]}$ (the subscript denotes the polarization direction of the incident linearly polarized light with respect to the crystalline axis) can be fitted by adjusting the depolarization factor (the L value in equation 1). An anisotropy in α also implies that the refractive index differs in these two directions. Therefore a system of preferentially uniaxial aligned NCs is birefringent [3].

5.7. DIELECTRIC ANISOTROPY OF THE REFRACTIVE INDEX OF PSI

A preferential uniaxial alignment of Si ellipsoids implies a different macroscopic refractive index for light linearly polarized with the polarization vector parallel or perpendicular to the long ellipsoidal axis. Here the refractive index is a maximum if the input polarization is parallel to the long axis (a similar dielectric screening problem is discussed in the Maxwell-Garnett effective medium session). The first observations of such alignment of Si NCs in PSi was observed in (100) PSi.

Birefringent (100) PSi

As evident from the measurements of the polarization phenomenon the NCs in (100) PSi are uniaxial aligned in growth direction [48]. Therefore (100) PSi is also birefringent under the appropriate geometrical conditions [74, 75]. Under normal incidence (100) PSi is birefringent in the cleaved edge geometry. However for practical applications it is first of all difficult to get optical access to the small stripe of the sample (in the range of a micrometer). Second, for birefringent elements it is essential to control the thickness and the parallelity of the front and back surface of the device, which is also problematic in such a geometrical arrangement. If (100) PSi is tilted in the top geometry (not normal incidence) it is also acting as a birefringent material because now all NCs are tilted in a uniaxial direction with respect to the incident light[4]. However, for practical applications normal incidence is desired. Therefore birefringent (110) PSi seems to be the more promising candidate.

Birefringent (110) PSi

Porosified (110) Si shows a pronounced in-plane birefringence under normal incidence [76,77] as a consequence of the uniaxial alignment of the NCs also seen in the high resolution TEM image at the beginning of this chapter (Fig. 5.4). An electron diffraction pattern of the porous Si ((110) PSi prepared from p-type, 1 mΩ cm substrate) layer is shown in Fig. 5.35.

---

[3] The measurements of ρ and α (Fig. 5.33 and Fig. 5.34) should be treated qualitatively because they are only correct for the polarization direction of the incident light parallel to one of the two main axis of the refractive index- the [001] and the $[1\bar{1}0]$ direction.

[4] Normal incidence for measurements of ρ is essential because otherwise the polarization state of the incident light is altered when traversing the sample because of its birefringence.

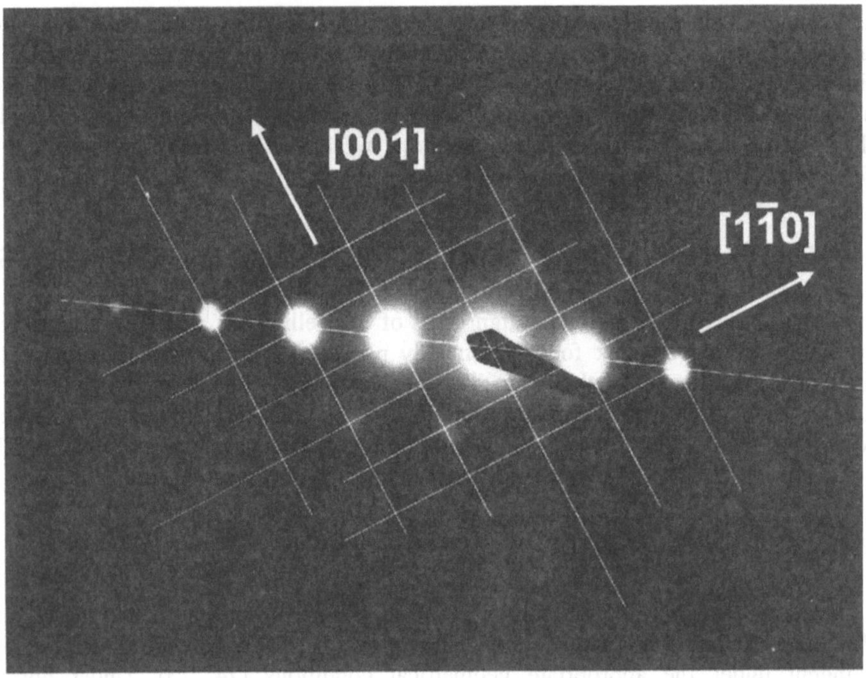

FIGURE 5.35. Electron diffraction pattern of (110) porous Si prepared from a p-type, 1 mΩ cm substrate.

The appearance of bright spots is a result of the crystalline structure of the layers. However, contrary to bulk Si or a (100) PSi layer the spots have an ellipsoidal shape along the [001] crystallographic direction. This can be understood in the following way: a completely regular crystal lattice structure results in spherical spots in the electron diffraction pattern. Considering the constriction of the crystal to a wire-like structure leads to sharp interference for electrons scattered along the wires. Perpendicular to the wire direction, however, the lattice has a reduced extension and consequently the interference is smeared and the spots get elongated (diffused) along this direction. This assures the preferential alignment of the Si wire-like structure along the crystallographic direction. To determine the anisotropy of the refractive index Δn ($\Delta n = n_{[1\bar{1}0]} - n_{[001]}$) transmission measurements in polarization-resolved mode are applied. The normally incident light of a 100 W halogen lamp was polarized 45^0 with respect to the two main optical axes (having refractive indices $n_{[1\bar{1}0]}$ and $n_{[001]}$) by a Glann prism. The polarization of the transmitted beam was analyzed by a second Glann prism and the detection of the transmitted beam was done by a Si and by a germanium photodiode following a single monochromator to cover the spectral range between 500 and 1800 nm. While the polarization of the incident light beam is kept fixed, the polarization of the analyzing Glann prism selects the components of the transmitted light being polarized parallel or perpendicular to the polarization of the incident beam, I_{par} or I_{per} respectively. The spectral dependencies of these two components were divided to each other and are shown in Fig. 5.36.

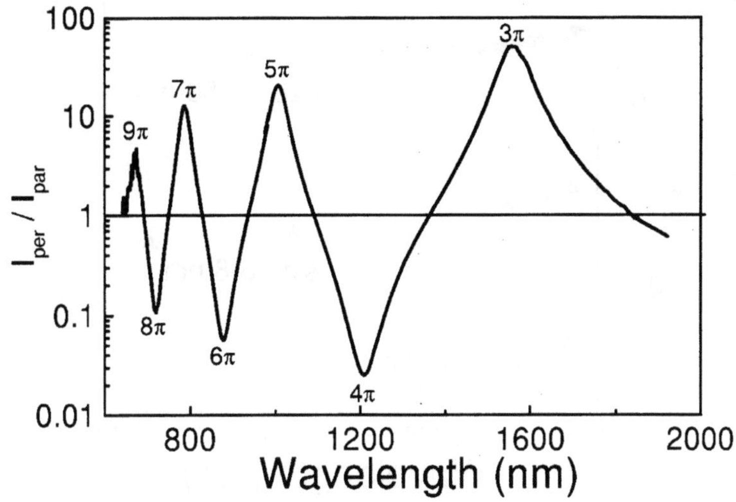

FIGURE 5.36. I_{per}/I_{par}. I_{par} (I_{per}) is the intensity of the light polarized parallel (perpendicular) to the polarization direction of the incident, linearly polarized, light after passing through the (110) PSi layer. The direction of the polarization direction of the incident light was chosen to be 45^0 tilted with respect to the [100] and $[1\bar{1}0]$ directions on the surface of the sample.

Maxima of the resulting curve belong to phase shifts corresponding to odd while minima to even orders of π between the two projections of the electric field vector along the main refractive indices axes. The difference in the refractive indices for the two main components Δn can be calculated according to the simple relation (equation 2):

$$\Delta n = \lambda / (2\pi d) \Delta\phi \tag{6.2}$$

where λ is the spectral position of a certain extremum in the I_{per}/I_{par} curve, $\Delta\phi$ is the corresponding phase shift and d denotes the thickness of the sample. To increase the number of experimental points a set of samples with thicknesses between 3 and 100 μm has been prepared.

Fig. 5.37 shows the values for Δn obtained from the procedure described above for a set of differently prepared samples. While for samples prepared from substrates with the same doping level but with different etching current densities only a small variation of Δn is observed, the doping level itself plays a crucial role for the magnitude of Δn. The samples prepared from a 1 mΩ cm substrate show the highest level of in-plane birefringence, which is even 2 times larger than that of the highest one found in natural crystals (Calcite and Rutile). The lowest level of birefringence is found for layers prepared from a 5 Ω cm substrate and those from 50 mΩ cm substrate have birefringence values in between.

The dependence of the values of Δn on the doping level and therefore on the typical structure size shows that the overall alignment of the nanowires is better for

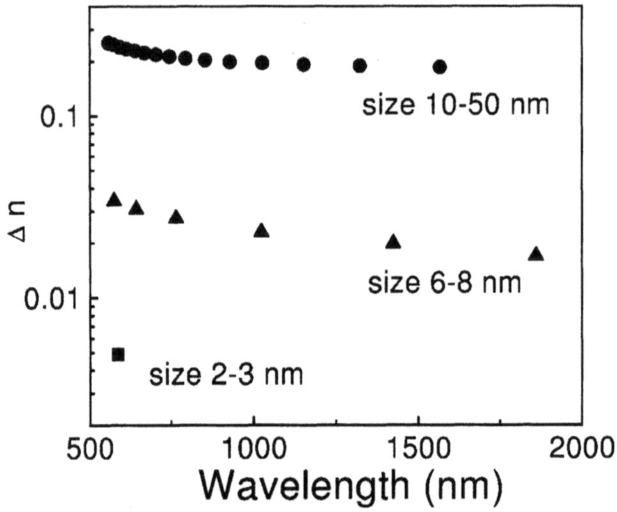

FIGURE 5.37. Δn ($\Delta n = n_{[1\bar{1}0]} - n_{[001]}$) for different doping level of the substrate (and therefore typical structural sizes). 1 mΩ cm (cirlcle), 50 mΩ cm (triangle) and 5 Ωcm (square).

higher doped substrates what is consistent with the behavior observed for (100) Si (see Fig. 5.3). At shorter wavelengths the spectral dispersion of Δn is caused by the spectral dispersion of the refractive index of bulk Si and is consistent with predictions of an effective medium approximation. Knowing Δn and the etching rate for different combinations of doping level of the Si substrate and etching current density is the basis of designing a variety of devices based on (110) PSi.

Devices based on birefringent (110) PSi

Dielectric nanostructuring of (110) Si opens a new way of realization of silicon-based retarders, which can for instance be integrated in optical communication lines. Lorentz has shown that for cubic lattices only a weak dependence of the refractive index value (n) on the light wave vector direction exists [78]. For instance, in bulk Si the maximum value of birefringence for light propagating along the [110] direction was found to be extremely small: $\Delta n = n_{[110]} - n_{[100]} = 5 \cdot 10^{-6}$, where subscripts denote the direction of the electric field vector [79]. Obviously this value does not allow construction of silicon-based controllers of the polarization state of light.

Optical retarders

To demonstrate the capability of (110) PSi layers to be efficient controllers of the polarization state of light propagating in optical fibers the thickness of layers was intentionally chosen to achieve desired retardation values ($\lambda/4$ and $\lambda/2$ conditions) in the transparency range of optical communication lines. Fig. 5.38 (left side) shows how linearly polarized incident light having wavelength 1.53 μm transforms into circularly polarized light.

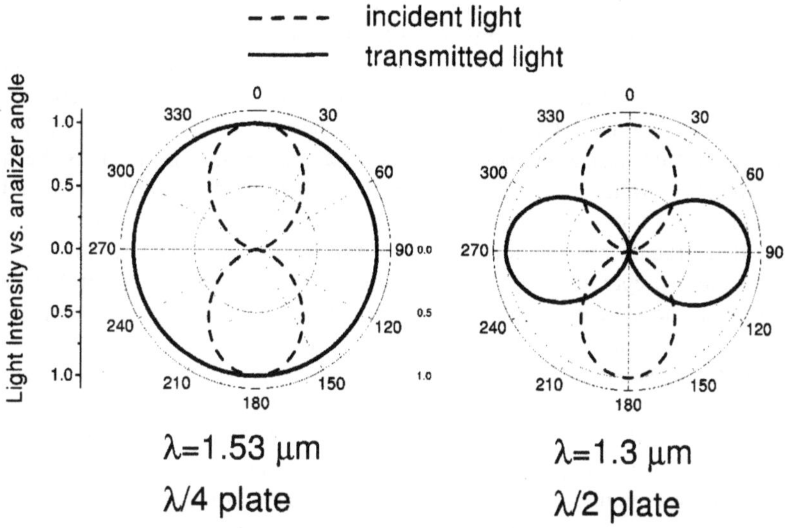

FIGURE 5.38. (110) PSi layers acting as a λ/4 (λ/2) optical retarder. The dashed line is the angular distribution of the intensity of the linearly polarized incident light and the solid line that of the light after traversing the sample.

The right side of Fig. 6.38 shows how a λ/2 retardation condition can be achieved at λ = 1.3 μm. The analysis of the angular polarization pattern of the transmitted light shows the high retardation accuracy: the deviation from the ideal circularly or linearly polarized light for λ/4 and λ/2 plates is of the order of 0.1 %, most likely limited by the accuracy of our experimental setup (see Fig. 5.38). This accuracy is comparable or even better than that of standard retardation plates produced from crystalline quartz. The observation of higher retardation orders is not possible due to Si substrate absorption. In similar experiments performed on free-standing layers partially transparent in the red spectral range retardation values as high as 4 λ have been observed. The accessible wavelength region for such devices is in the long wavelength region only limited by the self-absorption of infrared radiation in the 5 μm (Si-H species) and 10 μm range (Si-O compounds) [80]. By using free standing layers (mesoporous Si is actually an intrinsic material and does not show strong free carrier absorption [81]), free carrier absorption in a Si substrate can be avoided. Besides changing the etching parameters the effective dielectric constant of PSi can be modified by incorporating dielectric liquids having different ε than air in the pores of PSi (as already shown before for standard (100) PSi). In a similar manner Δn of (110) PSi is altered by dielectrics incorporated in the pores. Due to the rather high porosity such a layer can act as very sensitive gas/vapor sensor.

Gas/vapor sensors

The specifics of (110) PSi is that adsorption modifies the birefringent properties of the layers since an additional dielectric substance is incorporated in the pores. In accordance to the Bruggeman effective medium model this results in an increased average refractive index, which is accompanied by a reduction in the optical anisotropy parameter Δn (in the limit of the complete filling of the pores with a medium having a dielectric constant of Si the layers become fully isotropic). Therefore, as expected from Eg. 5.2, a spectral shift of the maxima (minima) to shorter wavelength, depending on the amount of adsorbed liquid and on its high frequency dielectric constant (ε) has been observed. This system has obviously a big potential for sensor applications: because of the large dynamic range of the ratio of the transmitted linearly polarized light components I_{par}/I_{per} (in our case up to 4 orders of magnitude), even a small spectral shift enables the detection of small concentrations of the adsorbent as well as low ε materials. Beside the simple preparation procedure of the freestanding layers several degrees of freedom allow the adjustment of their sensing properties to specific requirements.

The spectral position of a maximum (minimum) of I_{par}/I_{per} can be tuned to coincide with the wavelength of a laser diode used as illumination source by adjusting the thickness of the layer, i.e. by varying the etching time. The spectral shift of the maxima (minima) of the curve I_{par}/I_{per} depends on the amount of the condensed liquid in the pores and therefore can be controlled by a proper choice of the porosity. Fig. 5.39 (a) shows the I_{par}/I_{per} values vs. wavelength for a 42 μm thick layer (51% porosity) with initially empty pores (solid line) and when exposed to different organic vapors (acetone (ε=1.85), cyclohexane (ε=2.05), benzene (ε=2.25)) at their bulk saturation pressures. For all vapors a large spectral shift is observed being significantly different for the used substances and largest for that with the highest ε. The filling of the pores is accompanied by an increase of the I_{par}/I_{per} value at 800 nm up to a factor of 10^3 to 10^4 (a variation of the transmitted light intensity of about 100 can be achieved if only one polarization component (I_{par}) is measured). By plotting the spectral shift of a maximum of I_{par}/I_{per} observed at low acetone vapor pressure (see inset of Fig. 5.39) it is possible to show the high sensitivity of this method. Already at vapor pressures as low as 5 mbar a noticeable shift takes place. At the same time due to the large slope of the $I_{par}/I_{per}(\lambda)$ curve this spectral shift results in a large variation of the absolute value of I_{par}/I_{per}, and therefore a small amount of adsorbate can be detected.

The spectral shift is defined by the dielectric constant of the liquid incorporated in the porous layer and the filling factor of the pores. Therefore, the analysis of the spectral shift allows quantitative determination of the overall condensation behavior of liquids. The modified effective medium approach [24], taking into account the dielectric anisotropy of the layers is applied (two dimensional Bruggeman model). Since the birefringence of nanostructured (110) silicon arises from the preferential alignment of the nanocrystals and the pores, its macroscopic structure has to be taken into account by a modified model based on anisotropic cells. The potential effect of pore shape on the dielectric function is examined using an adjustable depolarization factor for ellipsoidal voids [24], being different for the two [001] and $[1\bar{1}0]$ in-plane directions. This model correlates the optical anisotropy parameters of our layers, i.e. the measured spectral shift with the pore filling factor. The L value in equation 1 is adjusted to fit the spectral position of the maximum in the I_{par}/I_{per} curve for the empty specimen. The spectral positions of the maxima for the filled case are calculated by changing the dielectric constant of the medium filling the pores. In Fig. 5.39 the predicted values for the spectral shift for completely filled pores for every liquid used are indicated by a triangle. They are found to be in a good agreement for high-ε liquids with our experimental results.

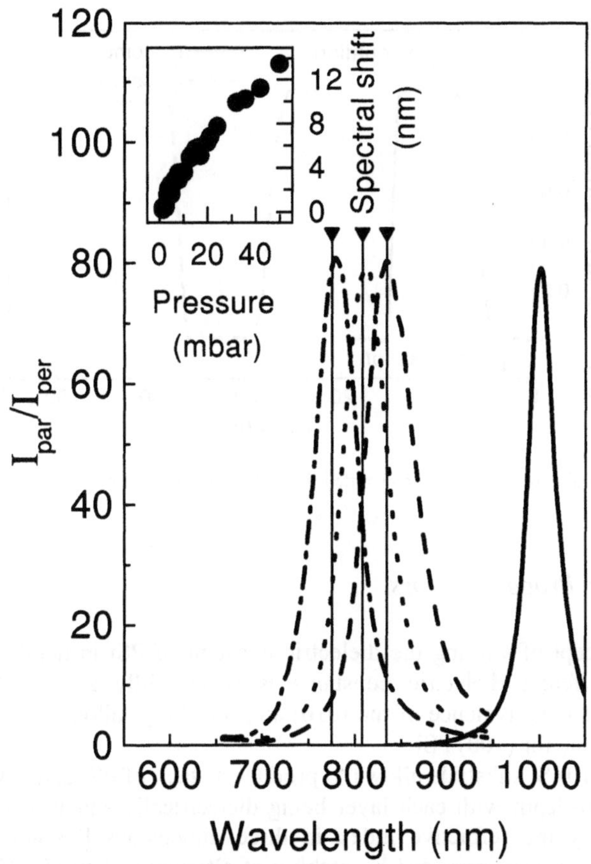

FIGURE 5.39. I_{per}/I_{par}. I_{par} (I_{per}) is the intensity of the light polarized parallel (perpendicular) to the polarization direction of the incident, linearly polarized, light after passing the (110) PSi layer. The direction of the polarization direction of the incident light was chosen to be 45^0 tilted with respect to the [100] and $[1\bar{1}0]$ directions on the surface of the sample. The solid line is the measurement for an empty sample. Dashed line is for a sample with pores filled with acetone, dotted line with cyclohexane and dash dotted line with benzene. The insert shows the spectral shift of the maximum measured for partially filled pores at low acetone vapor pressure. The triangles are a fit explained in the text.

Another important feature of any sensor is its response time. For characterizing such layers in regard to this parameter, the transients of the orthogonally polarized components of the transmitted light I_{par} and I_{per} at a fixed wavelength (λ = 800 nm) are monitored separately. The observed dynamic behavior is presented in Fig. 5.40, where a layer with initially evacuated pores was exposed to an ambient of acetone vapor at bulk saturation pressure. The condensation (filling) process takes place on a time scale of seconds, whereas draining proceeds in the ms range. This permits the fabrication of a fast reacting device. Both, the fast response time and high sensitivity makes such layers promising candidates for gas/vapor sensing [82].

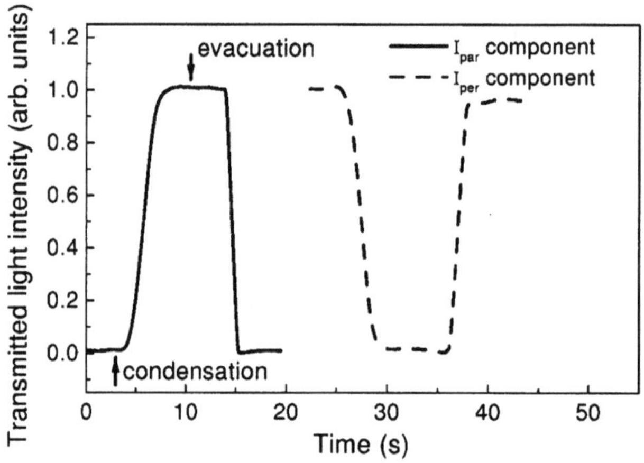

FIGURE 5.40. Temporal evolution of the transmitted light of (110) PSi before and after filling the pores with acetone.

Dichroic Bragg reflectors

The concept of varying the dielectric constant of PSi in depth and to produce layers having different dielectric constants is successfully applied for (100) PSi. However, due to the equivalence of the [010] and [001] crystallographic directions they are in plane isotropic for (100) PSi.

Using (110) Si substrate allows to produce a stack of PSi layers with modulated refractive index in depth with each layer being dielectrically anisotropic in the surface plane, *i.e.*, to vary the refractive index in three dimensions [83,84,85]. (110) PSi multilayer structures were prepared by etching at alternating low (J_1=22 mA/cm^2) and high (J_2=70 mA/cm^2) current density (10 repetitions) in order to produce a DBR. A standard DBR exhibits a high reflectivity band with the Bragg wavelength (λ_{Bragg}) approximately in its center. λ_{Bragg} depends on the thickness of the layers (d_1, d_2) and the corresponding refractive indices (n_1, n_2). The *m*th order of the Bragg peak is given by [83]:

$$m\lambda_{Bragg} = 2(d_1 n_1 + d_2 n_2)$$

However, for birefringent (110) PSi layers the values of n_1 (n_2) depend also on the polarization direction of the incident light ($n_{[1\bar{1}0]} \neq n_{[001]}$, with the subscript denoting the polarization direction of the incident light). Δn is varying with the current density but $n_{[1\bar{1}0]}$ and $n_{[001]}$ can be determined experimentally [86] for J_1 and J_2 ($n_{[001]}{}^{J}{}_{1}$=2.39, $n_{[1\bar{1}0]}{}^{J}{}_{1}$ = 2.55, $n_{[001]}{}^{J}{}_{2}$=1.55, $n_{[1\bar{1}0]}{}^{J}{}_{2}$ =1.75, with J_1 (J_2) indicating the refractive index at current density J_1 (J_2)).

Consequently a (110) PSi DBR exhibits different reflection bands for the polarization direction of the incident light parallel to the [001] or the direction $[1\bar{1}0]$:

$$m\lambda_{Bragg[001]} = 2\left\langle d_1 n^{J_1}_{[001]} + d_2 n^{J2}_{[001]} \right\rangle$$

$$m\lambda_{Bragg[1\bar{1}0]} = 2\left\langle d_1 n^{J_1}_{[1\bar{1}0]} + d_2 n^{J2}_{[1\bar{1}0]} \right\rangle$$

Fig. 5.41 shows the reflectivity of a (110) PSi DBR at almost normal incidence. Contrary to standard porous silicon DBRs [10, 30] the spectral position of $m\lambda_{Bragg}$ depends on the polarization direction of the incident light. For the polarization direction parallel to to the [001] direction (short dashed line) the first order band appears around 1060 nm but is significantly shifted towards longer wavelength (≈ 1170 nm) for light polarized parallel to the $[1\bar{1}0]$ direction (solid line). Based on the etching times and the measured etching rates for the different current densities the thickness of the layers is estimated to be approximately 90 nm for J_1 and 200 nm for J_2. The calculated λ_{Bragg} are 1050 nm for $\lambda_{Bragg\ [001]}$ and 1160 nm for $\lambda_{Bragg\ [1\bar{1}0]}$, which is in good agreement with the measurement. For a high quality dichroic device it is essential that the two reflection bands do not overlap, which is not the case for the first order Bragg bands in Fig. 5.41. However, the second order maxima in the visible spectral range are fully spectrally separated. The slightly reduced value of the reflectivity with respect to the first order is caused by self-absorption of visible light in the porous layers. This dichroic behavior can be easily seen as a change of color of the reflected light from green (560 nm) to red (620 nm) depending on the orientation of the polarization direction of the incident light. Compared to the first order Bragg bands in the near infrared region the second order bands are twice narrower than expected. The dispersion of the dielectric constant of bulk

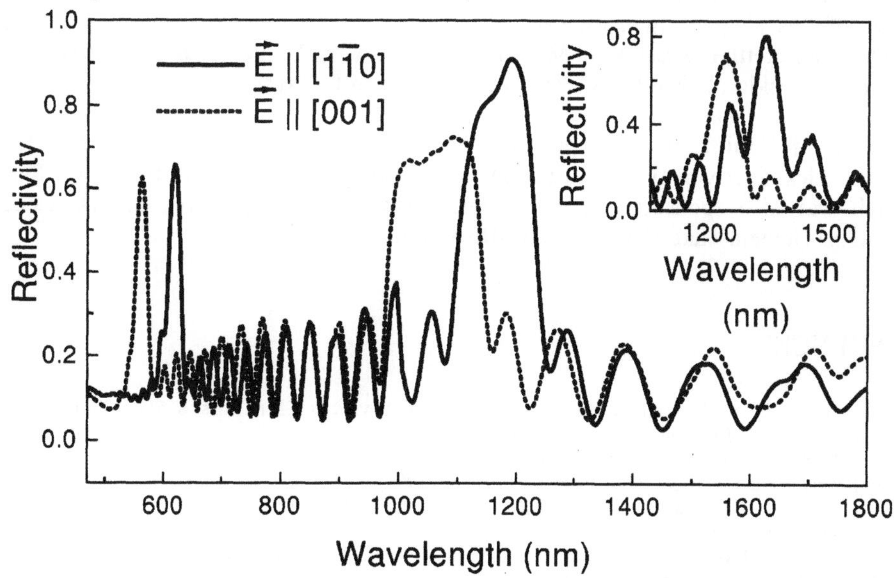

FIGURE 5.41. Spectrally and polarization-resolved reflection of a (110) distributed Bragg reflector. Inset: The same measurement for a (110) distributed Bragg reflector with thicker layers of low and high porosity.

Si leads to their spectral positions, which are not exactly half of the first order. For optical communication lines two separate spectral bands of high reflectivity in the infrared region are desired. Therefore a stack with thicker alternating layers of high and low refractive indices has been produced in order to achieve the second order Bragg bands in the IR spectral range (Fig. 5.41). However, due to the slight variation of etching rate with increased etching time and depth of the multilayers [30] the quality is not as good as for the DBR structure discussed above. In general, multilayers of anisotropically dielectric nanonostructured silicon can act as a dichroic, polarization-sensitive, distributed Bragg reflector. These structures can yield optical effects that are difficult to achieve with conventional Bragg reflector, *i.e.*, they can be used for sensing the color *and* the linear polarization of light.

5.8. SUMMARY

It has been shown how the shape of the Si nanocrystals (NCs) in porous silicon influences the photoluminescence (PL) properties of this material. The overall alignment of the aspheric NCs determines the degree of the preferential linear polarized PL response when excited with linear polarized light. Nonradiative Auger saturation processes can be applied to gain active control on the polarization properties of the PL. A preferential alignment of the aspheric NCs also affects the dielectric response of the porous layer. For porous silicon produced from (110) oriented silicon wafers such a uniaxial orientation leads to strong in-plane birefringence. Passive optical devices based on birefringent porous silicon were presented and their good optical quality was confirmed.

The authors want to thank many colleagues and coworkers who have contributed ideas, and theoretical and experimental work and who are not formally co-authors of this chapter. In particular: Al. L. Efros, M. Rosen, S. Tikhodeev, S. D. Ganichev, W. Prettl, I. Yassievich, M. Ben-Chorin, A. Kux, B. Averboukh, O. Sreseli, Y. R. Shen, H. Heckler, F. Müller, V. Yu. Timoshenko and L. Tsybeskov. We are especially grateful to V. Lehmann (Infineon) for his permission to include Fig. 5.2 and 5.3 from [20] and to M. Fujii for the TEM pictures (Fig. 5.4 and 5.35). J. Diener and D. Kovalev thank the Alexander von Humboldt Foundation for the partial support of this work. This research was also sponsored by the Siemens (Infineon) cooperation, the Deutsche Forschungs Gemeinschaft (DFG), the BMBF and Intas.

REFERENCES

1. A. I. Ekimov, A. A. Onushchenko and V. A. Tzekhomskii, Sov. Phys. Chem. Glass **6**, 511 (1980).
2. A. Henglein, Ber. Bunsenges. Phys. Chem. **88**, 301 (1982).
3. L. T. Canham, Appl. Phys. Lett. **57**, 1046 (1990).
4. V. Lehmann and U. Gösele, Appl. Phys. Lett. **58**, 856 (1991).
5. A. G. Cullis, L. T. Canham and P. D. J. Calcott, J. Appl. Phys. **82**, 909 (1997).
6. L. Canham, Nature **408**, 411 (2000).
7. L. Pavesi, L. Dal Negro, C. Mazzoleni, G. Franco and F. Priolo, Nature **408**, 440 (2000).
8. P. Fauchet, Physics World **14**, 19 (2001).
9. K. D. Hirschman, L. Tsybeskov, S. P. Duttagupta and P. M. Fauchet, Nature **384**, 338 (1996).
10. G. Vincent, Appl. Phys. Lett. **64**, 2367 (1994).
11. E. Yablonovitch, Phys. Rev. Lett. **58**, 2059 (1987).

12. E. Yablonovitch and T. J. Gmitter, Phys. Rev. Lett. **63**, 1950 (1989).
13. O. Bisi, S. Ossicini and L. Pavesi, Surface Science Reports **38**, 1 (2000).
14. D. Kovalev, H. Heckler, G. Polisski and F. Koch, phys. stat. sol. (b) **215**, 871 (1999).
15. A. Uhlir, Bell Syst. Tech. J. **35**, 333 (1956).
16. K. H. Jung, S. Shih and D. L. Kuong, J. Electrochem. Soc. **140**, 3046 (1993).
17. A. Halimaoui, in *Porous Silicon Science and Technology*, edited by J. C. Vial and J. Derrien (Springer, Berlin, 1995) p. 33.
18. L. T. Canham, A. G. Cullis, C. Pickering, O. D. Dosser, T. I. Cox and T. P. Lynch, Nature **368**, 133 (1994).
19. St. Frohnhoff, R. Arens-Fischer, T. Heinrich, J. Fricke, M. Arntzen and W. Theiss, Thin Solid Films **255**, 115 (1995).
20. V. Lehmann, R. Stengl and A. Luigart, Materials Science and Engineering **B69-70**, 11 (2000).
21. M. Christophersen, J. Carstensen, A. Feuerhake and H. Föll, Materials Science and Engineering **B69-70**, 194 (2000).
22. D. A. G. Bruggeman, Ann. Phys. (Paris) **24**, 636 (1935).
23. D. Kovalev, G. Polisski, M. Ben-Chorin, J. Diener and F. Koch, J. Appl. Phys. **80**, 5978 (1996).
24. J. E. Spanier and I. P. Herman, Phys. Rev. B **61**, 10437 (2000).
25. P. Menna, G. Di Francia and V. La Ferrara, Solar Energy Materials and Solar Cells **37**, 13 (1995).
26. H. F. Arrand, T. M. Benson, A. Loni, M. G. Krueger, M. Thönissen and H. Lüth, Electron. Lett. **33**, 1724 (1997).
27. M. Araki, H. Koyama and N. Koshida, Appl. Phys. Lett. **69**, 2956 (1996).
28. V. Mulloni and L. Pavesi, Appl. Phys. Lett. **76**, 2523 (2000).
29. M. G. Berger, C. Dieker, M. Thönissen, L. Vescan, H. Lüth, H. Munder, W. Theiß, M. Wernke and P. Grosse, J. Phys. D **27**, 1333 (1994).
30. C. Mazzoleni and L. Pavesi, Appl. Phys. Lett. **67**, 2983 (1995).
31. P. A. Snow, E. K. Squire, P. St. J. Russell and L. T. Canham, J. Appl. Phys. **86**, 1781 (1999).
32. J. Diener, D. Kovalev, G. Polisski and F. Koch, Phys. Stat. Sol. (a) **182**, 341 (2000).
33. J. Diener, D. Kovalev, G. Polisski, N. Künzner and F. Koch, Phys. Stat. Sol. (b) **224**, 297 (2000).
34. A. V. Andrianov, D. I. Kovalev, N. N. Zinov`ev and I. D. Yaroshetskii, JETP Lett. **58**, 427 (1993).
35. H. Koyama and N. Koshida, Phys. Rev. B **52**, 2649 (1995).
36. S. V. Gaponenko, E. P. Petrov, U. Waggon, O. Wind, C. Klingshirn, Y. H. Xie, I. N. Germanenkom and A. P. Stupka, J. Lumin. **70**, 364 (1996).
37. P. D. J. Calcott, K. J. Nash, L. T. Canham, M. J. Kane and D. Brumhead, J. Lumin. **57**, 257 (1993).
38. G. Fishman, R. Romestain and J. C. Vial, J. Lumin. **57**, 235 (1993).
39. J. Diener, M. Ben-Chorin, D. Kovalev, S. Ganichev and F. Koch, Thin Solid Films **276**, 116 (1996).
40. J. Diener, D. Kovalev, G. Polisski, H. Heckler and F. Koch, Phys. Stat. Sol.(b) **214**, R13 (1999).
41. for example G.E. Pikus and E.L. Ivchenko, *Excitons: Modern Problems in Con-densed Matter Sciences*, ed. by E.I. Rashba and M.D. Sturge, North-Holland, Amsterdam (1982).
42. F. Meier and B.P. Zakharchenya (Eds.), *Optical Orientation*, Modern Problems in Condensed Matter Sciences, North-Holland Publ. Co., Amsterdam (1984).
43. P. Lavallard and R.A. Suris, Solid State Commun. **25**, 267 (1995).
44. H. Koyama and P. Fauchet, Appl. Phys. Lett. **77**, 2316 (2000).
45. M. G. Bawendi, W.L. Wilson, L. Rothberg, P.J. Carrol, T.M. Jeddju, M.L. Stegerwald and L. Brus, Phys. Rev. Lett. **65**, 1623 (1990).
46. M. Chamarro, C. Gourdon and P. Lavallard, J. Lumin. **70**, 222 (1996).
47. D. Kovalev, M. Ben-Chorin, J. Diener, B. Averboukh, G. Polisski and F. Koch, Phys. Rev. Lett. **79**, 119 (1997).
48. D. Kovalev, M. Ben-Chorin, J. Diener, F. Koch, Al. Efros, M. Rosen, N. Gippius and S. Tikhodeev, Appl. Phys. Lett. **67**, 1585 (1995).
49. D. Kovalev, M. Ben-Chorin, J. Diener, F. Koch, A. Kux, Al. Efros, M. Rosen, N. Gippius and S. Tikhodeev, Thin Solid Films **276**, 120 (1996).
50. N. Gippius, S. Tikhodeev, Al. Efros, M. Rosen, D. Kovalev, M. Ben-Chorin, J. Diener and F. Koch, in *Surface/Interface and Stress Effects in Electronic Material Nanostructures*, Symposium. Mater. Res. Soc, Pittsburgh, PA, USA, 1996, p. 203.
51. L. D. Lanau, E. M. Lifshitz and L. P. Pitaevskii, *Electrodynamics of Continuous Media*, 2nd ed. (Pergamon, Oxford, 1984). Ellipsoids: E. C. Stoner, Phil. Mag. **36**, 803 (1945).
52. A. I. Ekimov and Al. L. Efros, Phys. Stat. sol.(b) **150**, 627 (1988).
53. D. Kovalev, H. Heckler, B. Averboukh, M. Ben-Chorin, M. Schwarzkopff and F. Koch, Phys. Rev. B **57**, 3741 (1998).
54. V. Grivickas, J. Linnros and JA. Tellefsen, Thin Solid Films **255**, 208 (1995).

55. D. Kovalev, B. Averboukh, M. Ben-Chorin, F. Koch, Al. L. Efros and M. Rosen, Phys. Rev. Lett. **77**, 2089 (1996).
56. Al. L. Efros, M. Rosen, B. Averboukh, D. Kovalev, M. Ben-Chorin and F. Koch, Phys. Rev. B **56**, 3875 (1997).
57. M. Nirmal, B. Dabbousi, M. Bawendi, J. Macklin, J. Trautman, T. Harris and L. Brus, Nature **383**, 802 (1996).
58. Al. L. Efros and M. Rosen, Phys. Rev. Lett. **78**, 1110 (1997).
59. S. Empedocles and M. Bawendi, Science **278**, 2114 (1997).
60. R. Neuhauser,K. Shimizu,W. Woo, S. Empedocles and M. Bawendi, Phys. Rev. Lett. **85**, 3301 (2000).
61. D. Kovalev, J. Diener, H. Heckler, G. Polisski, N. Künzner, F. Koch, Al. L. Efros and M. Rosen, Phys. Rev. B **61**, 15841 (2000).
62. M. E. Schmidt, S. A. Blanton, M. A. Hines and P. Guyot-Sionnest, Phys. Rev. B **53**, 12629 (1996).
63. J. Diener, D. Kovalev, G. Polisski and F. Koch, Appl. Phys. Lett. **74**, 3350 (1999).
64. J. Diener, Y. R. Shen, D. Kovalev, G. Polisski and F. Koch, Phys. Rev. B **58**, 12629 (1998).
65. Y. R. Shen *The Principles of Nonlinear Optics*, John Wiley \& Sons, Inc., 1984.
66. V. Lin, K. Motesharei, K. Dancil, M. Sailor and M. Ghadiri, Science **278**, 840 (1997).
67. Jun Gao, Ting Gao and M. Sailor, Appl. Phys. Lett. **77**, 901 (2000).
68. H. A. MacLeod, *Thin Optical Filters*, Adam Hilger, London (1969).
69. T. Kawazoe and Y. Masumoto, Phys. Rev. Lett. **77**, 4942 (1996).
70. Y. Masumoto, J. Lumin. **70**, 386 (1996).
71. D. Kovalev, H. Heckler, B. Averboukh, M. Ben-Chorin, M. Schwarzkopff and F. Koch, Phys. Rev. B **57**, 3741 (1998).
72. D. Kovalev, J. Diener, H. Heckler, G. Polisski, N. Künzner and F. Koch, Phys. Rev. B **61**, 4485 (2000).
73. G. Polisski, B. Averboukh, D. Kovalev and F. Koch, Appl. Phys. Lett. **70**, 1116 (1997).
74. S. Zangooie, R. Jansson and H. Arwin, J. Mater. Res **14**, 4167 (1999).
75. M. Kompan, J. Salonen, and I. Shabanov, Journal of Experimental and Theoretical Physics **90**, 324 (2000).
76. D. Kovalev, G. Polisski, J. Diener, H. Heckler, N. Künzner and F. Koch, Phys. Stat. Sol.(a) **180**, r8-r11 (2000).
77. D. Kovalev, G. Polisski, J. Diener, H. Heckler, N. Künzner, V. Yu. Timoshenko and F. Koch, Appl. Phys. Lett. **78**, 916 (2001).
78. H. A. Lorentz, Collected Papers, Martinus Nijhoff, The Hague 1936 (Vol. II, p. 79).
79. J. Pasternak and K. Vedam, Phys. Rev. B **3**, 2567 (1971).
80. H. D. Fuchs, M. Stutzmann, M. S. Brandt, M. Rosenbauer and J. Weber, Phys. Rev. B **48**, 8172 (1993).
81. V. Yu. Timoshenko, Th. Dittrich, V. Lysenko, M. G. Lisachenko and F. Koch, Phys. Rev. B **64**, (2001) 85314
82. E. Gross, D. Kovalev, N. Künzner, V. Yu. Timoshenko, J. Diener, F. Koch, J. Appl. Phys. **90**, (2001) 3529.
83. J. Diener, N. Künzner, D. Kovalev, E. Gross, V. Yu. Timoshenko, G.Polisski and F. Koch, Appl. Phys. Lett. **78**, (2001) 3887.
84. J. Diener, N. Künzner, D. Kovalev, E. Gross, F. Koch, J. Appl. Phys. **91**, (2002) 6704.
85. J. Diener, N. Künzner, D. Kovalev, E. Gross, F. Koch, M. Fujii, Phys. Stat. Sol. (a) **197**, (2003) 582
86. N. Künzner, D. Kovalev, J. Diener, E. Gross, V. Yu. Timoshenko, G.Polisski, F. Koch and M. Fujii, Optics Letters **26** (2001) 1265.

6

Nanocrystalline Silicon-Silicon Dioxide Superlattices: Structural and Optical Properties

Leonid Tsybeskov[1] and David J. Lockwood[2]

[1] *Department of Electrical and Computer Engineering, New Jersey Institute of Technology, Newark, NJ, USA*

[2] *Institute for Microstructural Sciences, National Research Council, Ottawa, Canada*

6.1. INTRODUCTION

This chapter is focused on the fabrication and characterization of layered Si-based nanostructures, which have been called nanocrystalline silicon (nc-Si)–silicon dioxide superlattices [1]. Among the many semiconducting materials, silicon is one of the most studied and definitely the most important material for commercial microelectronics. During the last several decades, the exponential growth of electronic chip complexity and drastic decrease of transistor dimensions has highlighted new directions in electronic device evolution and the potential applicability of Si nanocrystals for nanoelectronics and integrated light-emitters. The latter was stimulated by the discovery of efficient light emission in different forms of Si nanostructures [2] and by the demonstration of a Si-based light-emitting device prototype integrated into conventional microelectronic circuitry [3]. Hence, the interest in reliable fabrication of Si based nanostructures with control over the nanocrystal size, shape, and crystallographic orientation has been growing continuously over the last decade. Recently, the application of Si nanocrystals in electronic devices was suggested and proved by the demonstration of a Si nanocrystal non-volatile memory and other devices utilizing the Coulomb blockade effect [4, 5].

Despite the strong interest in different forms of nc-Si, the fabrication of Si nanocrystals with control over the nanocrystal size and shape was significantly less successful compared to other semiconducting materials. The latest progress in semiconductor nanocrystal fabrication has been based on advances in chemical synthesis in II-VI materials [6] and the successful application of the Stranski-Krastanow growth mode in molecular beam epitaxy (MBE) for III-V semiconductors (see the Chapter by Brunner and Zrenner). Both of these techniques are not applicable for Si/SiO_2 structures. In general, crystalline Si (c-Si) and amorphous SiO_2 are quite different materials with large mismatches between local crystallographic order and thermal expansion. Therefore, crystalline Si does not wet the SiO_2 surface, and standard heteroepitaxy based on MBE, chemical vapor deposition (CVD) or magnetron sputtering, produces highly disordered polycrystalline Si on SiO_2-covered substrates. Several attempts to introduce layered Si/SiO_2 nanoscale structures were reported in the mid and late 1990s, aiming for deep carrier confinement in Si nanocrystals due to high Si/SiO_2 barriers. Two major approaches were focused on (1) high quality amorphous Si (a-Si)/amorphous SiO_2 (a-SiO_2) structures [8] and (2) layers of Si nanocrystals sandwiched between tunnel transparent, nanometer-thin a-SiO_2 separating layers [1]. In the former case, involving a-Si layer thicknesses between 1 and 2 nm, the ultrathin Si layers could not be crystallized even after extended annealing at high temperatures owing to the strain at the Si/SiO_2 interface [9]. Nevertheless, clear optical evidence of carrier confinement in the Si layers was obtained [9, 10]. The other approach utilizes a unique property of the c-Si/SiO_2 interface, the same property that made the MOSFET the unique generic device for modern electronics. Since crystalline (or nanocrystalline) Si and amorphous SiO_2 belong to different thermodynamic classes of materials, they do not mix but rather segregate at high temperatures. Therefore, an atomically smooth interface and very homogenous growth of SiO_2 on c-Si can be produced by a simple technological procedure: Si thermal oxidation. In general, controlled crystallization of amorphous Si/SiO_2 superlattices follows the same rules with only one exception: the a-Si layer must be crystallized before interstitial oxygen diffuses into the Si layer. It was realized almost immediately that rapid thermal annealing, which currently is a standard fabrication procedure for many semiconductor processes, is able to provide very fast crystallization of nanometer thick a-Si layers. At the same time, an a-SiO_2 layer remains amorphous, because annealing at temperatures close to 1200°C merely improves its stoichiometry.

This technological process (full technical details are presented in Section II) is able to produce layers of Si nanocrystals sandwiched between a-SiO_2 layers or nc-Si/a-SiO_2 superlattices. In view of considerable prior research, the physical properties of these structures could be expected to be similar to that of grainy polycrystalline Si and should be affected by nanocrystal (or nanograin) boundaries. However, it turned out that many interesting and unique properties of nc-Si/a-SiO_2 superlattices stem from their vertical periodicity and nearly defect-free, atomically flat, and chemically abrupt nc-Si/SiO_2 interfaces [11]. In addition, by combining a less than 5% variation in a-Si layer thickness with control over the Si nanocrystal shape and crystallographic orientation [11, 12], a system of nearly identical Si nanocrystals can be produced and studied.

The aim of this chapter is to review in detail the fabrication of nc-Si/a-SiO_2 superlattices and their structural and optical (mainly Raman scattering and photoluminescence) properties. Recently, a number of surprising result (such as resonant tunneling and quantum carrier transport [13]) were obtained from investigations of the electronic properties of these structures. Carrier transport in layered structures of Si nanocrystals is a complex subject, and by itself could provide material for an entire monograph. Therefore, we only briefly mention here the most interesting and significant results on the electronic properties of nc-Si/a-SiO_2 superlattices.

6.2. FABRICATION TECHNIQUES AND POST-TREATMENT PROCEDURES

Amorphous layered structures or amorphous superlattices have distinct advantages when compared to crystalline materials: since long-range order is not preserved, virtually any two materials can be deposited in the form of a hetero-pair or layered structure. However, the observation of desired physical phenomena such as Brillouin zone folding, band structure modification, and quantum carrier transport requires minimization of charge carrier scattering, and potential sources of such scattering are structural defects. Another important issue is a leaky and/or rough barrier. To avoid this problem, chemically abrupt and defect free interfaces must be preserved. Since the great majority of studied amorphous superlattices were fabricated using plasma enhanced chemical vapor deposition (PE CVD) [14-16], the most significant problem affecting the quality of these structures (a-Si:H/a-SiN:H, a-Si/a-SiC:H, etc.) is their diffused interfaces. Realizing this fact, Lockwood et. al., proposed to use MBE growth of a-Si in combination with ozone oxidation to achieve a perfect control over the a-Si layer thickness and abrupt a-Si/a-SiO_2 interfaces [10]. Despite the technological complexity, they demonstrated that the wavelength of light emission in these structures, in general, can be controlled by the thickness of the a-Si layers in agreement with effective mass theory [8, 10]. This fact strongly suggests that luminescence in these structures, at least in part, is due to carrier confinement within a-Si layers sandwiched between a-SiO_2 barriers.

The use of an a-Si layer as a quantum well is limited due to intrinsic structural defects. Silicon dangling bonds are the most well known type of such defects [17], and they affect the quality of both the a-Si wells and a-Si/SiO_2 interfaces. Silicon dangling bonds in the a-Si matrix appear due to the absence of long-range order. The generally accepted model considers significant deviations in the Si bond angle and length and agrees with the fact that the a-Si density is typically 80% that of c-Si [18]. The model requires some critical thickness for the a-Si layer before the Si bonds are stretched enough to be broken. Therefore, the dangling bond density per unit volume is reduced in nanometer-thin a-Si layers compared to bulky, micron-thick layers of a-Si and the nanometer-thin a-Si layer density is 98% that of c-Si [19].

The fact that a-Si/a-SiO_2 superlattices are not defect-free structures stimulated a search for a post-treatment procedure that can reduce the defect density. Normally, structural defects that have been introduced by ion implantation or some other manufacturing steps can be eliminated using thermal annealing. Defect annealing utilizes the strong covalent nature of Si bonds and the ability of the crystal lattice to recover the initial, thermodynamically stable bond arrangement. Unfortunately, a-Si/a-SiO_2 superlattices cannot be treated the same way, because oxygen diffusion and the oxidation rate in a-Si is orders of magnitude faster at a given temperature compared to c-Si. Therefore, some intermixing of a-Si and a-SiO_2 may occur. Technically, an ideal solution is to have a stack of nanometer thick layers of c-Si sandwiched between SiO_2 barriers. However, current technology is not able to provide the required dimensions and the interface quality for such structures and many approaches (SIMOX, smart-cut, MBE growth of Si on dispersed SiO_2 layers, etc.) eventually failed to demonstrate the feasibility of reliable fabrication.

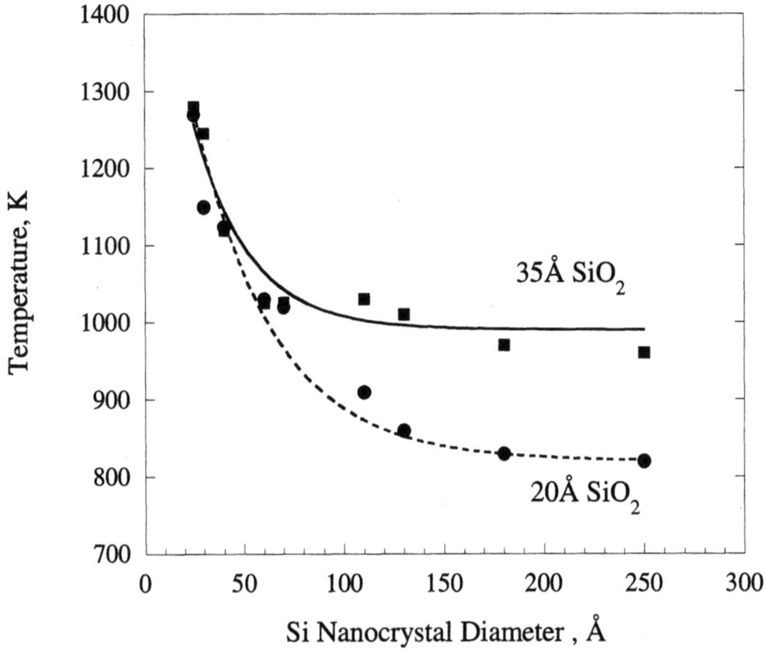

FIGURE 6.1. Crystallization temperature as a function of the Si nanocrystal vertical dimension in nc-Si superlattices with different thicknesses of a-SiO_2 separating layer.

Several independent reports proposed thermal crystallization of nanometer thick a-Si layers sandwiched between a-SiO_2 [20, 21], and they indicated that the crystallization temperature (normally $\leq 600^0$C for a micron-thick a-Si layer) increases significantly, and may reach 1000^0C or more for a-Si layers thinner than 2-3 nm. Later, detailed studies [22] supported the same conclusion, but they attributed such a strong increase in crystallization temperature to very strong strain within nanometer-thin a-Si layers sandwiched between layers of a-SiO_2. A model has been developed [23] that is able to describe the influence of strain at Si/SiO_2 interfaces, but it contains many empirical parameters. In fact, Fig. 1, which compares the crystallization temperature for two series of samples having different SiO_2 thickness fabricated by magnetron sputtering, proves that a thicker SiO_2 layer most likely introduces more strain into the layered Si/SiO_2 system. Therefore, the conclusion [22, 23] that the thickness of a partially disordered Si layer along such interfaces remains ~ 5 Å even for completely crystallized structures may not be exactly correct. Most likely, as proved by several unsuccessful attempts to crystallize a-Si/SiO_2 SLs fabricated by MBE and ozone oxidation [9], the values of the strain and crystallization parameters strongly depend on the local order of a

Si/SiO_2 structure, and such order may vary significantly for different fabrication techniques.

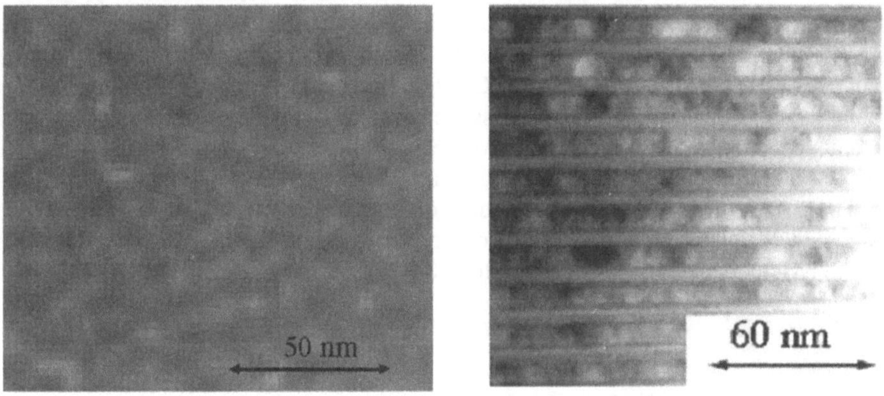

(a) (b)

FIGURE 6.2. TEM micrograph showing a result of thermal crystallization in an amorphous Si/SiO_2 superlattice with (a) initially diffuse (> 20A) and (b) abrupt (< 5A) Si/SiO2 interfaces. The interface abruptness has been estimated using Auger elemental microanalysis and TEM.

Compared to other a-Si/SiO_2 fabrication techniques, [e.g., radio-frequency (RF) sputtering and CVD], MBE in combination with ozone oxidation is the least desirable technique owing to its complexity. However, the alternatives to MBE fabrication are not able to produce abrupt a-Si/SiO_2 interfaces. Several experiments involving Auger microanalysis and transmission electron microscopy (TEM) showed the existence of a 20Å-thick interface layer at a-Si/SiO_2 interfaces in samples prepared by RF sputtering [24]. Detailed studies have shown that such diffuse Si/SiO_2 interfaces may result in strong intermixing and the total destruction of the defined layered structures or superlattices after annealing (see Fig. 2).

In addition to these interdiffused a-Si/SiO_2 interfaces produced by CVD or sputtering growth methods, the quality of CVD or sputtered SiO_2 is known to be poor. Detailed analysis had shown that the most of the SiO_2 structural defects originate from the formation of a columnar structure, which contains a large density of dangling Si bonds [25]. Such columns appear during growth due to the quasi-equilibrium (or low-rate) growth mode. To avoid the formation of the columnar structure, growth of amorphous materials (including a-Si and SiO_2) should be done under highly non-equilibrium conditions. To achieve these conditions, CVD requires a specific range of gas flow, reaction temperature and other growth parameters. Considering RF or magnetron sputtering, an increase in RF power generates highly energetic ions, which result in the destruction of the columnar structure. Figure 3 compares two samples of a-SiO_2 fabricated by RF sputtering using different RF power, and lower RF power visibly results in the formation of the unwanted columns.

The most basic figure of merit for these three growth techniques in a-Si/SiO_2 fabrication is the ability to control the growth rate and thus provide layers of both materials with angstrom precision. This is important for Si, because of the relatively large value of the reduced electron effective mass (~ 0.5 m_e), and a simple calculation shows that a desirable thickness for tunnel devices utilizing carrier confinement is of the order of 50 Å or less. It is even more important for SiO_2 barriers, because the large barrier offset of several electron volts constrains the barrier thickness to be thinner than 20 Å. Compared to the complex chemistry of CVD, RF or magnetron sputtering has the clear advantage of a physical deposition process, i.e., a linear dependence in the deposition rate as a function of time at a given RF power. Sputtering, as in any other plasma induced process, is a self-cleaning procedure, and it can be done in a moderate vacuum of ~ 10^{-7} Torr. The process allows the simple cleaning of a target and substrate as a pre-deposition step. A modern sputtering system also provides very good stabilization of the plasma and is able to bring the deposition rate down to 100 Å per minute or even less.

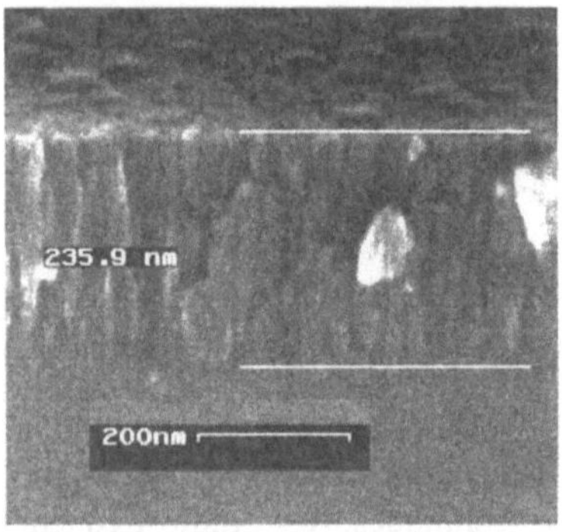

(a)

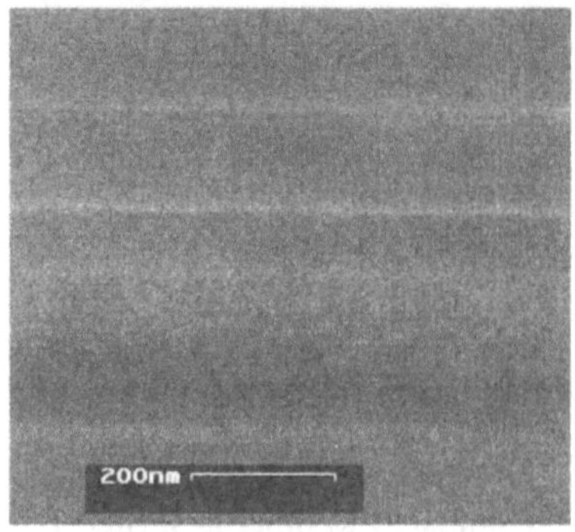

(b)

FIGURE 6.3. Two examples of sputtered a-SiO_2 layers (a) with and (b) without columnar microstructure. A sample with columnar microstructure has been prepared using a low (< 50W) RF power (provided by R. Krishnan).

Thermal crystallization of a-Si/SiO_2 superlattices can be carried out in a conventional, hot-wall furnace with control over the ambient conditions to avoid unwanted oxidation. At moderate temperatures (200-400°C) it was found that the interstitial oxygen may quickly diffuse into a-Si and initiate a-Si/SiO_2 intermixing [26]. It also was noticed that CVD or sputtered SiO_2 introduces less strain into layered a-Si/SiO_2 structures, and this is most likely due to the lower density of sputtered SiO_2 layers compared to SiO_2 fabricated by ozone or thermal oxidation. By considering these issues, it was suggested [27] that the best growth choice would be sputtered or CVD deposited a-Si/SiO_2 layers, but thermal crystallization must be performed using rapid thermal annealing (RTA). The RTA approach was found to provide very efficient crystallization of nanometer thick a-Si layers with thickness down to ~ 2 nm [27]. It was also found that the combination of RTA with subsequent furnace annealing results in the segregation of crystallized Si and amorphous SiO_2 layers, together with pronounced improvement in the nc-Si/SiO_2 interface abruptness and flatness. At the same time, thermal annealing at a temperature ≤ 1200°C improves the a-SiO_2 layer stoichiometry and density, and reduces the number of defects at the Si/SiO_2 interfaces.

To summarize this brief review of Si/SiO_2 layered structure fabrication, amorphous Si/SiO_2 superlattices can be prepared by several techniques, but only one post-treatment process can bring them to the level of device quality structures. This process is a properly controlled thermal annealing, and it results in (1) crystallization of a-Si, (2) enhancement of the Si/SiO_2 interface abruptness, and (3) reduction in the interface and intrinsic defect density (mostly Si dangling bonds).

6.3. STRUCTURAL CHARACTERIZATION OF NANOCRYSTALLINE SILICON – SILICON DIOXIDE SUPERLATTICES: TRANSMISSION ELECTRON MICROSCOPY AND X-RAY DIFFRACTION

Ever since the first report on the successful fabrication of nanocrystalline Si superlattices [28], a wide variety of techniques have been proposed and employed to characterize these structures. The obvious initial focus was on transmission electron microscopy (TEM), which despite its complexity, remains the most important technique to study nanocrystal geometry (size, shape, etc.) and crystallography. TEM is particularly difficult to perform in Si/SiO_2 structures, because of the relatively low contrast and significant probability of charging the SiO_2 layers. However, due to the high interest in Si-based nanostructures, considerable experience has been developed in the last decades, and a number of useful methods for sample preparation and image enhancement [29] finally allowed TEM to become a routine characterization method. In nc-Si/a-SiO_2 superlattices, TEM images have provided critical information and a number of important results will be presented below.

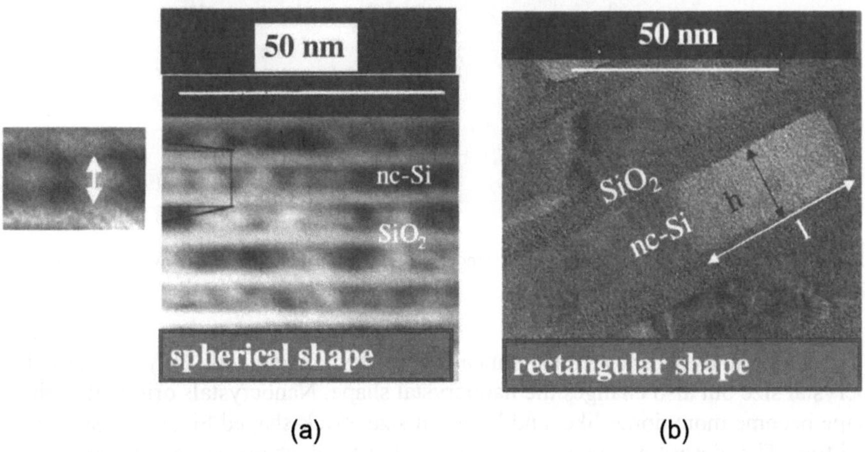

FIGURE 6.4. TEM micrograph showing examples of (a) a spherical and (b) rectangular shape Si nanocrystals named "quantum dots" and "quantum bricks

The TEM images in Fig. 4 show a clear size-dependent trend in the Si nanocrystal shape: smaller Si nanocrystals with vertical sizes ≤ 6 nm are spherical or slightly elliptical but larger nanocrystals (≥ 10 nm) clearly have a rectangular (or brick-like) shape. For the brick-shaped nanocrystals, the ratio between lateral and vertical dimensions is close to 2:1. A high-resolution TEM image of an individual, brick-shaped Si nanocrystal (Fig. 5) proves that the nc-Si/SiO_2 interfaces are flat and abrupt. This observation highlights at least two issues: (1) brick-shaped Si nanocrystals should have a better nc-Si/SiO_2 interface than spherically shaped nanocrystals; and (2) it is important to fabricate laterally-extended Si nanocrystals with an even higher aspect ratio. Laterally extended Si nanocrystals can be considered and used as a Si quasi-well sandwiched between SiO_2 barriers. Figure 6 shows that by utilizing a series of annealing cycles in different ambient conditions (N_2, diluted O_2, etc. [12]) the aspect ratio in the Si nanocrystal lateral and vertical dimensions can be extended to $\geq$ 3:1. Finally, nearly continuous Si layers with a thickness of ~ 25 nm have been produced (Fig. 7), but some of them clearly show fluctuations in the brightness of the reflected electron beam indicating possible intrinsic defects.

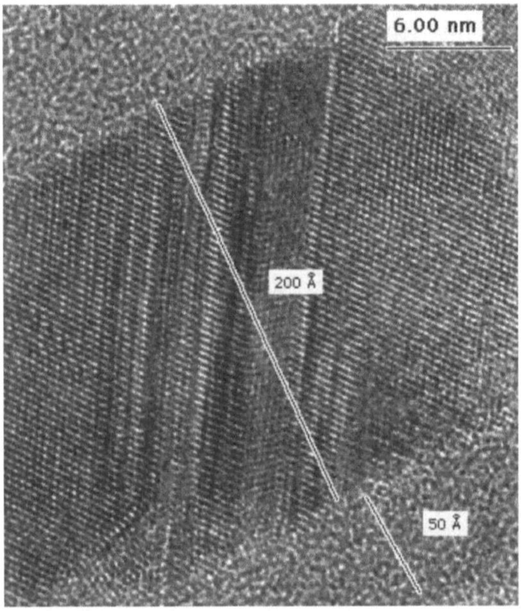

FIGURE 6.5. A lattice image of a rectangular Si nanocrystal (i. e., "quantum brick") showing nearly atomically flat interfaces with a-SiO_2.

TEM analysis shows that annealing in dry oxygen not only reduces the Si nanocrystal size but also changes the nanocrystal shape. Nanocrystals originally spherical in shape become more dome-like, and larger in size, brick-shaped Si nanocrystals change very little. The described modification in nanocrystal shape is extremely slow and requires hours of oxidation at a temperature ~ 1000°C confirming the strong resistance of Si nanocrystals to oxidation in dry oxygen. In contrast, wet oxidation is extremely aggressive and destroys Si nanocrystals very effectively. The precise mechanism of Si nanostructure oxidation is not understood, but most likely it is connected with different oxidation rates in different crystallographic directions. Figure 8 confirms this preliminary conclusion and shows that the originally brick-shaped Si nanocrystals after several minutes exposure to N-diluted H_2O vapor at 950°C have been transformed into pyramids of different sizes but having the same angles.

Another interesting feature has been found in nc-Si/SiO_2 superlattices with a large number of periods (>> 20). In some of parts of the initial layers of nanocrystals

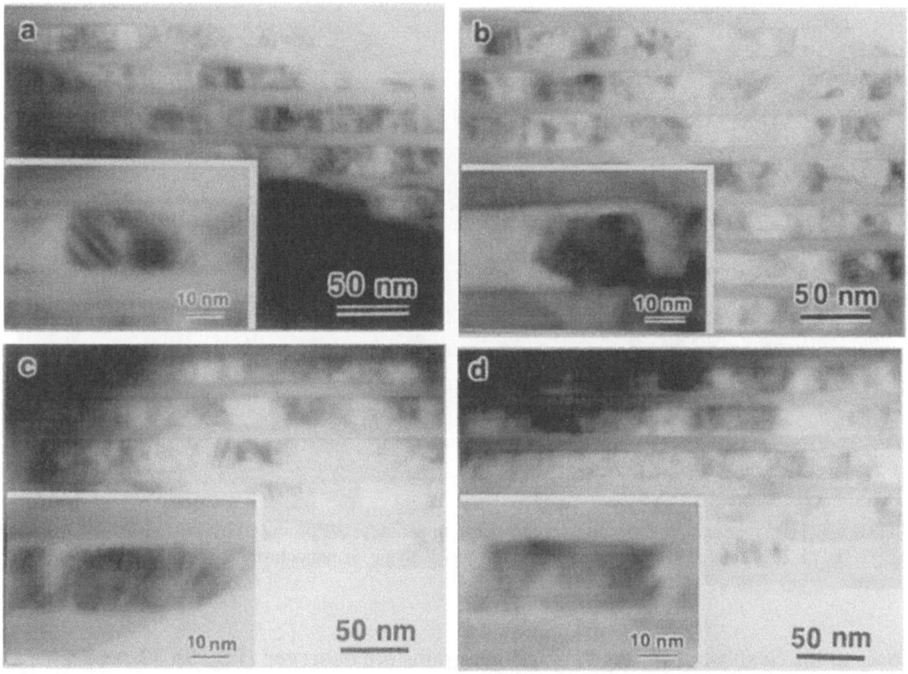

FIGURE 6.6. TEM images of a nc-Si/a-SiO_2 superlattice subjected to (a) initial two hours of annealing at 950°C in a dry nitrogen; (b) 15 minutes annealing at 750°C in ~ 15% nitrogen diluted water vapor followed by (a); (c) 30 minutes annealing at 750°C in ~ 15% nitrogen diluted water vapor followed by (a); (d) 45 minutes annealing at 750°C in ~ 15% nitrogen diluted water vapor followed by (a). From (a) to (d), the Si nanocrystal size slightly reduces in vertical dimension and significantly increases laterally.

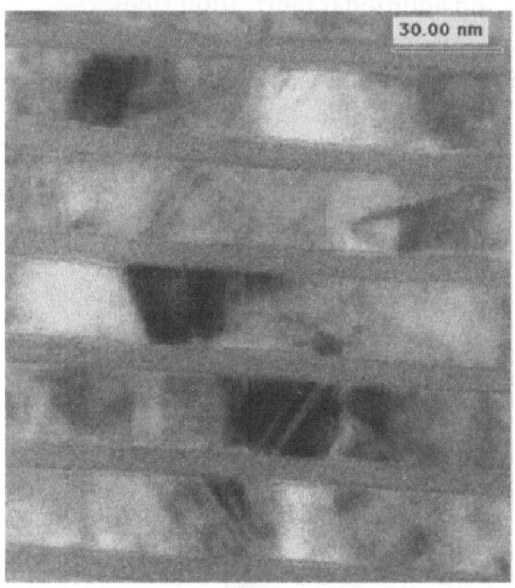

FIGURE 6.7. Nearly continuous, nanometer-thin Si layer produced by RTA followed by 30 minutes wet oxidation at 800°C and 2 hour annealing in dry N_2 at 1100°C

FIGURE 6.8. TEM micrograph showing pyramid-shape Si nanocrystals produced by 15 minutes wet oxidation at 950°C in 15-20% diluted water vapor of initially brick-shape Si nanoclusters. The pyramid locations are indicated.

(closer to the Si substrate), large crystalline grains are observed (Fig. 9). These grains are seen as bright spots, which is due to a specific crystallographic orientation of these grains. These 'broken' layers most likely appear due to strain forces. This strain is built in during crystallization, because of the significant difference in thermal expansion between Si and SiO_2. Interestingly, this local distortion does not propagate throughout the nc-Si superlatices, and the rest of the structure is perfectly ordered. This observation proves that the ordered structure in nc-Si superlatices is robust and can survive during post-crystallization treatments and additional fabrication steps.

Among the many other characterization techniques, X-ray diffraction (XRD) shows reasonable agreement with TEM results (Fig. 10) using conventional analysis for the grain sizes [30, 31]. However, the quantitative application of XRD is limited to

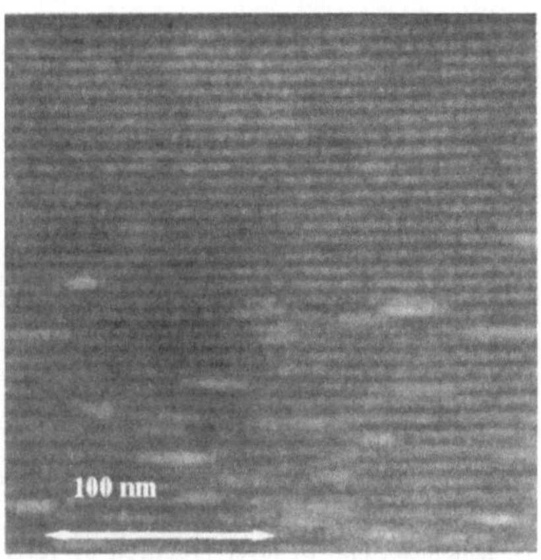

FIGURE 6.9. Multi-period nc-Si/a-SiO_2 with areas of "broken" Si layers shown as bright spots (provided by M. Zacharias).

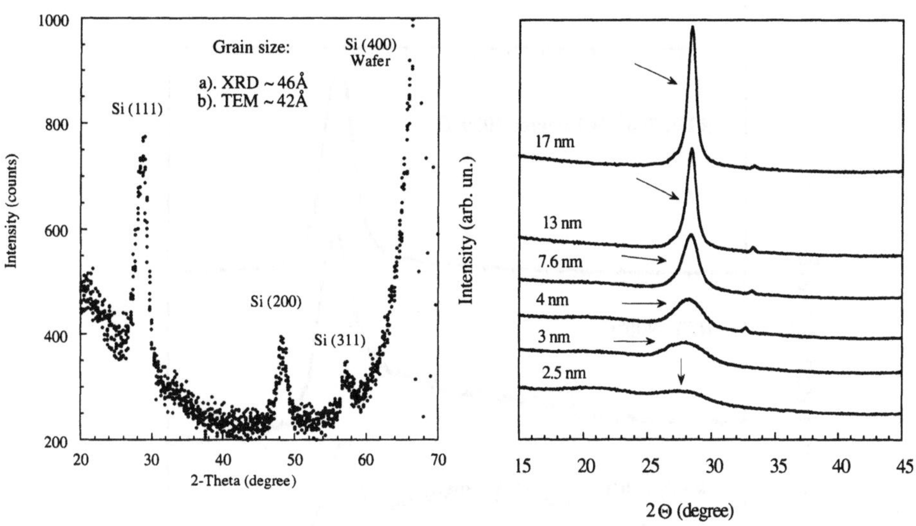

(a) (b)

FIGURE 6.10. XRD data showing (a) peaks attributed to different crystallographic directions in Si nanocrystals and (b) a progression of Si <111> diffraction peak for continuously decreasing in size Si nanocrystals (provided by T. Blanton and M. Zacharias).

moderate sizes of Si nanocrystals (> 2.5 nm). It is not clear if the reason for this limitation arises from a distortion of the lattice structure in small size Si nanocrystals, the presence of residual a-Si tissue, or something else.

6.4. RAMAN SCATTERING FROM NANOCRYSTALLINE SILICON-SILICON DIOXIDE SUPERLATTICES: OPTICAL AND ACOUSTIC PHONONS AND POLARIZATION RAMAN MEASUREMENTS

Raman scattering in nanocrystals is a well-developed subject, starting with the work of [32] and [33] where a semi-quantitative model considering a partial modification of the phonon selection rules was developed. Controlled crystallization is a fabrication method where Raman scattering can be an even more informative characterization method, due to the easily observed difference between the spectra of amorphous and nc-Si phases. Figure 11 compares the Raman spectra of several samples. In the as prepared a-Si (11 nm)/a-SiO_2 superlattice, a strong Raman peak at ~ 470 cm^{-1} clearly indicates a purely amorphous phase. Rapid (~ 30 s) thermal annealing at 700°C partially crystallizes the samples and a Raman contribution from both amorphous and nc-Si phases is shown by the two peaks at 485 and 520 cm^{-1}, respectively. Rapid (~ 30 s) thermal annealing at 900°C further reduces the contribution from the amorphous phase. Finally, furnace annealing for 30 minutes at 1050°C completes the crystallization, and that can be verified from the Raman spectrum as a perfectly symmetric, slightly broader peak at 520 cm^{-1}.

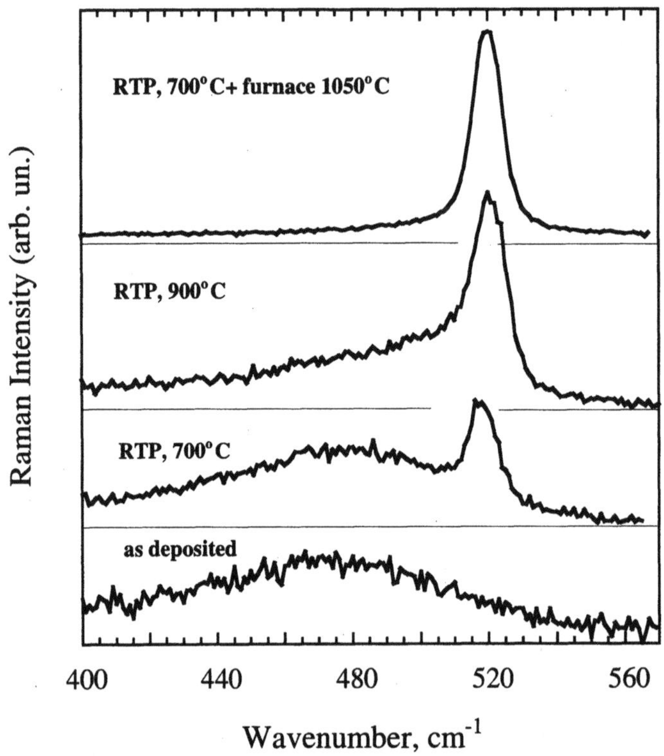

FIGURE 6.11. Raman scattering in several samples of a nc-Si/a-SiO_2 superlattice annealed at different (indicated) conditions.

More detailed Raman measurements (Fig. 12) show that in smaller Si nanocrystals (from 14 down to 4 nm), several features associated with the disordered Si phase can be seen. The major feature related to small Si nanocrystals is a first-order, transverse-optical (TO) Raman peak at 520-500 cm^{-1} (depending on the Si nanocrystal size) with a broad tail to the lower wavenumbers. Qualitatively similar broadening is observed for overtones related to combinations of acoustic and optical phonons at 150 and 300 cm^{-1} as well as for the second-order Raman scattering at ~ 1000 cm^{-1}. The Raman data indicate not only the modification of phonon selection rules in small Si nanocrystals but also the presence of partial disorder, most likely due to the higher surface-to-volume ratio and stronger contribution from the nc-Si/SiO_2 interface. Following a semi-quantitative analysis developed for a-Si [34, 35], we can consider that the degree of disorder is indicated by the intensity ratio between the major Raman peak at 520 cm^{-1} and overtones (~ 300 cm^{-1}) that are forbidden in perfect c-Si [36]. Figure 12 shows that this ratio between the Raman peaks at 300 and 520 cm^{-1} increases as the Si nanocrystal size decreases.

The Campbell-Fauchet and other models that describe the changes in the phonon selection rules in nanometer-size Si nanocrystals, have been criticized for being both a simplified theoretical approach and inconsistent with experiments [37-39]. However, it is not possible to combine all desirable considerations (nanocrystal size reduction, shape influence, strain, specific interface, etc.) in one model. It is also impossible to separate such effects as, for example, partial disorder due to a strained interface and phonon selection rule modification due to a nanocrystal size reduction. All these complications have made quantitative analysis of the optical phonon Raman spectra in Si nanocrystals difficult, if not impossible.

The fact that a nc-Si/a-SiO_2 superlattice is a layered, periodic structure with flat and abrupt interfaces stimulated interest in Raman measurements in the wavenumber range of acoustic phonons. We follow here the original Rytov model [40], which has been successfully applied to conventional, MBE-grown superlattices [41]. In a more simplified picture, a superlattice made of alternating layers of nc-Si having thickness d_1 and a-SiO_2 having thickness d_2, induces a new periodicity $d = d_1+d_2$ and leads to a reduction of the bulk crystal Brillouin zone in the superlattice growth direction. A new minizone is created with a maximum wavevector $q_{mz} = \pi/d$. Since the nc-Si/a-SiO_2 superlattice contains more than 10 atomic layers per period, it can be treated as an elastic medium and the elastic continuum model of acoustic vibrations can be used to calculate the superlattice phonon dispersion [40, 41]. According to this theory, the acoustic phonon dispersion in the superlattice is obtained by folding the acoustic phonon branches of the average compound into the superlattice minizone. For bulk acoustic phonons with frequencies less than 100 cm^{-1}, the dispersion is almost linear, and in the superlattice away from the Brillouin minizone center ($q = 0$) and edge ($q = q_{mz}$) it can be described by $\omega = 2mv_{SL}\pi/d \pm qv_{SL}$, where $m = 0,1,2,...$ is the Brillouin zone folding index and q is the phonon wavevector. The superlattice sound velocity v_{SL} is given by

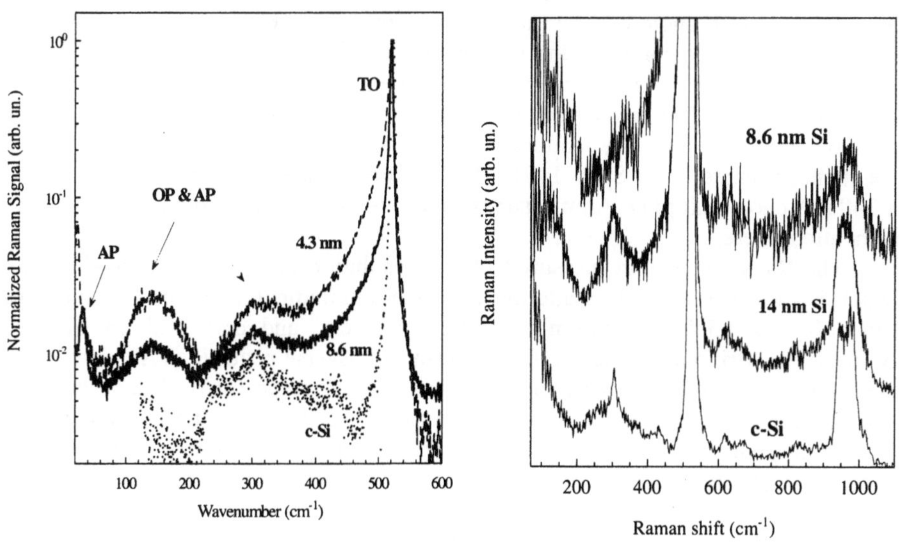

(a) (b)

FIGURE 6.12. (a) A high resolution Raman spectra and (b) details of second order Raman scattering (~ 1000 cm^{-1}) and overtones (~ 300 cm^{-1}) for c-Si and Si nanocrystals of different sizes.

$$v_{SL} = d/[(d_1/v_1)^2+(d_2/v_2)^2+(\kappa+1/\kappa)\cdot(d_1d_2/v_1v_2)] \quad (6.1)$$

where $\kappa = \rho_1v_1/\rho_2v_2$, and v_1 and v_2, ρ_1 and ρ_2 are the sound velocities and densities, respectively, in the nc-Si and a-SiO_2 layers. In the backscattering geometry, Raman scattering from phonons with wavevector $q_p = 4\pi n_{SL}/\lambda$ is observed, where n_{SL} is the superlattice effective refractive index at the laser wavelength λ, defined as

$$n_{SL} = \xi n_1^2+(1-\xi)n_2^2, \quad (6.2)$$

where n_1 and n_2 are the corresponding refractive indices for the nc-Si and a-SiO_2 layers, respectively, and $\xi = d_1/d$. The case of m = 0 corresponds to conventional Brillouin scattering from acoustic phonons ($\omega_{Br} = q_pv_{SL}$). Each value of the folding index m ($m \neq 0$) gives rise to a doublet ($\omega_{\pm m} = 2mv_{SL}\pi/d \pm \omega_{Br}$), which is the signature of the folded phonon modes in Raman scattering. In the exact continuum theory [42], small energy gaps appear in the phonon dispersion at the minizone center ($q_p/q_{mz} = 0$) and edge ($q_p/q_{mz} = 1$) due to a mismatch between the acoustic impedances ρv of the nc-Si and a-SiO_2 layers. The gaps can be ignored when q_p/q_{mz} is not close to 0 or 1.

The investigated a-Si/a-SiO_2 superlattices were grown on a Si wafer by RF magnetron sputtering and plasma oxidation. The layer thicknesses for both a-Si and a-SiO_2 were kept constant throughout the superlattice. The a-Si layers were crystallized by RTA at 900°C for 60 s followed by quasi-equilibrium furnace annealing starting at 750°C with a 10°C/min temperature ramp-up to 1100°C (see Ref. 11, 12 for details). Rapid thermal and furnace annealing were performed in a nitrogen atmosphere. The layer thicknesses were determined using TEM.

Raman spectra were measured in a 90° scattering geometry with the sample surface inclined at an angle of 12.3° to the incident light (which is near to Brewster's angle in c-Si [41]). This configuration is close to exact backscattering inside the sample since the superlattice has a quite large refractive index ($n_{SL} > 3$ as calculated below). The excitation source was 300 mW of argon ion laser light at wavelengths of 457.9, 476.5, 488.0, 496.5, and 514.5 nm. The scattered light was analyzed with a SOPRA DMDP2000 double monochromator or a Spex 14018 double spectrometer equipped with a cooled RCA 34031A photomultiplier. All measurements were carried out at room temperature in a helium atmosphere.

The Raman spectrum measured with a resolution of 0.06 cm^{-1} and 457.9 nm excitation is shown in Fig. 13. Scattering from the longitudinal acoustic (LA) phonons with m = 0 ($\omega_{Br} = 3.4$ cm^{-1}) and m = ±1 ($\omega_{-1} = 17.4$ cm^{-1} and $\omega_{+1} = 24.3$ cm^{-1}) can be easily identified. Brillouin scattering from LA and transverse acoustic (TA) phonons in the c-Si (100) substrate is also observed at 5.6 and 3.9 cm^{-1} respectively. The observation of weak scattering from TA phonons is consistent with the small deviation from the exact backscattering geometry, for which Raman scattering from TA phonons in (100) Si is forbidden. The observation of Brillouin scattering is possible only in the case of a very low laser light scattering background, proving the case of a very smooth sample surface and flat nc-Si/a-SiO_2 interfaces. The m = ±1 doublet was also resolved in Raman spectra measured with a Spex 14018 double spectrometer with resolution better than 0.1 cm^{-1} and excitation by 457.9 nm, 476.5 nm, 488.0 nm, 496.5 nm and 514.5 nm lines of an Ar^+ laser.

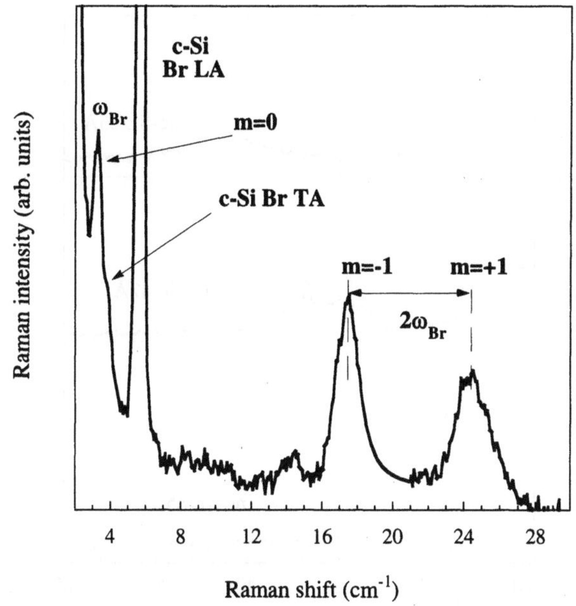

FIGURE 6.13. A high resolution Raman spectrum in the vicinity of folded acoustic phonons (indexes are shown) in a fully crystallized nc-Si/a-SiO_2 superlattice comprised of brick-shape Si nanocrystals.

Using the experimentally determined d and $\omega_{\pm 1}$, it is possible to estimate the nc-Si/a-SiO_2 superlattice sound velocity. We found that $v_{SL} = d(\omega_{+m} + \omega_{-m})/4\pi$, averaged over the data obtained with different laser excitation lines is 7.5×10^5 cm/s. The reduced wavevectors of phonons participating in the scattering were calculated from $q_p/q_{mz} = d(\omega_{+m} - \omega_{-m})/2\pi v_{SL}$. The use of different excitation wavelengths allows the observation of scattering from the LA phonons in the 0.26-0.31 reduced wavevector range of the Brillouin minizone. The assumption that the phonon dispersion is linear (i.e., the gaps can be ignored) is indeed justified by the above result: the light scattering involved phonons far from the minizone center or edge. Several measurements over the sample area of 2-3 cm^2 show very small deviation in the difference of the position of folded phonon peaks, and this difference indicates less than 5 % variation in the superlattice thickness. Also, assuming that the sound velocity in the a-SiO_2 layers is identical to that in bulk a-SiO_2 (6.0×10^5 cm/s), the sound velocity in the Si nanocrystals is calculated as 8.6×10^5 cm/s. This calculated sound velocity in nc-Si is close to the sound velocity in c-Si) along [100] direction (8.5×10^5 cm/s), compare with 9.2×10^5 cm/s and 9.4×10^5 cm/s along the [110] and [111] directions, respectively [43, 44, 45]. This result indicates that the Si nanocrystals might have some degree of preferential [100] orientation. This conclusion is not unreasonable, since under certain experimental conditions crystallized poly-Si films have [100] preferential orientation due to the fact that [100] is the fastest growing crystallographic direction [46]. The effective refractive index of the SL for different λ is calculated from $n_{SL} = \lambda(\omega_{+m} - \omega_{-m})/8\pi v_{SL}$. The refractive index in Si nanocrystals, n_1, then can be derived assuming that the refractive index of a-SiO_2, n_2, is 1.46. The refractive index of nc-Si with an average nanocrystal size of 8.6 nm is noticeably lower than that of c-Si or a-Si.

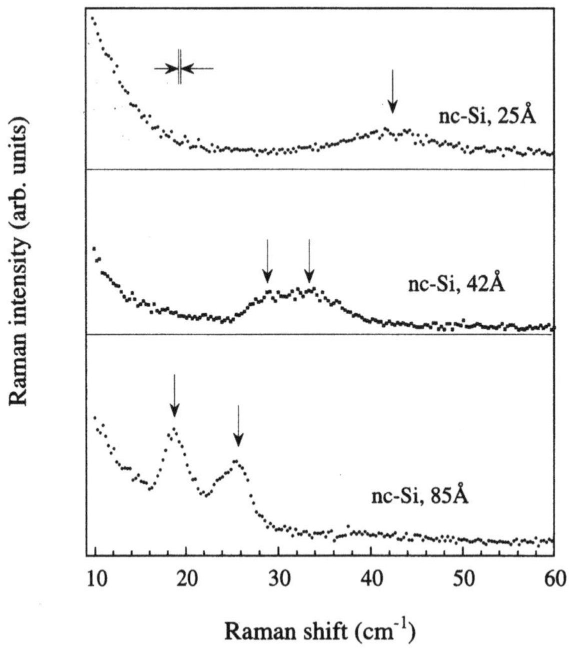

FIGURE 6.14. Moderate resolution low frequency Raman spectra for samples of nc-Si/a-SiO_2 superlattices with different sizes of Si nanocrystals.

Figure 14 shows moderate resolution (~0.2 cm^{-1}) Raman spectra in the range of acoustic phonons for three samples of nc-Si/a-SiO_2 superlattices with different vertical sizes of the Si nanocrystals: 2.5, 4.2 and 8.5 nm. In agreement with the simplified theory of acoustic phonon dispersion in a layered structure, the Raman signal associated with folded acoustic phonons shifts to greater wavenumbers as the Si nanocrystal size decreases: from ~ 20 cm^{-1} for 8.5 nm Si nanocrystals to ~ 45 cm^{-1} for 2.5 nm Si nanocrystals. However, the structure of the observed Raman signal is then not clearly resolved, and that could be due to a decrease in quality of the nc-Si/a-SiO_2 interfaces. This decrease of the interface flatness is expected for the observed changes in Si nanocrystal shape: Si nanocrystals smaller than 7 nm are more spherical compared to larger, brick-shaped in size Si nanocrystals. In addition, these uncertainties in Si nanocrystal shape may be accompanied by some changes in other structural properties; for example, their crystallographic orientation.

For many applications, it is very important to know and control the crystallographic orientations of Si nanocrystals. In particular, many physical effects (e.g., energy quantization) can only be observed for nanocrystals with similar crystallographic orientations, because of a six-fold symmetry of the electron energy dispersion in Si. Among other nondestructive and relatively fast analytical techniques, polarized Raman spectroscopy can be used directly to determine the crystallographic orientations of a Si film. We briefly review the theoretical approach developed in [47, 48] for the Raman polarization analysis and apply it to characterize nc-Si films.

Raman scattering is a two-photon process and the scattering intensity is determined by second-rank Raman tensors (Rij, with i, j, = x, y, z). As a result, the Raman intensity depends on the orientations of polarization vectors of the incident and scattered light relative to the crystallographic axes of the sample. This means that the crystallographic orientation of the sample can be determined by examining the anisotropy of the Raman scattering intensity. Crystalline Si belongs to the O_h point group of the 32 crystal classes, and the Raman-active optical phonon has F_{2g} symmetry, which is triply degenerate. Raman tensors for the F_{2g} mode in a crystal axis coordinate system are represented by

$$R_{yz} = \begin{pmatrix} 0 & 0 & 0 \\ 0 & 0 & d \\ 0 & d & 0 \end{pmatrix}, \ R_{zx} = \begin{pmatrix} 0 & 0 & d \\ 0 & 0 & 0 \\ d & 0 & 0 \end{pmatrix}, \text{ and } R_{xy} = \begin{pmatrix} 0 & d & 0 \\ d & 0 & 0 \\ 0 & 0 & 0 \end{pmatrix}. \quad (6.3)$$

Raman polarization experiments are performed in the exact backscattering geometry, where the wave vectors of the incident (k_i) and scattered (k_s) light are normal to the crystal surface. In this configuration, changing the incident polarization direction has no effect on the percentage of incident laser light being absorbed by the sample. The Raman scattering intensity is given by

$$S = \eta \sum_j (e_i R_{ij} e_s)^2 , \quad (6.4)$$

where η is a constant of proportionality related to the standard Raman cross section, and depends on the frequency of the scattered light and the intensity of the incident laser light, e_i and e_s are unit polarization vectors of the incident and scattered light, respectively, expressed in the crystal axis coordinate system using θ, ϕ, ψ_0, ψ and ψ'. Angles θ, ϕ and ψ_0 relate the laboratory coordinate system (fixed with respect to the sample) to the crystal axis coordinate system and are required to be known to identify the crystallographic orientation of the sample. By definition, θ is the angle made by the <001> axis and k_s, ϕ is the angle between the <100> axis and the projection of k_s onto the (001) plane, and ψ_o is the angle between the x axis of the laboratory coordinate system and the projection of the <001> axis onto the x-y plane. In addition, ψ (ψ') is introduced as the angle between the x axis and e_i (e_s). Equation (4) can be rewritten in terms of the angle ψ as

$$I(\theta, \phi, \psi_0; \psi, \psi') = S/\eta d^2 =$$
$$A(\theta, \phi, \psi_0; \psi') + B(\theta, \phi, \psi_0; \psi')\cos 2\psi + C(\theta, \phi, \psi_0; \psi')\sin 2\psi. \quad (6.5)$$

The determination of the crystallographic orientation of the sample requires the measurement of its Raman intensity as a function of the angle ψ for at least two linear polarization directions of the scattered light. Usually, these two directions are chosen to be orthogonal to each other. When the polarization vector of the scattered light is parallel to the x axis ($\psi' = 0$) or y axis ($\psi' = 90$) in the laboratory coordinate system, Eq. (6.5) becomes

$$I_0 = A_0(\theta, \phi, \psi_0) + B_0(\theta, \phi, \psi_0)\cos 2\psi + C_0(\theta, \phi, \psi_0)\sin 2\psi \quad (6.5a)$$

$$I_{90} = A_{90}(\theta, \phi, \psi_0) + B_{90}(\theta, \phi, \psi_0)\cos 2\psi + C_{90}(\theta, \phi, \psi_0)\sin 2\psi \quad (6.5b)$$

By combining coefficients A_0, B_0, C_0, A_{90}, B_{90}, and C_{90}, and eliminating ψ_0, we obtain quantities that are functions of only θ and ϕ:

$$G(\theta, \phi) \equiv (\beta_+^2 + \gamma_+^2)/\alpha_+^2 = (k_1^2 + k_2^2)/ k_0^2, \quad (6.6)$$

and

$$H(\theta, \phi) \equiv [(2\beta_- - \alpha_+ + 1)^2 + 4\gamma_-^2]/\alpha_+^2 = [(k_3 - k_4)^2 + 4 k_5^2]/ k_0^2, \quad (6.7)$$

where $\alpha_+ = A_0 + A_{90}$, $\beta_+ = B_0 + B_{90}$, $\gamma_+ = C_0 + C_{90}$, $\alpha_- = A_0 - A_{90}$, $\beta_- = B_0 - B_{90}$, $\gamma_- = C_0 - C_{90}$;

$k_0 = [4x(1 - x) + 2 + (x - 1)^2 y]/2$,
$k_1 = [-4x + (x + 1)y](x - 1)/2$,
$k_2 = (1 - x)[xy(1 - y)]^{1/2}$,
$k_3 = [4x(1 - x) - 2 + (x + 1)^2 y]/2$,
$k_4 = 1 - 2xy$,
$k_5 = (1 + x)[xy(1 - y)]^{1/2}$,

with $x = \cos^2\theta$, $y = \sin^2 2\phi$.

Angles θ and ϕ are found by solving simultaneously Eqs. (6) and (7). The angle ψ_0 can be found by solving

$$\alpha_- = k_1\cos 2\psi_0 - k_2\sin 2\psi_0. \quad (6.8)$$

With the restrictions $\theta \geq 0$ and $\phi \leq 90°$ six sets of (θ, ϕ) satisfy Eqs. (6) - (8). Determination of the angle ψ_0 reduces the six sets (θ, ϕ) into three equivalent orientations (θ, ϕ, ψ_0) that are obtained by the rotation of the surface normal by 120° about the <111> axis. For the (001) and (111) surfaces the variation of Raman intensity is given by

$$I_{001} = \sin^2(\psi + \psi' + 2\psi_0) \quad (6.9)$$

$$I_{111} = [\cos^2(\psi - \psi') + 2]/3 \quad (6.10)$$

Any oblique incidence and scattering will have a negligible effect on the Raman polarization analysis of Si when one uses an objective lenses with numerical aperture (NA) smaller than 0.5. This is because the refractive index of Si at 514.5 nm is 4.22 and the collection cone within the sample is only 6.6°.

The experimental set-up for these measurements is quite standard. The 514.5 nm line of an Ar ion laser was focused on a sample surface by a lens of NA = 0.1. The scattered light was collected by an objective lens of NA = 0.5 and focused onto the entrance slit of a Jobin-Yvon 1.0 m double spectrometer. The signal was detected by a photomultiplier with photon-counting electronics. The laser power at the sample surface was about 250 mW. The temperature of the sample was calculated from $I_S/I_{AS} = \exp(h\nu/kT)$, where I_S and I_{AS} are the intensities of the Stokes and anti-Stokes Raman components, respectively and $h\nu$ is the energy of the phonon (62 meV for the Brillouin zone center optical phonon in Si). Practically no heating of the sample was observed. A Glan-Thompson beam-splitting prism is used to produce linearly polarized light. A $\lambda/2$ wave plate was used to rotate the polarization of the incident light. The sample was mounted on a movable X-Y stage and adjusted to be normal to the incident light. A thin-film polarizer was used as an analyzer of the scattered light. A scrambler that compensates for the polarization dependent transmission of the spectrometer was placed between the analyzer and the spectrometer entrance slit.

Raman polarization measurements were performed by varying the polarization of the incident light (ψ) in 10° steps. The polarization of the scattered light is fixed at either perpendicular or parallel to the entrance slit of the spectrometer, *i e.*, H (horizontal) or V (vertical) configurations, respectively. For each of the polarization configurations the spectrum in the 505-535 cm^{-1} range was measured, the intensity of the background at 505 cm^{-1} was subtracted and the spectrum was numerically integrated. The spectral slit width of the spectrometer was ~1 cm^{-1} and the Raman intensity was measured in 0.5 cm^{-1} increments.

For Raman polarization measurements, (100) and (111) oriented single crystal Si wafers were used as test samples. Figure 15 shows their integrated Raman intensity in the radial direction as a function of the rotation angle of the polarizer (ψ). Two sets of data (H and V) were collected for two orthogonal polarizations of the scattered light (along the x and y axes, respectively). The dots are experimental points fitted by Eqs. (9-10) (solid lines). The deviation between the calculated and nominal crystallographic orientations of Si (100) and (111) single crystals is found to be less than 1°.

For nc-Si/a-SiO_2 superlattices grown on c-Si substrates, a quantitative Raman polarization analysis is possible when the condition $d_{Si}N\alpha(\lambda) \geq 1$ is satisfied (d_{Si} is a nc-Si layer thickness, N is a number of periods, and $\alpha(\lambda)$ is the absorption coefficient of nc-Si). This condition ensures that the excitation light is completely absorbed within the nc-Si layers and no signal from the substrate will be detected: For a sample with 60 periods and the layer thickness of d_{Si} = 14 nm, $\alpha_{\lambda = 514.5\ nm} = 1.465\cdot10^4\ cm^{-1}$, where α is the absorption coefficient of c-Si. Indeed, no difference in the shapes of normalized Raman spectra measured for orthogonal polarizations of the analyzer is observed. This proves that the Raman scattering is entirely determined by the nc-Si layers. Figure 15 shows the measured integrated Raman intensity as a function of a rotation angle of the polarizer (dots) for a sample having d_{Si} = 20 nm. The solid lines are the curves obtained from a least-squares fit of the experimental data to Eq. (6.10). Using Eqs. (6.6)-(6.8) we find that $\theta = 52°$, $\phi = 45°$, $\psi_0 = 13°$. This set of (θ, ϕ) angles is very close to the one defined by a (111) surface: $\theta = \arctan(\ \ ) \approx 54.8°$, $\phi = \arctan(1) = 45°$. This result suggests that the majority of Si nanocrystals have their <111> axis normal to the sample surface or, in other words, they have preferred <111> crystallographic orientation.

The same Raman polarization analysis has been applied to samples with smaller Si nanocrystals (~ 8.6 mn in vertical dimension). Figure 15d shows the result, and our immediate impression (see Fig. 15) is that the Raman signal from the <100> single crystal Si substrate is interfering with the signal from Si nanocrystals. The result of the numerical deconvolution of the Raman polarization signal is not accurate enough for any solid conclusions. However, a recent luminescence polarization analysis and additional structural measurements [49] indicate that Si nanocrystals smaller than 10 nm most likely have random crystallographic orientation. We will discuss the crystallographic orientation of smaller Si nanocrystals in a later paragraph on luminescence in nc-Si superlattices.

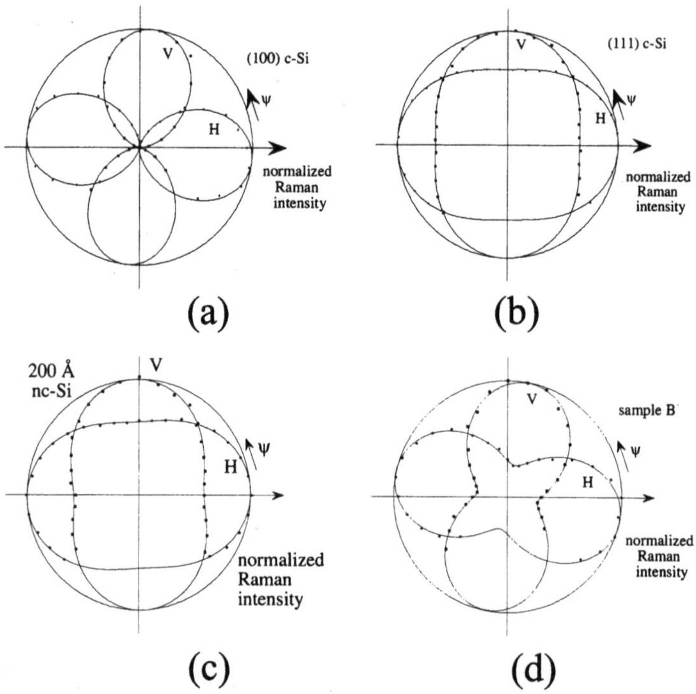

FIGURE 6.15. Polarization Raman diagram for (a) <100> and (b) <111> single crystal Si and (c) and (d) samples of crystallized nc-Si/a-SiO_2 superlattices.

During studies of solid phase crystallization of a-Si films on amorphous substrates such as a-SiO_2, it was been found that crystallized a-Si films may exhibit a preferred crystallographic orientation only in cases when an orientation dependent crystallization process (e.g., laminar crystallization along the a-Si/a-SiO2 interfaces) takes place [50]. In c-Si, growth rates along <113>, <110>, <112> and <111> crystallographic directions are 1, 1.5, 3, 5 and 24 times slower than the growth along the <100> direction [51]. That difference is related to the number of Si atoms that are necessary to complete a crystal plane during the crystallization process. The crystallographic orientation of the crystallized a-Si film is strongly dependent on the density of nucleation centers, their location in the film, and the thickness of the film.

Normally, crystallization of a-Si films on a-SiO_2 starts near the Si/SiO_2 interface where a large strain is present due to a difference in the thermal expansion coefficients of Si and SiO_2. In the case of interface nucleation, crystallization of a-Si proceeds in columnar or laminar fashion, depending on the ratio between the film thickness h and the mean distance d between nuclei. When $d \ll h$, the crystallites grow in a three-dimensional way until they impinge on each other, then the growth continues in a one-dimensional way along a direction perpendicular to the Si/SiO_2 interface. In this case, the structure of the film is columnar and has a (110) or (112) crystallographic orientation, since the fastest growth occurs by the formation of twins along <110> and <112> axes. If $d \gg h$, crystallites grow in a three-dimensional way until they reach the surface and then the growth proceeds in a two-dimensional way. Nuclei with an orientation permitting a fast lateral rate of growth extend at the expense of other less favorably oriented nuclei. The resultant structure is laminar and has a (111) orientation. When $d \approx h$, no preferred crystallographic orientation is expected because crystallites have the same chance to grow in any direction. It is quite possible that this particular scenario is realized in the case of Si nanocrystals smaller than 10 nm.

Our results show that the initial thickness of the a-Si film during crystallization controls not only the Si nanocrystal size but also the nanocrystal shape and crystallographic orientation. The shape of relatively large Si nanocrystals (> 10-20 nm) is rectangular and expanded laterally (i.e., bricks). In contrast, smaller Si nanocrystals (< 10 nm) are nearly spherical. Also, our results have proved that brick-shape Si nanocrystals have preferred (111) orientations while smaller, spherical nanocrystals are suspected to have random crystallographic orientation.

6.5. PHOTOLUMINESCENCE SPECTROSCOPY IN NANOCRYSTALLINE SILICON – SILICON DIOXIDE SUPERLATTICES

Strong interest in the optical properties of Si nanocrystals was stimulated by the initial report of a high-efficiency, room-temperature photoluminescence (PL) in porous Si [52], and by the later discussions of the PL origin [53-57]. This subject has already been covered by many papers, reviews and book chapters [2, 58, 59]. The aim of our review is (1) to compare PL properties in nc-Si/a-SiO_2 superlattices with similar data obtained in other materials containing Si nanocrystals (porous Si, Si nanocrystals prepared by CVD, ion implantation, etc.), and (2) to find a correlation between Si nanocrystal structural (i.e., size, shape and crystallographic orientation) and light-emitting properties. Since the beginning of these studies, it was obvious that we should start with relatively large size (> 20 nm) Si nanocrystals of well-defined, brick-like shape and <111> crystallographic orientation, and should continue toward much smaller, but less controlled in shape, Si nanograins. This research strategy provided a unique opportunity to observe the transformation of phonon-assisted carrier recombination controlled by well-understood selection rules to a much less ordered process where phonons are still involved in carrier recombination but all selection rules are significantly relaxed. Figure 16 perfectly illustrates this statement where a set of narrow PL lines in a bulk Si sample with the dominant TO-phonon line practically does not move, but broadens and slightly changes the ratio between TO-, TA-, and second TO-phonon lines in Si nanocrystals with sizes down to 8.6 nm in vertical dimension. Continuing to decrease the Si nanocrystal size down to ~ 6 nm, we observe a blue (toward higher photon energy) shift in the PL peak (which is at ~ 1.32 eV) and significant broadening up to ~ 100 meV in the PL full width at half maximum (FWHM). The phonon involvement can be recognized by the much less pronounced but still observable structure, with the main PL peak separated by ~ 60 meV (TO-phonon) from a shoulder at lower photon energy (Fig. 16). Further decrease in Si nanocrystal size down to 4.2 nm results in a shift of the PL peak to 1.4 eV and an almost 200 meV FWHM. This PL looks quite different compared to PL from bulk Si and, at the same time, is similar to PL spectra observed in porous Si and Si nanocrystals prepared by ion-implantation [58, 60]. The dependence of the PL peak position and FWHM as a function of Si nanocrystal size is summarized in Figure 17, showing a clear correlation between the PL broadening and PL peak blue shift. Note that the PL peak in smaller Si nanocrystals shifts to higher photon energy in a just slightly superlinear fashion, but the PL FWHM increases exponentially as the Si nanocrystal size decreases. To our surprise, we have found that the observed selection rule relaxation in PL from small Si nanocrystals is very similar to the conclusions that we were able to draw out of Raman studies.

Considering the origin of this broad PL band, we concentrate on two major possibilities. During more than a decade of searching for a convincing experiment on the origin of a broad and featureless PL in Si nanocrystals, the technique of resonant PL excitation has been widely recognized as the most useful one [58, 61]. The PL has been linked to carrier recombination in Si nanocrystals due to a clear indication of the phonon-

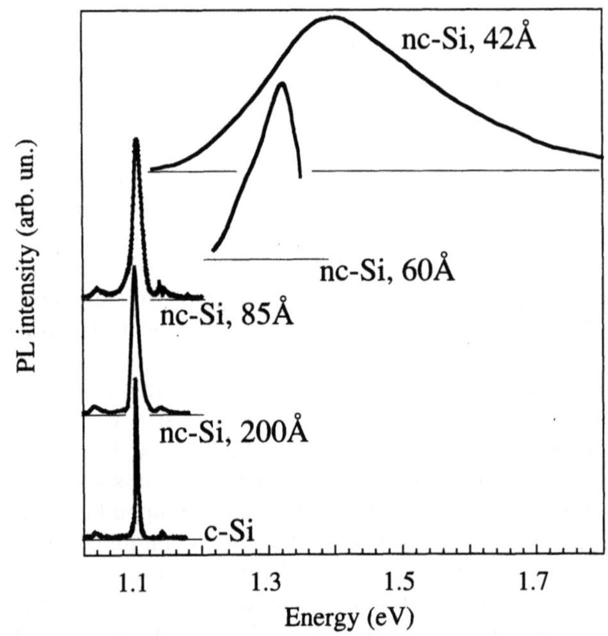

FIGURE 6.16. PL spectra from samples of nc-Si/a-SiO_2 superlattices with different (indicated) sizes of Si nanocrystals.

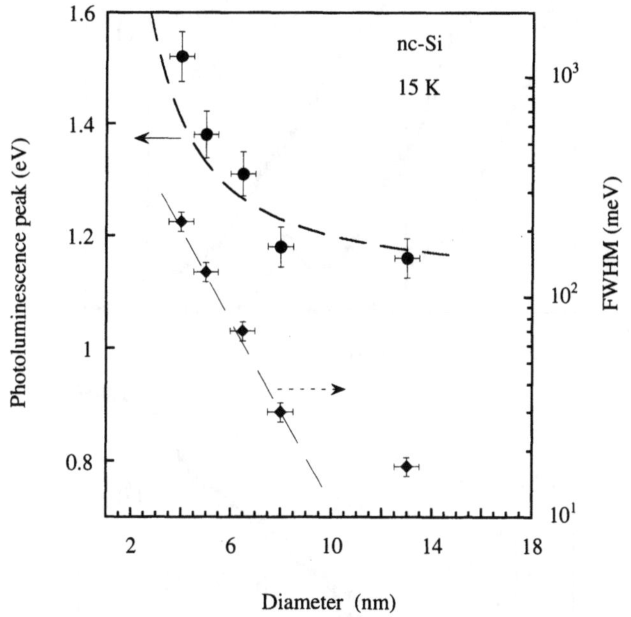

FIGURE 6.17. The PL peak and full width of half maximum as a function of Si nanocrystal size.

assisted PL mechanism and the observation of structure in PL spectra where almost all characteristic Si phonons and their combinations are found [58]. Tuning the wavelength of optical excitation toward the PL band, the inhomogeneous PL broadening due to Si nanocrystal size distribution is reduced [58]. In addition, resonant PL excitation may require phonon absorption due to the indirect nature of optical transitions in Si nanocrystals. These measurements have been described and performed in porous Si and other systems containing Si nanocrystals [58, 62]. In the case of nc-Si/a-SiO_2 superlattices, an additional contribution to the inhomogeneous broadening is, most likely, coming from the previously mentioned uncertainties in the shape and crystallographic orientation of small-size Si nanocrystals. This is particularly important for Si due to its indirect band structure and strong dispersion in the effective mass along different crystallographic directions.

Technically, the experiment requires a source of tunable excitation, which is usually a tunable laser (e.g., Ti:Al_2O_3, etc.) because the desired condition is to bring the excitation wavelength as close as possible to the PL line. This is demonstrated in Fig. 18a, where a decrease of the excitation photon energy from 1.66 to 1.42 eV depicts a step-like structure more clearly. Figure 18 b shows a higher resolution PL spectrum under resonant excitation, and contributions from specific Si phonons and their combination are indicated.

However, the fact that the PL lifetime has a strong temperature dependence (Fig. 19) indicates that at least two competing recombination mechanisms (most likely radiative and non-radiative) may contribute to the overall PL signal. Interestingly, the PL kinetics look single exponential compared to the more complex, usually stretched

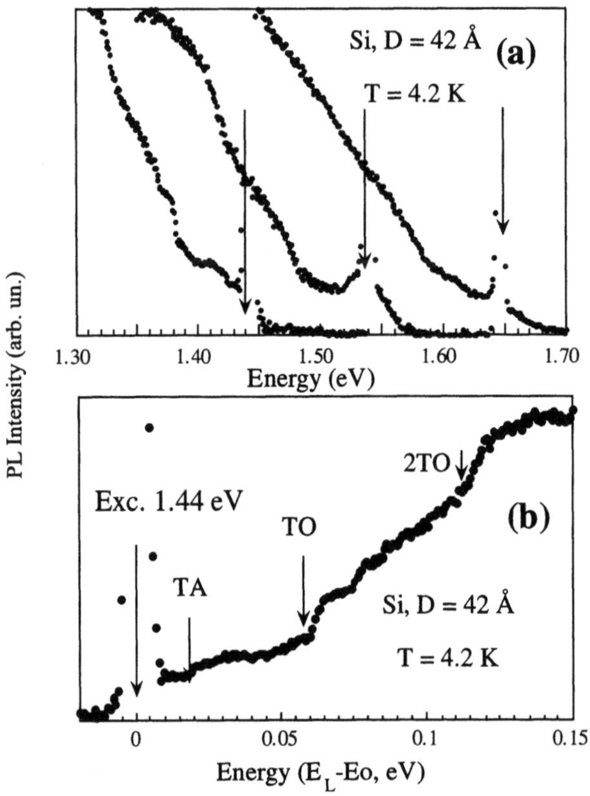

FIGURE 6.18. (a) An example of resonantly excited PL spectra in nc-Si/a-SiO2 superlattice with 42A diameter of Si nanocrystal showing better resolved PL structure as the excitation photon energy decreases. (b) A high resolution resonantly excited PL spectrum with indicated energies of characteristic Si phonons.

exponential behavior reported in Si nanocrystals prepared by other techniques [63]. Another indication that the PL may have a more complex rather than single origin comes from the data in Fig. 20, where we compare PL spectra in a nc-Si superlatice with an average Si nanocrystal size of ~ 4.2 nm. At low excitation power (0.05 W/cm^2), the PL peak is at ~ 1.4 eV but with the increase of the excitation up to 10 W/cm^2, the PL spectrum broadens and the PL peak shifts to > 1.7 eV. The observed broadening of the PL spectra has triggered a search for another technique that is able to separate different possible contributions to luminescence in nc-Si superlattices.

To our surprise, the most interesting results with a very clear interpretation have been found in studies of the PL spectral response as a function of external magnetic and electric fields. Since the initial report that an external magnetic field increases the gap between singlet and triplet states in Si nanocrystals, the fact that the PL lifetime increases and intensity decreases in Si nanocrystals was quite well understood [64-66]. However, our studies show that in nc-Si superlattices this PL quenching is selective and affects only PL at wavelengths longer than ~ 750 nm. The PL intensity at shorter wavelength does not show any quenching under an applied magnetic field as high as 9 T (Fig. 21). The

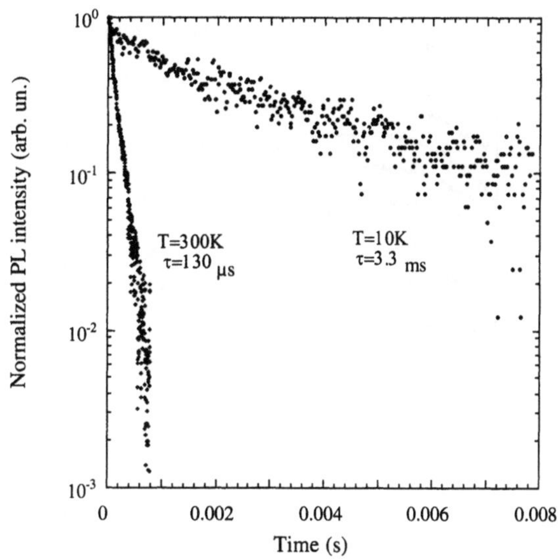

FIGURE 6.19. PL decay in fully crystallized nc-Si/a-SiO_2 superlattice with 42A diameter Si nanocrystals showing a single exponential decay and strong PL lifetime temperature dependence.

observed selective quenching of PL in nc-Si superlattices proves that the PL origin is more complex than just a single mechanism based on the quantum confinement effect in Si nanocrystals.

The application of an external electric field is also an informative experiment due to quite different mechanisms of PL quenching, including direct exciton deformation and dissociation under an external electric field, exciton impact ionization, etc. Different electric field induced PL quenching mechanisms have very different thresholds and field dependences. Therefore, we studied the PL dependence as a function of the applied electric field in nc-Si/a-SiO_2 superlattices. Our technique is based on the application of an AC electric field and detection of a modulated PL component using a lock-in amplifier. This technique is more sensitive compared to the standard technique where CW PL is influenced by an applied DC electric field. We have found that an electric field dependent PL spectral component is strongly correlated with magneto PL measurements: the modulated PL component is red-shifted compared to CW PL at zero electric field, and the electric field modulates only the PL at wavelengths longer than 750 nm (Fig. 22). This data is in complete agreement with the magneto-PL measurements. Therefore, we can conclude that a portion of the PL in nc-Si superlattices at shorter wavelengths does not depend on electric or magnetic field and most likely has a different origin than PL associated with confined excitons in Si nanocrystals. Additional support for this conclusion can be found in our early work on porous Si showing that the PL peak follows the ratio between the numbers of Si-O and Si-H bonds in samples with different degrees of oxidation [67].

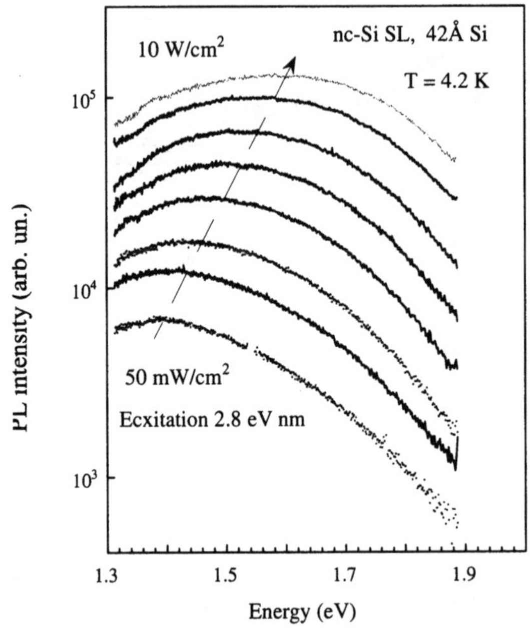

FIGURE 6.20. PL spectra in nc-Si/a-SiO_2 superlattice with 42A diameter Si nanocrystals under different intensity of excitation.

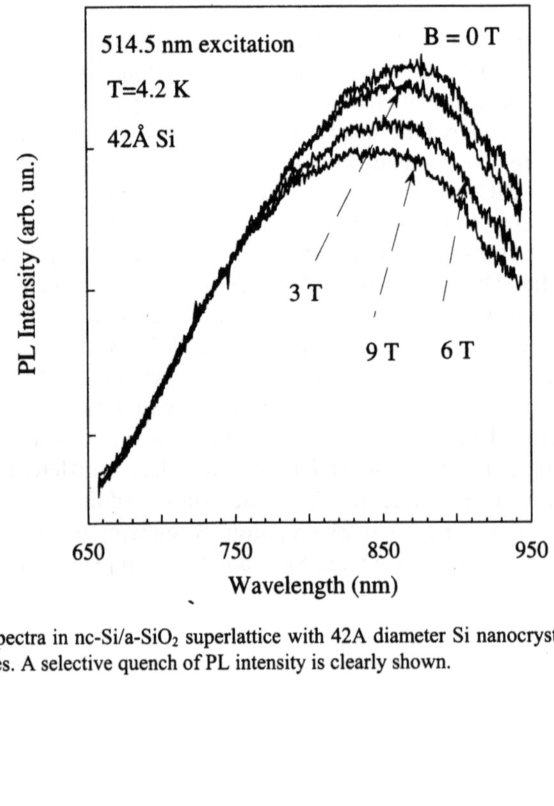

FIGURE 6.21. PL spectra in nc-Si/a-SiO_2 superlattice with 42A diameter Si nanocrystals under magnetic field of different intensities. A selective quench of PL intensity is clearly shown.

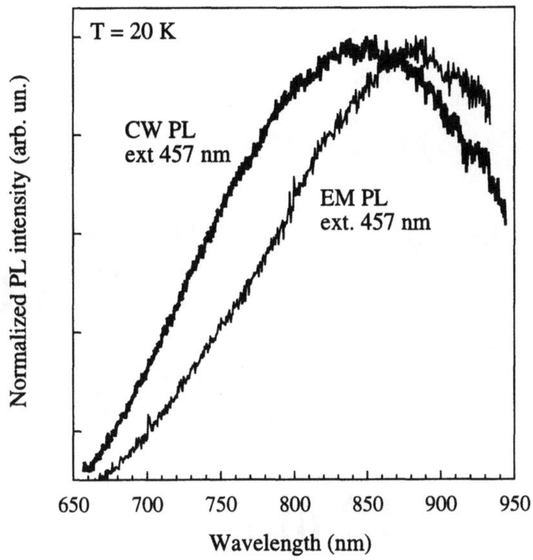

FIGURE 6.22. Strong red shift of the electric field modulated (EM) PL spectrum in 20 period nc-Si/a-SiO_2 superlattice with 42A diameter Si nanocrystals measured using a lock-in amplifier and ac voltage of 15 V.

6.6. SUMMARY AND FUTURE DIRECTIONS IN NANOCRYSTALLINE SILICON - AMORPHOUS SILICON DIOXIDE BASED NANOSTRUCTURES

As has been mentioned earlier, the initial interest in structures based on Si nanocrystals was raised mostly due to their unusual optical properties, e.g., visible PL with high quantum efficiency as reported in 1990 [52]. The experimental fact that PL in Si nanocrystals as small as 2-3 nm remains slow and requires phonon assistance to conserve momentum due to the indirect nature of the quasi-band in Si nanocrystals prevents their use in practical light-emitters utilizing intrinsic radiative transitions [68]. In contrast, recent work on complex or doped (i.e., Er+ ions) Si nanocrystals and nc-Si superlattices is important due to the observation of anti-correlation between intrinsic PL and PL related to dopants [69]. However, at least two critical steps still need to be made: (i) a better understanding of the charge-transfer mechanism and (ii) a significant increase in the PL and EL quantum efficiency.

Following the first paper reporting the charging effect and possible applications of Si nanocrystals in memory devices [70], additional interest in their electronic properties has started to grow quickly. The fact that Si nanocrystals can be prepared in the form of a layered structure with precise control over their size, shape and crystallography has also stimulated general interest in carrier transport in such systems. Despite obvious imperfections, i.e., grain boundaries separating Si nanocrystals laterally, the periodic nature of nc-Si/a-SiO_2 superlattices with tunnel transparent SiO_2 controls

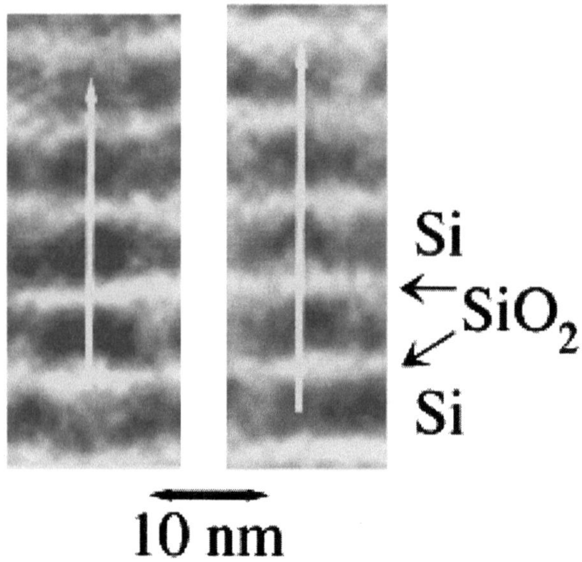

FIGURE 6.23. TEM micrograph showing vertical self-organization in a nc-Si/a-SiO_2 superlatice. Note a nearly pyramidal shape of Si nanocrystals.

carrier transport at low temperature and many interesting and potentially important features such as sequential resonant tunneling, unusual magnetoresistance and sustained oscillations [13] can be observed. In addition, it has been predicted that vertical quantum carrier transport can significantly be enhanced in structures where their nanocrystals display vertical self-alignment due to a uniform strain propagation. So far, this work has been done in epitaxial crystal growth (MBE, MOCVD or CVD of III-V and IV-IV semiconductors, etc. The first successful attempt to achieve similar results in nc-Si/a-SiO_2 superlattices with focus on the vertical self-alignment of Si nanocrystals is shown in Figure 23. In addition, our preliminary data indicates that quantum carrier transport in these structures is enhanced, and these structures can find applications in practical electron devices. Therefore, it is quite possible that the most interesting results in the development of novel electron devices based on nc-Si superlattices are yet to come.

ACKNOWLEDGEMENT

We would like to thank our many colleagues from the University of Rochester (P. M. Fauchet, L. Rothberg, and D. G. Hall), National Research Council of Canada (J. P. McCaffrey, J.-M. Baribeu, G. I. Sproule, H. J. Labbe,), the Technical University of Munich (F. Koch, M. Stutzmann, D. Kovalev, J. Diener, V. Timoshenko), Motorola (B. E. White, S. Zollner), Kodak (T. Blanton), NJIT and other collaborators (M. Zacharias, G. Allan, C. Delerue) for their help and contribution. LT specifically thanks his former graduate students at the University of Rochester, Dr. G. F. Grom and R. Krishnan. He also thanks the German Academic Exchange Service (DAAD) for a 1999 summer research program in Germany and the National Science Foundation, Semiconductor Research Corporation, Motorola, Foundation at NJIT, and Army Research Office for financial support.

REFERENCES

1. L. Tsybeskov, K. D. Hirschman, S. P. Duttagupta, M. Zacharias, P. M. Fauchet, J. P. McCoffrey and D. J. Lockwood, Appl. Phys. Lett. **72**, 43 (1998).
2. Light emission from Silicon: From Physics to Devices, Edited by D. J. Lockwood, Academic Press, San Diego, 1998.
3. K. D. Hirschman, L. Tsybeskov, S. P. Duttagupta, and P. M. Fauchet, Silicon-based light-emitting devices integrated into microelectronic circuits, Nature, **384**, 338 (1996).
4. T. Baron, P. Gentile, N. Magnea, and P. Mur, Appl. Phys. Lett. **79**, 1175 (2001).
5. N. Takahashi, H. Ishikuro, and T. Hiramoto, Appl. Phys. Lett. **76**, 209 (2000).
6. M. G. Bawendi, P. J. Carroll, William L. Wilson, and L. E. Brus, J. Chem. Phys. **96**, 946 (1992).
7. Z.-H. Lu, D. J. Lockwood, and J.-M. Baribeau, Nature, **378,** 258 (1995).
8. Z.-H. Lu, D. J. Lockwood, and J.-M. Baribeau, Solid-State Electron. 40, 197 (1996).
9. D. J. Lockwood, Z. H. Lu, and J.-M. Baribeau, Phys. Rev. Lett. 76, 539 (1996).
10. G. F. Grom, D. J. Lockwood, J. P. McCaffrey, H. J. Labbe, L. Tsybeskov, P. M. Fauchet, B. White, J. Diener, H. Heckler, D. Kovalev, and F. Koch, **Nature 407**, 358 (2000).
11. G. F. Grom, PhD Thesis, University of Rochester Library, pp. 134, 2001.
12. L. Tsybeskov, G. F. Grom, R. Krishnan, L. Montes, P. M. Fauchet, D. Kovalev, J. Diener, V. Timoshenko, F. Koch, J. P. McCaffrey, H. J. Labbe, G. I. Sproule, D. J. Lockwood, Y. M. Niquet, C. Delerue and G. Allan, Europhysics Letters 55, (4), 2001, p.552-558.
13. P. D. Persans, A. F. Ruppert, B. Abeles, and T. Tiedje, Phys. Rev. **B32**, 5558 (1985).
14. P. V. Santos, M. Hundhausen, L. Ley, and C. Viczian, J. Appl. Phys. **69**, 778 (1991).
15. J. Kakalios, H. Fritzche, N. Ibaraki, and S. R. Ovshinsky, J. Non-Cryst. Solids **66**, 339 (1984).
16. T. Sakurai and T. Sugano, J. Appl. Phys. **52**, 2889 (1981).
17. W. Y. Ching, Chun C. Lin, and Lester Guttman, Phys. Rev. B **16**, 5488 (1977).
18. A. F. Ruppert, P. D. Persans, G. J. Hughes, K. S. Liang, B. Abeles, and W. Lanford, Phys. Rev. B **44**, 11381 (1991).
19. P.D. Persans, A. Ruppert, B. Abeles, J. Non-Cryst. Solids **102,** 130 (1988).
20. Z.H. Lu, Solid State Electron. **40**,191 (1996).
21. J. Gonzalez-Hernandez, R. Tsu, Appl. Phys. Lett. **42,** 90 (1983).
22. M. Zacharias , J. Heitmann, M. Schmidt, P. Streitenberger, Physica **E11**, 245 ((2001).
23. J. P. McCaffrey, private communication.
24. P. Wickboldt, D. Pang, W. Paul, J. H. Chen, F. Zhong, J. D. Cohen, D. L. Williamson and Y. Chen, Journal of Non-Crystalline Solids, **198-200**, 567 (1996).
25. L. Tsybeskov, not published.
26. L. Tsybeskov, K. D. Hirschman, S. P. Duttagupta, M. Zacharias, P. M. Fauchet, J. P. McCoffrey and D. J. Lockwood, Physica Status Solidi **165**, 69 (1998).
27. L. L. Chang, L. Esaki, W. E. Howard, R. Ludeke, and G. Schul, J. Vac. Sci. Technol. **10**, 655 (1973).
28. S. D. Walck and J. P. McCaffrey, Thin Solid Films, **308-309**, 399 (1997).
29. S. T. Ngiam, K. F. Jensen, and K. D. Kolenbrander, J. Appl. Phys. **76**, 8201 (1994).
30. D. Reznik, C. H. Olk, D. A. Neumann, and J. R. D. Copley, Phys. Rev. B **52**, 116 (1995).
31. H. Richter, Z. P. Wang and L. Ley, Sol. State Commun. **39**, 626 (1981).
32. I. H. Campbel, P. M. Fauchet, Sol. State Commun. **58**, 739 (1986).
33. S. T. Kshirsagar and J. S. Lannin, Phys. Rev. B **25**, 2916 (1982).
34. J. S. Lannin, L. J. Pilione, S. T. Kshirsagar, R. Messier, and R. C. Ross, Phys. Rev. B **26**, 3506 (1982).
35. P. A. Temple and C. E. Hathway, Phys. Rev. B **7**, 3685 (1973).
36. Y. Sasaki and M. Kitahara, J. Appl. Phys. **76**, 4344 (1994).
37. G. Mariotto, F. Ziglio and F. L. Jr. Freire, Journal of Non-Crystalline Solids, **192-193,** 253 (1995).
38. N. Brunetto and G. Amato, Thin Solid Films, **297**, 122 (1997).
39. S. M. Rytov, Akust. Zh. **2**, 71 (1956).

40. D. J. Lockwood, M. W. C. Dharma–wardana, J. –M. Baribeau, and D. C. Houghton, Phys. Rev. B **35**, 2243–2251 (1987).
41. C. Colvard, R. Merlin, and M. V. Klein, Phys. Rev. Lett. **45**, 298 (1980).
42. D. Royer and E. Dieulesaint, J. Acoust. Soc. Am. **76**, 1438 (1984).
43. M. A. Tamor and J. P. Wolfe, Phys. Rev. B **26**, 5743 (1982) and references therein.
44. I. R. Cox-Smith, H. C. Liang, and R. O. Dillon, J. Vac. Sci. Technol. A **3**, 674 (1985)
45. G. D. Ivlev and E. I. Gatskevich, Applied Surface Science, **143**, 265 (1999)
46. J. B. Hopkins and L. A. Farrow, J. Appl. Phys. **59**, 1103 (1986)
47. Kohji Mizoguchi and Shin-ichi Nakashima, J. Appl. Phys. **65**, 2583 (1989)
48. L. Tsybeskov, not published
49. R. A. Street, Technology and Applications of Amorphous Silicon, Springer Series in Materials Science, vol. 37, Springer, Berlin, 2000, pp. 94–146 and the references therein
50. H. F. Wolf, Silicon Semiconductor Data, Pergamon Press, NY 1969, pp. 648
51. L. T. Canham, APL **37**, 1046 (1990)
52. A. G. Cullis and L. T. Canham, Nature, **353**, 335 (1991)
53. L. Tsybeskov, Ju. v. Vandyshev, and P. M. Fauchet, Phys. Rev. B **49**, 7821 (1994)
54. M. Brandt *et al.*, Solid State Commun. **81**, 307 (1992)
55. Y. Kanemitsu, Phys. Rep. **263**, 1 (1995)
56. D. Kovalev, H. Heckler, B. Averboukh, M. Ben-Chorin, M. Schwartzkopff, and F. Koch, Phys. Rev. B **57**, 3741 (1998)
57. A. G. Cullis, L. T. Canham, and P. D. J. Calcott, J. Appl. Phys. **82**, 909 (1997)
58. Reuben T. Collins, Philippe M. Fauchet, and Michael A. Tischler, Phys. Today **50**, 83 (1997)
59. D. Kovalev, H. Heckler, G. Polisski, J. Diener and F. Koch, Optical Materials **17**, 35 (2001)
60. D. Kovalev, H. Heckler, M. Ben-Chorin, G. Polisski, M. Schwartzkopff, and F. Koch, Phys. Rev. Lett. **81**, 2803 (1998)
61. M. Fujii, D. Kovalev, J. Diener, F. Koch, S. Takeoka, and S. Hayashi, J. Appl. Phys. **88**, 5772 (2000)
62. L. Pavesi and M. Ceschini, Phys. Rev. B **48**, 17625 (1993)
63. H. Heckler, D. Kovalev, G. Polisski, N. N. Zinov'ev, and F. Koch, Phys. Rev. B **60**, 7718 (1999)
64. C. H. Perry, F. Lu, F. Namavar, N. M. Kalkhoran, and R. A. Soref, Appl. Phys. Lett. **60**, 3117 (1992)
65. J. Diener, D. Kovalev, H. Heckler, G. Polisski, and F. Koch, Phys. Rev. B **63**, 73302 (2001)
66. L. Tsybeskov and P. M. Fauchet, Appl. Phys. Lett. **64**, 1983 (1994)
67. L. Tsybeskov MRS Bulletin **23** (4), 1998
68. J. Heitmann, R. Scholz, M. Zacharias, J. Non-Crys. Solids, 299-302 (2002), 678-682
69. S. Tiwari, F. Rana, H. Hanafi, A. Hartstein, E. F. Crabbé, and K. Chan, Appl. Phys. Lett. **68**, 1377 (1996)

7

Quantum Dot Photonic Crystals

David J. Norris[1] and Yurii A. Vlasov[2]

[1]Department of Chemical Engineering and Materials Science, University of Minnesota

[2]Physical Sciences Department, Thin Films and Optical Physics, IBM T. J. Watson Research Center

7.1. INTRODUCTION

7.1.1. Electronic Confinement

An early goal of research in semiconductor quantum dots was to utilize the finite size of these materials to modify the electronic properties of the semiconductor. In particular, researchers wished to modify its electronic density of states, defined as the number of electronic states per unit energy per unit volume. In a bulk semiconductor the density of states, ρ_e, can be described as a smooth function near the valence and conduction band edges, as depicted in Fig. 7.1a.[1] However, in a quantum dot, where the continuous bands of the bulk crystal evolve into a series of atomic-like levels due to quantum confinement, ρ_e is dramatically altered.[2-4] Indeed, ρ_e can be concentrated into a series of individual features, as shown in Fig. 7.1b.[5]

But why is such a change in the electronic density of states desirable? The reason is two-fold. First, many electronic processes strongly depend on ρ_e, which describes how many states are "available" to participate. For example, the spontaneous emission rate of an optical transition, which is described by Fermi's golden rule, is proportional to the electronic density of states.[6] Thus, due to the enhancement of ρ_e, the finite size of quantum dots can lead to more efficient emission.[7,8] This effect has been utilized to obtain novel light emitting semiconductor devices.[9-11]

The second motivation for altering the electronic density of states is that it can, in principle, provide a means to externally control electronic processes.[5] For example, if ρ_e consists of a series of sharp peaks, as in Fig. 7.1b, these peaks can be manipulated by external fields. By modulating the position of a feature at energy E, one can cause significant changes in the optical properties of the semiconductor at this energy (*e.g.*, absorption, index of refraction, *etc.*). Some effort has focused on using these effects in quantum dots to create optical switches and other optical control devices.

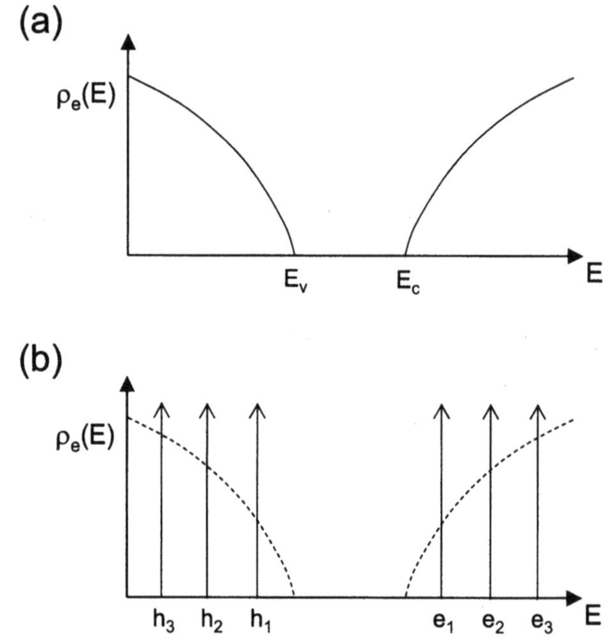

FIGURE 7.1. (a) The electronic density of states at the valence and conduction band edges in a typical bulk semiconductor. (b) In semiconductor quantum dots, the bulk density of states (dashed line) is concentrated into a series of sharp features (vertical arrows).

7.1.2. Photonic Confinement

However, it is interesting to note that during these studies on the influence of electronic confinement, the optical microcavity community was simultaneously investigating the implications of photonic confinement.[12-16] Just as quantum wells, quantum wires, and quantum dots, were used to confine electrons; parallel plates, cylinders, and spheres were used to confine photons in one, two, and three dimensions. The confinement could be achieved either by metallic coatings (*i.e.*, mirrors) or total internal reflection (TIR). For example, in a dielectric sphere photons can be trapped[17] via TIR in a *whispering gallery mode* in which light circulates just inside the surface of the sphere (see Fig. 7.2).[16]

As we would expect from our knowledge of electronic confinement, such photon confinement can lead to a modification of the *photonic* density of states, defined as the number of photonic modes per unit frequency per unit volume. In free space (*i.e.*, no confinement), the photonic density of states (ρ_{ph}) is a smooth function proportional to ν^2, where ν is the frequency. This is depicted by the dashed line in Fig. 7.3. In contrast, if the photons are confined in a cavity, only certain wavelengths of light will satisfy the boundary conditions. When the light is in resonance with one of these modes, ρ_{ph} can be higher than the free space value. Indeed, Purcell showed that the increase in the density of states can be approximated by:

$$\frac{3DQ\lambda^3}{4\pi^2 V} \qquad (1),$$

where D, Q, and V, are the degeneracy, quality factor, and modal volume of the cavity resonance, respectively.[13] Therefore, in small, micrometer-scale optical cavities, with high-Q (*i.e.*, low loss) resonances, ρ_{ph} can be significantly increased. Conversely, if the frequency of the light is off resonance with a cavity mode, ρ_{ph} can be lower than the free space value. Both situations are present in the solid line in Fig. 7.3, which is a qualitative description of ρ_{ph} for a planar cavity. Clearly, ρ_{ph} goes both above and below the free space density of states (dashed line) due to the influence of the cavity.

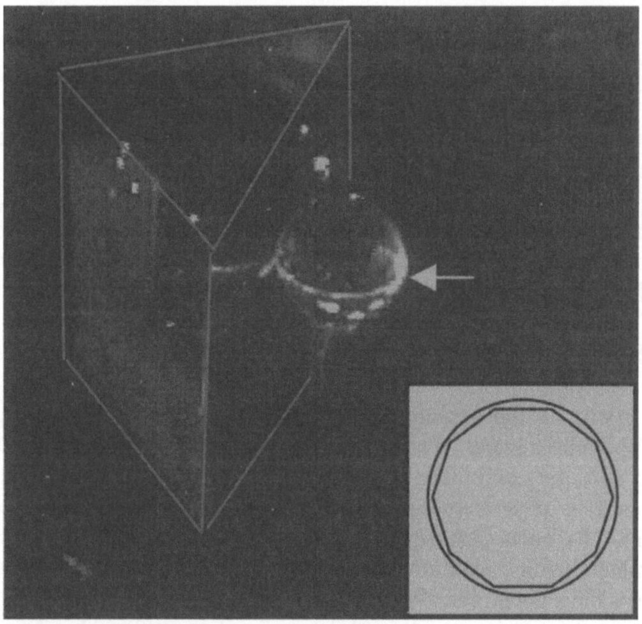

FIGURE 7.2. Photograph of an optical resonance in a 1 centimeter glass sphere. Scattering from a whispering gallery mode (inset) is seen as a bright horizontal band at the equator of the sphere (arrow). This mode is excited by coupling resonant laser light into the sphere via a prism (outline shown).

In analogy with the electronic case, two motivations exist for manipulating the photonic density of states. First, the light-matter interaction depends on how many optical modes are "available". Indeed, above we described how Fermi's golden rule depends on the electronic density of states. But it also depends on ρ_{ph}. In fact, one of the primary implications of Eq. 1, is that the spontaneous emission rate of a transition, a parameter that is commonly considered to be fixed for a particular material, can be modified by an optical microcavity. This effect, known as the Purcell effect, can lead to the enhancement or inhibition of spontaneous emission.

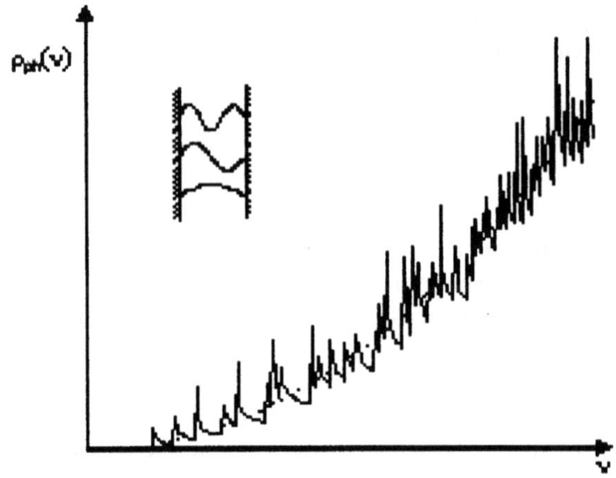

FIGURE 7.3. Cartoon of the photonic density of states for a linear optical cavity. The photonic density of states for free space is proportional to ν^2 (dashed line). See Ref. 14 for a quantitative treatment.

The second motivation is that if one has narrow optical resonances in ρ_{ph}, these can be manipulated to control the optical properties of the material. For example, the frequency of a high-Q cavity resonance is strongly dependent on the cavity parameters, *e.g.*, the index of refraction, physical size, *etc.* Thus, by slight time-dependent modifications of the microcavity, peaks and valleys in the density of states can be shifted. Eq. 1 implies that this could be used to control the light-matter interaction.

7.1.3. Combining Electronic and Photonic Confinement

As a result of the research described above, the ability to manipulate the electronic and photonic density of states has become quite sophisticated. Of course, it is also possible to consider combining both electronic and photonic confinement to control the optical properties of semiconductors in an optimal way. To achieve this, a large amount of work has been performed on one-dimensional optical microcavities with embedded semiconductor quantum wells or quantum dots. Consequently, many interesting effects have been recently demonstrated.[18]

However, the purpose of this chapter is not to review this prior microcavity work. For this subject, the reader can find extensive discussions elsewhere. Rather, here we will discuss the possibility of using colloidal semiconductor quantum dots (nanocrystals) and photonic crystals to obtain both electronic and photonic confinement. Photonic crystals (described further below) can provide the ultimate control over the photonic density of states. Therefore, the combination of quantum dots and photonic crystals is a potentially useful approach. In particular, we will focus on self-assembly methods that can provide a simple and inexpensive route to these materials.

7.2. PHOTONIC CRYSTALS

A photonic crystal is a material that has a periodic index of refraction, n.[19-22] An example is a solid in which bubbles have been embedded. If the bubbles are periodically placed on a three dimensional lattice, the material is referred to as a 3D photonic crystal. Obviously, the refractive index is periodic because the index alternates between $n=1$ (bubbles) and $n>1$ (surrounding solid) as the structure is traversed. Photonic crystals have been studied recently since they exhibit interesting properties when the lattice constant of the crystal becomes comparable to the wavelength of light. In this case, the propagation of photons through the photonic crystal can be dramatically altered due to optical Bragg diffraction. In particular, Yablonovitch and John showed that this effect can lead to the photonic equivalent of a semiconductor.[23,24] If properly tailored, a photonic crystal can possess a photonic band gap (PBG) in analogy to the electronic band gap. The existence of the PBG can be utilized to confine, manipulate, and guide photons in these materials. Thus, PBG crystals hold great promise for providing the fundamental building block for photonic devices.

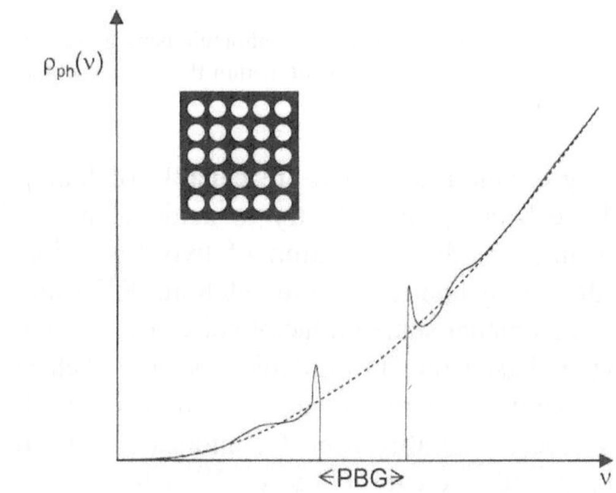

Figure 7.4. Cartoon of the photonic density of states for a material with a photonic band gap (PBG). The photonic density of states for free space is depicted by the dashed line.

The key property which is responsible for this great potential is the ability of PBG crystals to manipulate the photonic density of states. The PBG can be defined as a range of frequencies where the photon cannot exist inside the crystal. In other words, ρ_{ph} is, in principle, zero for frequencies within the gap (see Fig. 7.4). Indeed, in his first paper on the PBG, Yablonovitch envisioned using such a ρ_{ph} to completely inhibit spontaneous emission.[23] Furthermore, it was also realized that by adding a structural defect to the photonic crystal, a defect mode could be obtained with a frequency within the PBG (see Fig. 7.5). Photons in resonance with this mode can become trapped or confined at the defect site. Such a defect mode can have an extremely high Q since light would not be able to "leak" away due to the band gap in the surrounding material. In addition, since the modal volume of a defect-induced microcavity can be extremely small (*i.e.*, diffraction-limited), it can have an extremely large ρ_{ph}. Thus, structural defects can act as ideal optical microcavities with extremely large Purcell effects.

It is this ability of PBG materials, to tailor ρ_{ph} between zero and extremely high values that make them ideal for controlling light. One of their drawbacks, however, is that they have been extremely challenging to fabricate. To exhibit a PBG at optical frequencies, the structure must be extremely porous (typically ~80% air) with a very specific sub-micrometer, 3D periodicity (*i.e.*, a specific lattice symmetry). Conventional

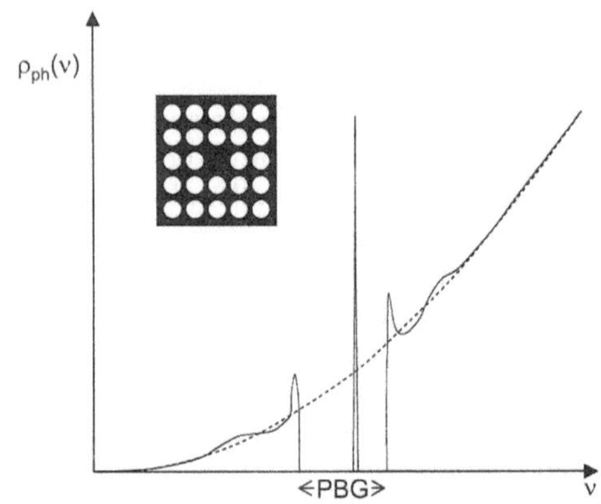

FIGURE 7.5. Cartoon of the photonic density of states for a photonic band gap crystal with a structural defect. If properly tailored, such a defect adds an additional mode within the gap. The photonic density of states for free space is depicted by the dashed line.

semiconductor processing techniques, which are capable of fabricating structures with nanometer precision, have been adapted to try to achieve this. The most promising advances have been realized in the fabrication of two-dimensional photonic crystals, where electron-beam lithography and reactive ion etching (RIE) can drill a periodic array of sub-micrometer holes in a planar semiconductor substrate.[25-27] Using these techniques, the first prototype devices based on 2D photonic crystals, such as nano-cavity lasers, have recently been demonstrated.[28] Unfortunately, the realization of 3D PBG crystals has been more challenging since nanofabrication techniques are intrinsically planar. One approach that has been pursued is to carefully stack 2D patterned structures in a layer-by-layer fashion. Indeed, 3D PBG crystals with a band gap at optical wavelengths have recently been obtained using this technique.[29,30]

However, a potentially simpler and more economical approach is to utilize the self-assembly of colloidal microspheres.[31] Sub-micrometer spheres (silica or polymer) can be induced to spontaneously organize on a crystalline lattice. In nature, this process leads to gemstone opals.[32,33] In analogy, sub-micron spheres assembled in the laboratory are sometimes referred to as *synthetic opals* (see Fig. 7.6). Although these structures do not have a photonic band gap, they may be used to obtain a PBG material. As shown in Fig. 7.7, the interstitial spaces between the spheres are filled with another material and then the spheres are selectively removed by chemical etching or calcination. The final structure, which is a replica of the original opal, consists of air spheres arranged on a face-centered cubic (*fcc*) lattice. This photonic crystal is often referred to as an *inverted opal* or *inverse opal*.

In recent years, research has demonstrated that inverted opals can be made from a variety of materials, including metal oxides,[34-38] graphitic carbon,[39] polymers,[40-42] metals,[43-46] and semiconductors.[47-53] Further, theory predicts that inverted opals can satisfy all the requirements for a PBG at optical wavelengths.[54,55] The band gap arises because this structure has the appropriate porosity, lattice spacing, and lattice symmetry. On the other hand, the band gap will occur only if we can utilize a material with a refractive index greater than 2.85. Unfortunately, this requirement has been satisfied by only a handful of the inverted opals that have been prepared to date.[50,51,53] And even in these cases, it is not clear *a priori*, as we will discuss below, if they have a PBG.

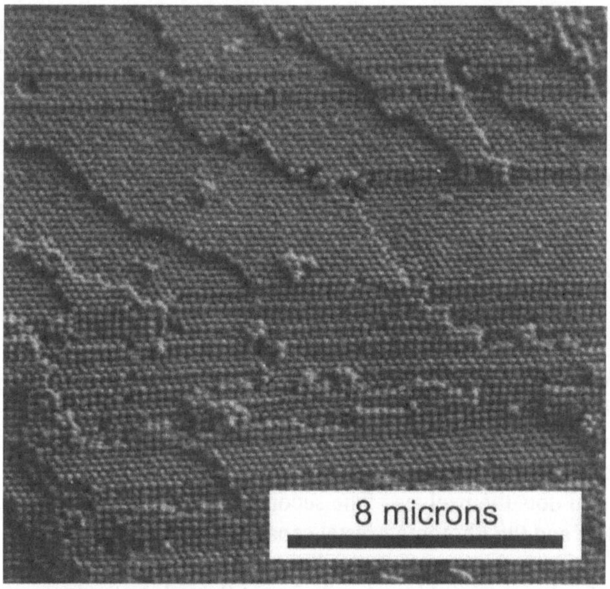

FIGURE 7.6. Scanning electron micrograph of a synthetic opal. A colloidal suspension of sub-micron silica spheres was allowed to slowly sediment under gravity. The spheres naturally organized on a face-centered cubic lattice.

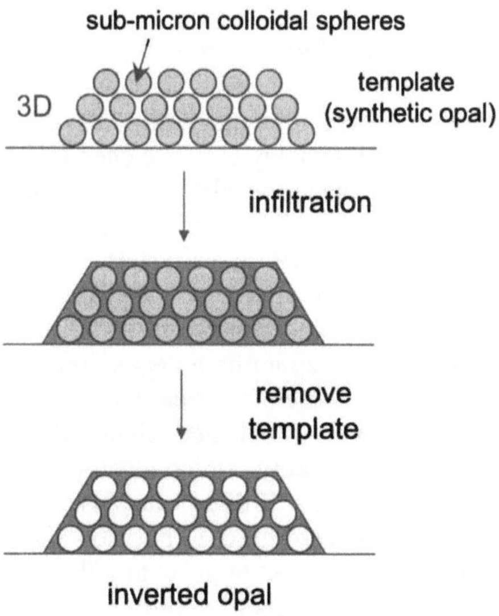

FIGURE 7.7. General procedure for the formation of inverted opals. Colloidal spheres are induced to organize on a crystalline lattice, which acts as a template. The space between the spheres is infiltrated with a material of interest. Subsequently, the spheres are selectively removed, typically by dilute hydrofluoric acid or calcination

Thus, in this chapter we will divide our photonic crystals into two classes: those that have a PBG and those that do not. When they have a PBG, they provide the optimal control over the photonic density of states and are the best choice for combining with nanocrystals. However, even without a PBG, a photonic crystal can be very interesting. In this case, the structure possesses a partial band gap, a frequency range where light can propagate only in a limited number of directions. Such photonic crystals can provide a novel electro-magnetic environment. For frequencies just outside the partial band gap, the photonic density of states for light traveling along forbidden directions can be higher than in free space. This increase indicates that more optical modes are available to interact with electronic excitations. Thus, if quantum dots can be incorporated into such a structure, they could utilize this enhanced light-matter interaction to further improve optoelectronic properties.

7.3. QUANTUM DOT PHOTONIC CRYSTALS

We have been exploring two approaches to combine quantum dots with 3D photonic crystals. The first, described in section 7.3.2, is to fabricate photonic crystals from the quantum dots themselves. The second, described in section 7.3.3 is to fabricate the quantum dots and the photonic crystal separately and then combine. At the outset, we note that our work in both areas is still in progress. Accordingly, this chapter will indicate both what has been achieved and what still needs to be done. Also, the interested reader should be aware of other related work, *e.g.*, semiconductor nanostructures have been incorporated into 2D photonic crystals[26,28] and both fluorescent dyes[56,57] and rare earth ions[58] have been embedded into 3D photonic crystals.

7.3.1. Colloidal Semiconductor Nanocrystals

Before proceeding, however, the quantum dots that have been utilized are briefly described. Since our general strategy is to use chemical-assembly to fabricate our structures, colloidal semiconductor nanocrystals[59-62] are employed as the quantum dots. Such nanocrystals can be easily manipulated and assembled into solids, as we discuss below. Another advantage of colloidal nanocrystals is that they can be prepared separately. Consequently, one can optimize their properties before incorporating them into the photonic crystal. In contrast, early efforts to infiltrate photonic crystals with CdS[31,63] and InP[64,65] quantum dots by vapor phase deposition suffered from an inability to control the size of the quantum dots, which is critical for determining their behavior.

By following the now well-established chemical synthesis of Murray and Bawendi,[66] or its variations,[67-72] high quality semiconductor nanocrystals can be prepared. Organometallic precursors (*e.g.*, for CdSe one uses dimethyl cadmium and trioctylphosphine selenide) are injected into a hot solvent, typically above 300°C. Nanocrystals then nucleate and are allowed to grow to a desired size. Agglomeration of the nanocrystals is prevented by surfactant molecules that coat the surface of the particles. Often the solvent is chosen such that it can play the role of surfactant. From such a synthesis, highly crystalline, nearly monodisperse (<5% rms deviation in size) nanocrystals can be obtained with surfaces that are electronically passivated by the surfactant molecules. Furthermore, by controlling the growth conditions, such samples can be easily obtained from ~0.8 to ~6 nm in mean radius.

The most heavily studied nanocrystal system has been CdSe. Consequently, the basic optical properties of this system are fairly well understood.[73-76] One particularly nice property is that they exhibit strong luminescence. As prepared, they exhibit fluorescence quantum yields at room temperature of ~10%. However, by encapsulating the CdSe nanocrystals in a higher band gap semiconductor, such as ZnS or CdS, the quantum yield can be further improved.[77-79] Emission efficiencies greater than 0.5 at room temperature have been reported. Due to quantum-confinement, the wavelength of this emission can be tuned by changing the size of the nanocrystals. Therefore, nanocrystals act as extremely robust chromophores that are convenient for a variety of optical applications. Indeed, devices, such as light-emitting diodes[80-82] and lasers[11,83] have been recently demonstrated.

In general, such devices rely, at least in part, on the ability to assemble the nanocrystals into densely-packed solids, where the nanocrystals are touching but have not fused. These materials are typically referred to as *quantum-dot solids*.[84] Two general types have been considered: 1) quantum dot crystals, where the nanocrystals act as "atoms" placed on a periodic lattice and 2) quantum dot glasses, where the nanocrystals are randomly, but densely packed as a solid with no long-range order. Note, that in either case, the weak van der Waals forces between the nanocrystals are responsible for holding the solid together. In many applications, a quantum dot glass is sufficient, as the long-range order of the nanocrystals is not necessary. However, further research into the behavior of quantum dot crystals is still necessary to explore the special properties of these materials.

Although CdSe has been the most heavily studied nanocrystal system, a large variety of other high-quality colloidal quantum dots can now be prepared. These include not just group II-VI semiconductors[66,70] such as CdS, CdTe, ZnS, ZnSe, and ZnTe, but also III-V semiconductors[68,69] such as GaP and InAs and group IV semiconductors such as Si.[85] In addition, significant progress has been made in obtaining other materials such as noble metals[86,87] and magnetic nanocrystals.[88]

7.3.2. *Periodic Quantum Dot Solids*

For optical device applications, it should be advantageous to pattern these quantum dot solids into photonic crystals. The structure of the solid can provide further control over the propagation of light inside the material. To explore this issue, we synthesized photonic crystals made from quantum dot glasses.[47] Our approach combines colloidal chemistry with two steps of hierarchical self-assembly. Colloidal silica particles were first used to construct an opal template. This template then served as a three-dimensional scaffolding for the self-assembly of colloidal nanocrystals, as shown in Fig. 7.8. The opal was immersed in a concentrated solution of nanocrystals. By slowly evaporating the solvent, a quantum dot glass was formed both outside and within the interstitial spaces of the opal. The silica was then selectively removed by etching with dilute hydrofluoric acid.

Figure 7.9a shows an example of the macroporous quantum dot solid that can be obtained with this approach. Spherical voids, caused by the removal of the silica spheres, are clearly seen. While in this image these voids may appear to be disordered, they are actually on a *fcc* lattice, as confirmed by the Fourier transform of the image (shown in the inset). The apparent disorder is caused by the "soft" (*i.e.*, van der Waals) nature of the quantum dot glass. This leads to an extremely rough fracture surface when the crystal is cleaved. Figure 7.9b shows a transmission electron micrograph of a small portion of the crystal. Arcs representing the initial locations of the silica spheres are observed. Between these arcs are the interstitial spaces, which are clearly filled with small grains of crystalline material. At higher resolution (Fig. 7.10), these grains can be verified to be ~4nm CdSe nanocrystals that have assembled into a densely packed quantum dot glass. Thus, by using this simple self-assembly approach, inverted opals that are comprised solely of semiconductor nanocrystals, which are stuck together by van der Waals forces, can be prepared.

In this photonic crystal, several important parameters that govern its final behavior can be controlled. First, we control the individual nanocrystals. As described above, semiconductor nanocrystals have interesting optical properties, such as discrete optical transitions, efficient luminescence, and optical nonlinearities that are tunable with the size of the nanocrystals. By changing the size of the nanocrystalline building blocks, we can dramatically alter the properties of the final material. Second, we control how the individual nanocrystals interact with each other. The collective electronic and optical behavior of the solid strongly depends on the spacing and arrangement of the nanocrystals. By modifying this secondary structure of the nanocrystals, we can optimize this collective behavior in the photonic crystal. Finally, we control the structure of the silica template. By adjusting the size, spacing, and arrangement of the silica particles, we can, in principle, determine how the quantum-dot solid is patterned on an optical-length scale. This tertiary structure modifies the internal electro-magnetic environment and influences the propagation of photons through the material.

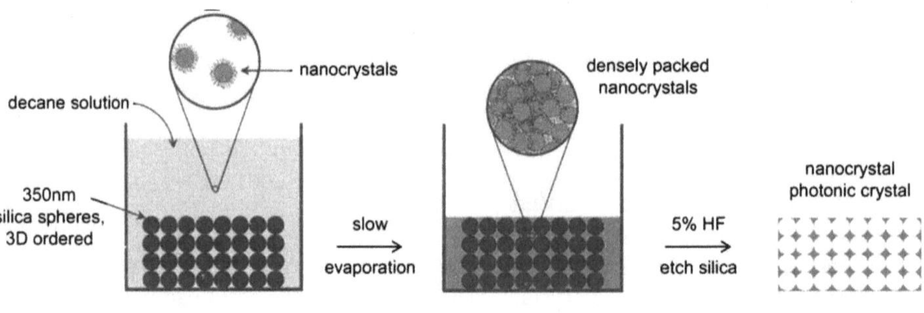

Figure 7.8. General procedure for the formation of quantum dot photonic crystals made from nanocrystals. A synthetic opal is immersed in a concentrated solution of nanocrystals. The solvent is allowed to slowly evaporate. The spheres are selectively removed by dilute hydrofluoric acid.

Therefore, these materials have great potential and their optical properties need to be studied. However, one drawback of these materials is that they cannot exhibit a photonic band gap. To have a PBG, the interstitials must be filled with a material with a refractive index above 2.85. A typical quantum dot glass made from CdSe nanocrystals has a refractive index around 2. One could start with nanocrystals from another, higher refractive index semiconductor material (*e.g.*, Si). Unfortunately, even in this case the index requirement cannot be satisfied. The index will always be lowered by voids left between the densely-packed nanocrystals.

Consequently, these materials will have only a partial band gap. As discussed above, this can still lead to enhanced light-matter interactions. These enhancements, particularly in relation to improved optical gain and optical nonlinearities, need to be explored. Further, because the approach outlined in Fig. 7.8 is both very simple and

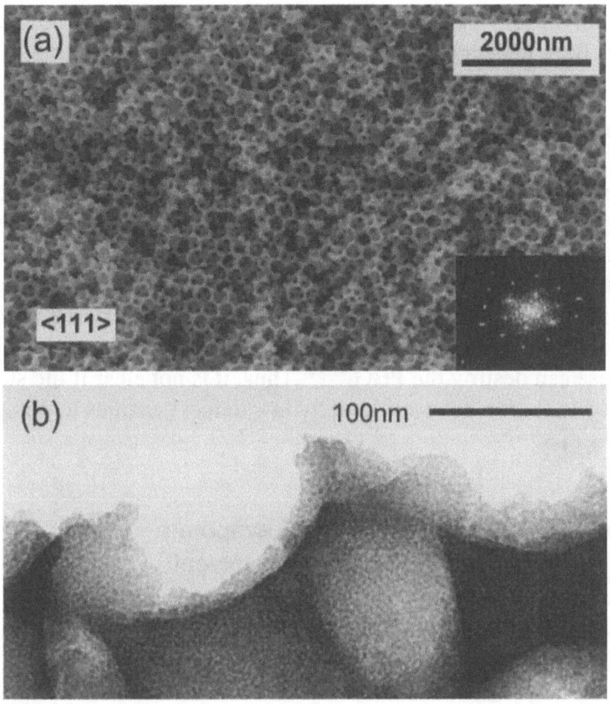

Figure 7.9. (a) Scanning electron micrograph of a quantum dot photonic crystal made from CdSe nanocrystals. The sample is fractured to expose the internal structure of the solid. Due to surface roughness, 10 monolayers of the original opal are exposed. Inset shows the Fourier transform of the image, illustrating long-range order in the [111] direction. (b) Transmission electron micrograph of a small fragment of the sample in (a). Arcs represent the positions of silica spheres prior to etching. The solid is composed of many small grains. Adapted with permission from Ref. 47. Copyright ©1999 John Wiley & Sons.

FIGURE 7.10. High resolution transmission electron micrograph of the sample in Fig. 7. 9. The image shows the atomic lattice fringes and moiré patterns typical of densely-packed CdSe nanocrystals. Adapted with permission from Ref. 47. Copyright ©1999 John Wiley & Sons.

completely general to any type of nanocrystalline material, the possibilities for new optical properties from nanocrystals are extensive.

7.3.3. Silicon Photonic Band Gap Crystals

To combine quantum dots with a structure that has a complete 3D PBG, a different strategy must be used. A reasonable route is to first self-assemble the PBG crystal and then incorporate the quantum dots. However, while inverted opals can theoretically satisfy all of the requirements to exhibit a PBG, in practice it is more problematic. The main difficulty is that self-assembly is not a perfect process. Defects, such as stacking faults, point defects, *etc.*, can be added to the crystal lattice during assembly and these can destroy the PBG.[89,90] Thus, it is not clear if the structures that are actually fabricated in the lab have sufficiently low defect densities to possess a band gap.

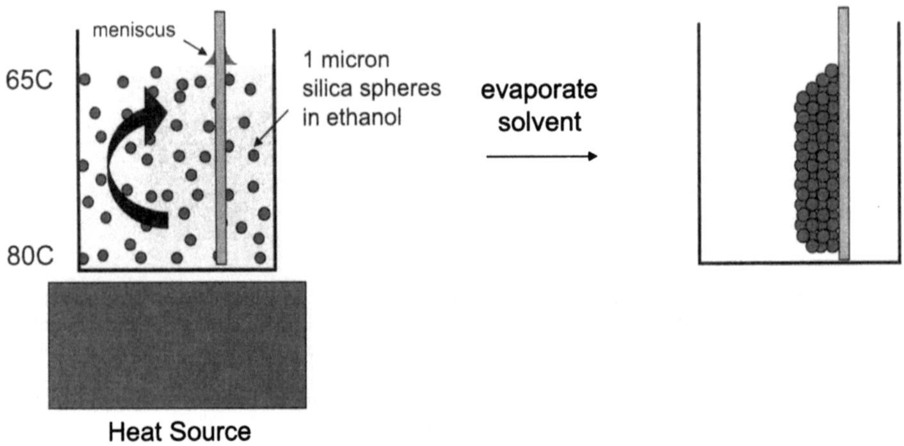

FIGURE 7.11. General procedure for the formation of thin planar opals made from silica spheres with diameters larger than 400nm. A small temperature gradient is added to a vial containing an ethanolic suspension of silica. As the ethanol evaporates a thin opal is deposited on a vertically placed substrate.

Consequently, before incorporating the quantum dots, one must first investigate if self-assembled photonic crystals can realistically provide a PBG. This is an important issue that we have recently addressed.[53] Due to prior work,[50] we focused on silicon inverted opals as the best candidate for forming self-assembled PBG crystals. To decrease the final defect density in the photonic crystal, we strived to dramatically improve the quality of the opal templates. This was achieved by growing thin opals on a planar substrate as shown in Fig. 7.11. It had previously been shown that colloidal crystallization could be induced within a meniscus between a substrate and a colloidal suspension.[91-93] Further, by slowly sweeping this meniscus across a vertically placed substrate, thin opals could be deposited.[94,95] Unfortunately, this method was commonly believed to be ill-suited to spheres larger than 400nm. Larger spheres sediment away from the meniscus due to gravity before deposition can take place. To make opals from larger spheres (*e.g.*, 800-1000nm), which are required to obtain a PBG at wavelengths

FIGURE 7.12. A cross-sectional scanning electron micrograph of a thin opal fabricated as shown in Fig. 7.10. The diameter of the silica spheres is 1 micrometer.

longer than the absorption edge of Si, we had to solve this problem. Unexpectedly, we found that by adding a small temperature gradient to the silica suspension the unwanted sedimentation could be avoided. Convective flows are established that gently replenish the silica spheres to the meniscus region (see Fig. 7.11). Thus, very nice, large sphere opals could be obtained, as demonstrated in Fig. 7.12. In addition, since these opals could be grown directly on silicon wafers, they are much more amenable to any potential applications.

To fabricate a PBG material, such templates were then infiltrated with silicon using a commercially available low-pressure chemical vapor deposition (LP-CVD) furnace. As illustrated in Fig. 7.13, the interstitial spaces were first filled with amorphous silicon. The silicon was then converted into polycrystalline silicon with an additional annealing step.

This procedure avoided undesirable surface roughness that occurred if grains of polycrystalline silicon were deposited directly. After infiltration, the opal could be fractured to reveal internal surfaces such as that shown in Fig. 7.14. Here silica spheres can be seen with silicon in between. If the silica spheres were then removed by a wet etch, Si inverted opals of various thickness could be obtained, as shown in Fig. 7.15.

As a self-assembled PBG material, such structures have several key advantages. First, as they are grown directly on a Si wafer, they are well-integrated onto the "chip". Indeed, even with close inspection of the images in Fig. 7.15, it is difficult to find the interface between the photonic crystal and the wafer. Second, as the structures are made

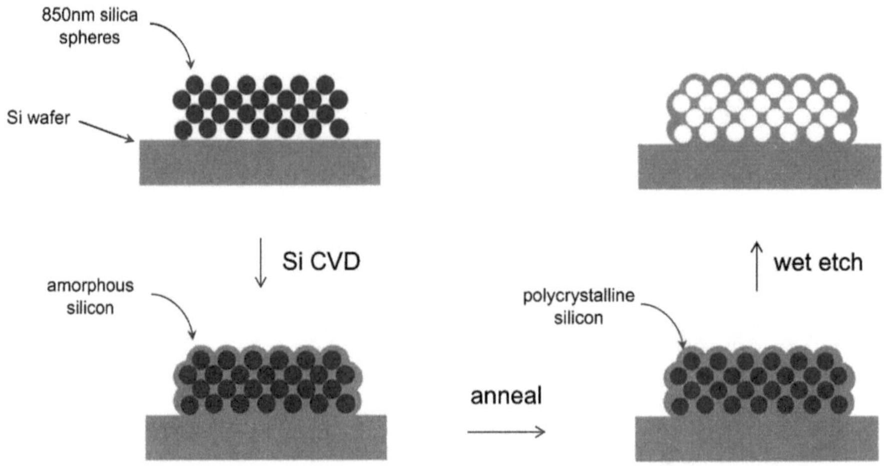

FIGURE 7.13. General procedure for the formation of self-assembled silicon photonic band gap crystals. A thin planar opal is assembled onto a silicon wafer and infiltrated with amorphous silicon via low-pressure chemical vapor deposition (LP-CVD). The silicon is then converted to polycrystalline silicon by annealing and the silica spheres are selectively removed by buffered oxide etchant.

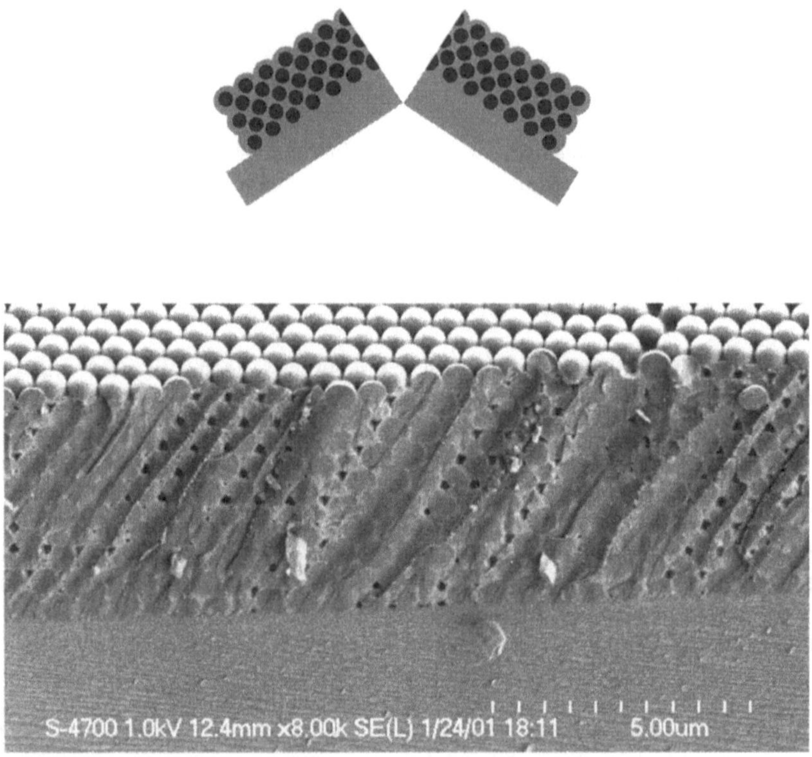

FIGURE 7.14. Scanning electron micrograph of a sample fractured after silicon infiltration.

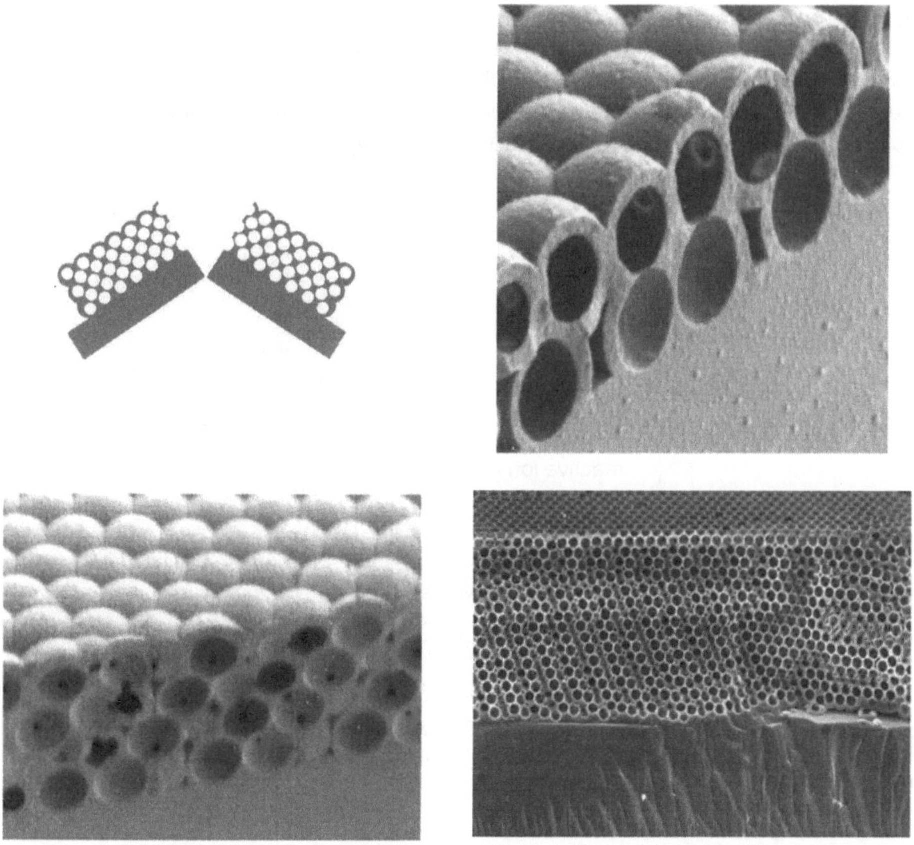

FIGURE 7.15. Scanning electron micrographs of silicon inverted opals as a function of the thickness of the initial opal template for 2, 4, and 16 layers. The silicon substrate is snapped and the fracture propagates up through the photonic crystal. Close examination of the lowest layer reveals that the photonic crystals are completely integrated into the wafer. Apparent disorder in the 16-layer sample is due to a ripple in the fracture surface that occurs in thicker photonic crystals. Adapted with permission from Ref. 53. Copyright © 2001 Macmillan Magazines Limited.

from polycrystalline Si, the photonic crystal is extremely robust. Attempts to scratch them (*e.g.*, with tweezers) showed that they are mechanically quite strong and able to survive the application of moderate pressure. In the future, these mechanical properties must be quantified. Third, since the initial opal template is flat (*i.e.*, it can have the same number of layers over centimeter length scales), the corresponding photonic crystal is in an intrinsically planar format. While this is obviously useful for potential applications, it also makes these samples ideal for probing the existence of the photonic band gap.

However, before optical measurements can be performed, one additional processing step is required. As depicted in Fig. 7.16, excess silicon is typically deposited on top of the opal template. If this extra silicon remains in the inverted structure, it can introduce unwanted photonic surface states in the optical data and complicate the analysis. To avoid this, we used reactive ion etching (RIE) to remove the excess silicon prior to the removal of the template. Under the proper conditions, the top silicon coating can be selectively removed by RIE while the silica spheres act as an etch stop. In other words, this process will automatically terminate when the uppermost layer of silica spheres has been exposed. Subsequently, the silica spheres can be removed by wet etching to obtain an "open top" structure, as shown in the electron micrograph of Fig. 7.16. Therefore, by following the simple steps outlined in Figs. 7.11, 7.13, and 7.16, extremely

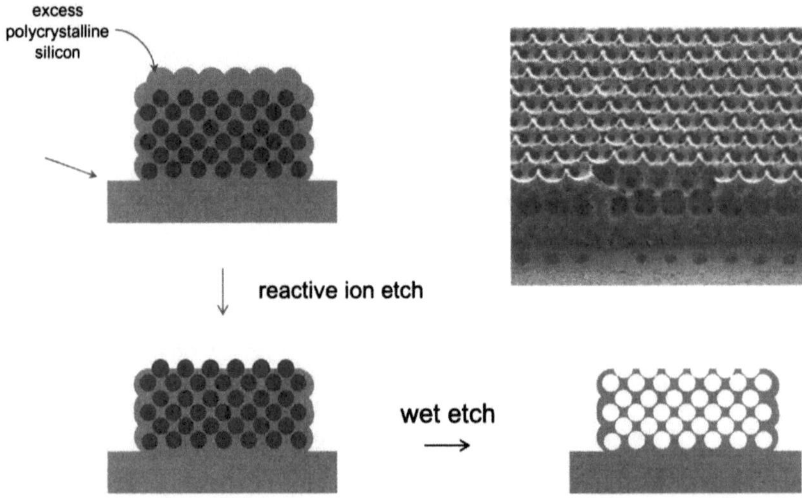

FIGURE 7.16. General procedure for the removal of excess silicon which is deposited on top of the opal. Reactive ion etching is used to expose the top layer of silica spheres, which act as an effective etch stop. The spheres are then removed by wet etching to obtain an "open top" silicon inverted opal, as demonstrated in the scanning electron micrograph.

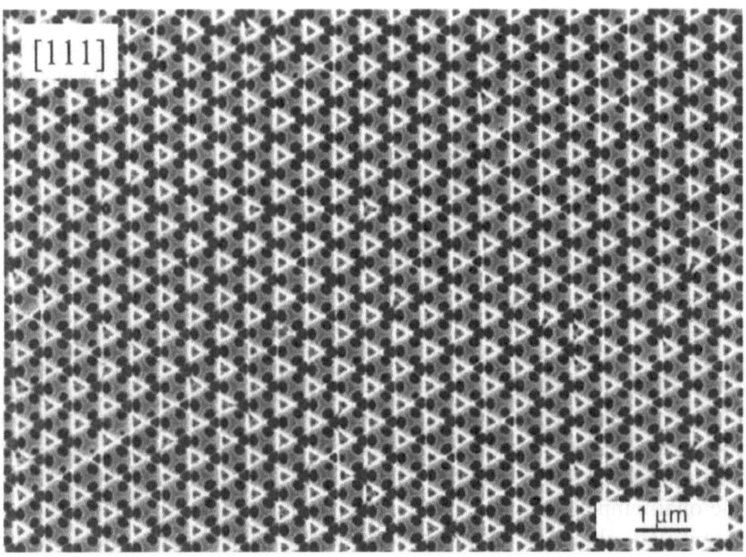

FIGURE 7.17. Scanning electron micrograph of the (111) surface of a silicon inverted opal. Adapted with permission from Ref. 53. Copyright ©2001 Macmillan Magazines Limited.

high quality silicon inverted opals can be obtained. For example, Fig. 7.17 shows a typical scanning electron micrograph of structures that have been prepared. This image looks down on the wafer and probes the [111] direction of the *fcc* photonic crystal lattice.

Once such structures were available, we then examined their photonic properties. In principle, one only needs to perform simple reflection (or transmission) measurements to determine if a crystal exhibits a PBG. For example, if a sample is completely reflective for any direction of internal light propagation in the frequency range expected by theory, this would be strong evidence that the band gap exists. In practice, it is more complicated. Here, we consider two issues.

First, as discussed previously,[63,96-98] it is not trivial to probe different directions of propagation in a photonic crystal due to the breakdown of Snell's law near a band gap. This implies that one cannot simply rotate the sample to change the direction of propagation. If this is attempted, Snell's law would be required to convert an external angle of incidence to an internal angle of propagation. Instead, one must always probe at normal incidence. For each measurement, a sample surface must be prepared with a different lattice direction exposed.

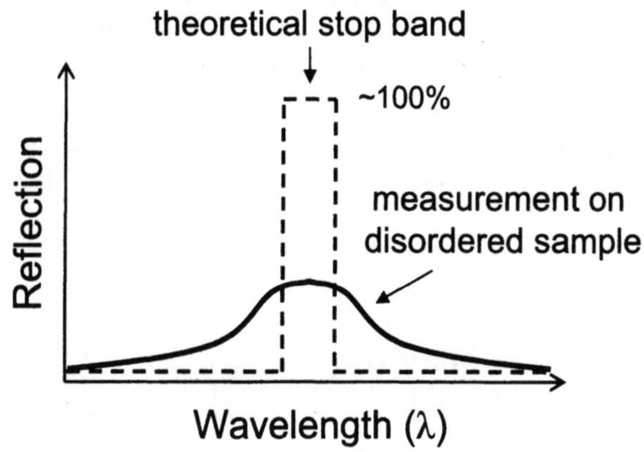

FIGURE 7.18. Cartoon depicting the effect of disorder on the optical reflectance spectrum of a photonic crystal.

Second, and more importantly, one always has to be concerned about the effects of disorder on the experimental results. While in theory one might expect a "square top" reflection peak with near unity reflectance, in experiment one can observe a less intense peak with a larger linewidth, as shown in Fig. 7.18. Some increase in linewidth will occur even in a perfect sample due to the finite size of the crystal.[98,99] However, significant broadening can also occur due to averaging over the disorder in the sample (*i.e.*, inhomogeneous broadening). Therefore, the presence of disorder should always be the experimentalist's first concern if a spectrum reveals a broad reflectance feature which is lower than unity in the band gap region.

Previously, we have demonstrated that an optical microscope can be a powerful tool to eliminate the effects of disorder in self-assembled photonic crystals and obtain homogeneous spectra which describe their intrinsic optical behavior.[100] The microscope can probe a specific local area of the crystal and greatly reduce the effects of inhomogeneities. We utilized this technique to investigate the existence of the PBG in our planar Si photonic crystals. A sample, such as that shown in Fig. 7.19, was carefully measured in two crystal orientations, [111] and [100]. Although the planar opals typically assemble with their [111] axis perpendicular to the substrate, small "defect" regions that are [100]-oriented can also be found. These regions are small, but provide another crystalline direction that can be probed with the microscope. The results are shown in Fig. 7.20 and compared with theoretical calculations. In the top panel, the photonic band diagram is shown. The hashed region indicates the expected position of the photonic band gap. For our samples, the band gap should be around 1.3 micrometers. In the lower two panels, experimental results are shown. These spectra revealed for the first time that self-assembled photonic crystals can exhibit optical properties consistent with a PBG. Unity reflectance (within experimental error) was observed in the band gap region for both directions of propagation. While further work needs to be done, *e.g.*, to examine more directions, these measurements are extremely encouraging since they indicate that self-assembled materials which are prepared in a straightforward manner can have defect densities sufficiently low that the PBG survives. Furthermore, since the crystals are integrated directly onto a Si wafer, one can imagine using these materials to mass produce low-cost Si self-assembled PBG devices or circuits. Indeed, we have already demonstrated that these materials can be easily patterned with standard techniques used to fabricate conventional microelectronics.[53]

FIGURE 7.19. Scanning electron micrograph of a typical silicon inverted opal that was used in optical measurements. Adapted with permission from Ref. 53. Copyright ©2001 Macmillan Magazines Limited.

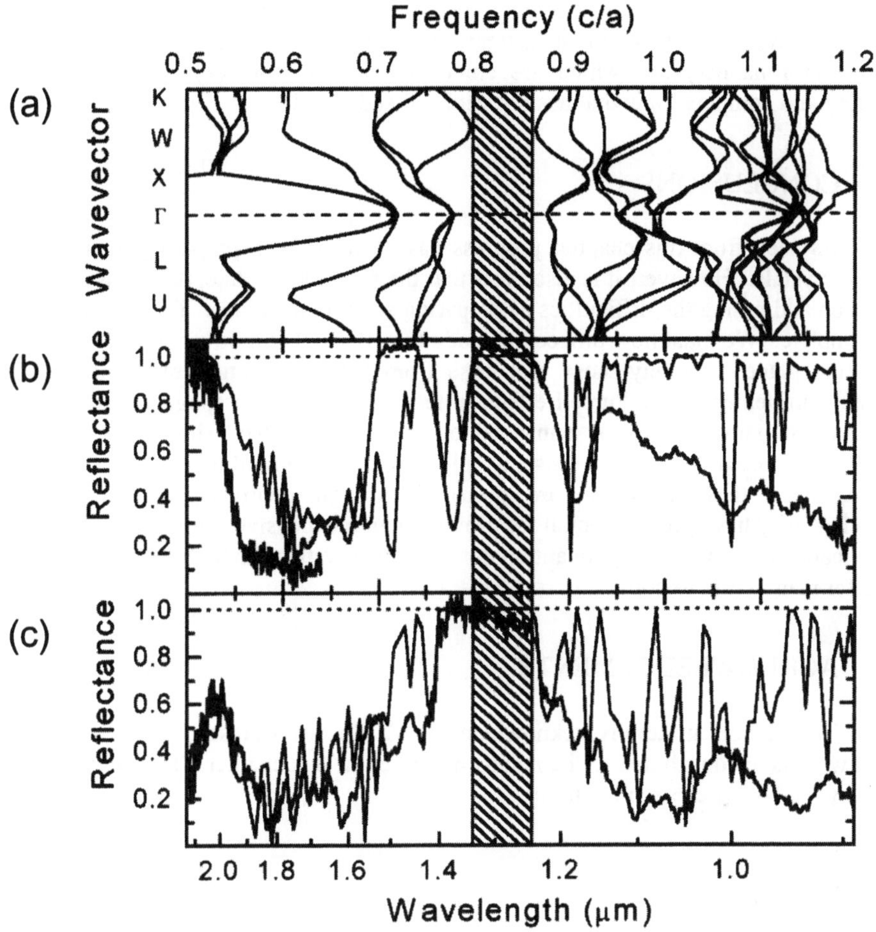

FIGURE 7.20. Probing the photonic band gap in silicon inverted opals. (a) The calculated photonic band diagram. The blue-hashed region denotes the expected frequency range of the bandgap. The frequency is plotted in units of c/a, where c is the speed of light and a is the lattice constant. (b) and (c) show experimental reflection spectra for amorphous Si inverted opals measured normal to the (111) and (100) planes, respectively. Data from two samples, with a equal to 1070nm (red) and 841nm (blue), are combined. The wavelength scale corresponds to the 1070nm sample. The black curve represents the best theoretical fit. Adapted with permission from Ref. 53. Copyright ©2001 Macmillan Magazines Limited.

However, because these self-assembled PBG materials have only recently been made, they have not yet been combined with quantum dots. For the reasons outlined above, it should be very interesting to embed fluorescent nanocrystals into the photonic crystal to observe changes in the light-matter interaction induced by the band gap. Due to the absorption edge of silicon, these inverted opals can only have a band gap in theinfrared. In particular, our structures are designed to have a PBG at either 1.3 or 1.55 micrometers, the primary wavelengths used in telecommunication. Thus, to explore the effects on emission, infrared emitting nanocrystals will be required. One possibility is to use recently developed PbSe nanocrystals,[101,102] which should exhibit size-tunable fluorescence between 1.2 and 2.0 micrometers. This is an area we are currently exploring. We also plan to investigate the enhanced optical nonlinearities of semiconductor nanocrystals which are embedded in our PBG crystals. This is an interesting and complex topic that has not yet been addressed.

7.4. CONCLUSIONS

As seen from this chapter, progress has been made in preparing quantum dot photonic crystals. However, it is also clear that the research, to date, has been primarily focused on addressing the difficulties of preparing suitable materials. The research is far from complete. We must now proceed to study the effects that motivated these materials in the first place – namely those that arise from the combination of electronic and photonic confinement. The Purcell effect, in terms of changes in spontaneous emission and enhanced gain, as well as enhanced optical nonlinearities, should be explored. Therefore, this research, in some sense, is only at the beginning of an exciting future. In addition, since these materials are prepared through relatively simple chemical approaches, they have great potential for providing an inexpensive semiconductor system for applications. Consequently, quantum dot photonic crystals should be a fruitful area to pursue for many years to come.

7.5. ACKNOWLEDGEMENTS

The authors gratefully acknowledge X.-Z. Bo, M. Deutsch, J. C. Sturm, and N. Yao who have contributed to the results described in this chapter. This research was funded by the NEC Research Institute.

REFERENCES

1. N. W. Ashcroft and N. D. Mermin, *Solid State Physics* (W. B. Saunders, Orlando, 1976), Ch. 28.
2. A. L. Efros and A. L. Efros, Sov. Phys. Semicond. **16**, 772 (1982).
3. L. E. Brus, J. Chem. Phys. **79**, 5566 (1983).
4. L. E. Brus, J. Lumin. **32**, 381 (1984).
5. S. Schmitt-Rink, D. A. B. Miller, and D. S. Chemla, Phys. Rev. B **35**, 8113 (1987).
6. P. Y. Yu and M. Cardona, *Fundamentals of Semiconductors* (Springer-Verlag, Berlin, 1996), p.460.
7. M. Asada, Y. Miyamoto, and Y. Suematsu, IEEE J. Quant. Elect. **QE-22**, 1915 (1986).
8. Y. Arakawa and H. Sakaki, Appl. Phys. Lett. **40**, 939 (1982).
9. C. Weisbuch and B. Vinter, *Quantum Semiconductor Structures, Fundamentals and Applications* (Academic, San Diego, 1991).
10. M. Grundmann, Physica E **5**, 167 (2000).
11. V. I. Klimov, A. A. Mikhailovsky, S. Xu, A. Malko, J. A. Hollingsworth, C. A. Leatherdale, H.-J. Eisler, and M. G. Bawendi, Science **290**, 314 (2000).
12. *Cavity Quantum Electrodynamics*, edited by P. R. Berman (Academic Press, San Diego, 1994).
13. E. M. Purcell, Phys. Rev. **69**, 681 (1946).
14. D. Kleppner, Phys. Rev. Lett. **47**, 233 (1981).
15. S. Haroche and D. Kleppner, Physics Today **42**, 24 (January 1989).
16. *Optical Processes in Microcavities*, edited by R. K. Chang and A. J. Campillo, (World Scientific, Singapore, 1996).
17. Since whispering gallery modes are "leaky", the photon can never be trapped indefinitely. See Ref. 16.

18. See, *e.g.*, *Confined Electrons and Photons*, edited by E. Burstein and C. Weisbuch (Plenum Press, New York, 1995).
19. J. D. Joannopoulos, R. D. Meade, and J. N. Winn, *Photonic Crystals* (Princeton University Press, Princeton, 1995).
20. J. D. Joannopoulos, P. R. Villeneuve, and S. Fan, Nature **386**, 143 (1997).
21. E. Yablonovitch, Sci. Am. **285**, 34 (2001).
22. For recent reviews, see articles in *Photonic Band Gap Materials*, edited by C. M. Soukoulis (Kluwer, Dordrecht 1996) and *Photonic Crystals and Light Localization*, edited by C. M. Soukoulis (Kluwer, Dordrecht 2001).
23. E. Yablonovitch, Phys. Rev. Lett. **58**, 2059 (1987).
24. S. John, Phys. Rev. Lett. **58**, 2486 (1987).
25. T. F. Krauss, R. M. De La Rue, and S. Brand, Nature **383**, 699 (1996).
26. C. J. M. Smith, H. Benisty, S. Olivier, M. Rattier, C. Weisbuch, T. F. Krauss, R. M. De La Rue, R. Houdré, and U. Oesterle, Appl. Phys. Lett. **77**, 2813 (2000).
27. E. Chow, S. Y. Lin, S. G. Johnson, P. R. Villeneuve, J. D. Joannopoulos, J. R. Wendt, G. A. Vawter, W. Zubrzycki, H. Hou, and A. Alleman, Nature **407**, 983 (2000).
28. O. Painter, R. K. Lee, A. Scherer, A. Yariv, J. D. O'Brien, P. D. Dapkus, and I. Kim, Science **284**, 1819 (1999).
29. S. Y. Lin, J. G. Fleming, D. L. Hetherington, B. K. Smith, R. Biswas, K. M. Ho, M. M. Sigalas, W. Zubrzycki, S. R. Kurtz, and J. Bur, Nature **394**, 251 (1998).
30. S. Noda, K. Tomoda, N. Yamamoto, and A. Chutinan, Science **289**, 604 (2000).
31. V. N. Astratov, V. N. Bogomolov, A. A. Kaplyanskii, A. V. Prokofiev, L. A. Samoilovich, S. M. Samoilovich, and Y. A. Vlasov, Nuovo Cimento D **17**, 1349 (1995).
32. J. V. Sanders, Acta Cryst. A **24**, 427 (1968).
33. J. V. Sanders, Nature **204**, 1151 (1964).
34. O. D. Velev, T. A. Jede, R. F. Lobo, and A. M. Lenhoff, Nature **389**, 447 (1997).
35. B. T. Holland, C. F. Blanford, and A. Stein, Science **281**, 538 (1998).
36. J. E. G. J. Wijnhoven and W. L. Vos, Science **281**, 802 (1998).
37. G. Subramania, K. Constant, R. Biswas, M. M. Sigalas, and K.-M. Ho, Appl. Phys. Lett. **74**, 3933 (1999).
38. G. Subramanian, V. N. Manoharan, J. D. Thorne, and D. J. Pine, Adv. Mater. **11**, 1261 (1999).
39. A. A. Zakhidov, R. H. Baughman, Z. Iqbal, C. Cui, I. Khayrullin, S. O. Dantas, J. Marti, and V. G. Ralchenko, Science **282**, 897 (1998).
40. S. H. Park and Y. Xia, Adv. Mater. **10**, 1045 (1998).
41. P. Jiang, K. S. Hwang, D. M. Mittleman, J. F. Bertone, and V. L. Colvin, J. Am. Chem. Soc. **121**, 11630 (1999).
42. M. Deutsch, Y. A. Vlasov, and D. J. Norris, Adv. Mater. **12**, 1176 (2000).
43. O. D. Velev, P. M. Tessier, A. M. Lenhoff, and E. W. Kaler, Nature **401**, 548 (1999).
44. K. M. Kulinowski, P. Jiang, H. Vaswani, and V. L. Colvin, Adv. Mater. **12**, 833 (2000).
45. J. E. G. J. Wijnhoven, S. J. M. Zevenhuizen, M. A. Hendriks, D. Vanmaekelbergh, J. J. Kelly, and W. L. Vos, Adv. Mater. **12**, 888 (2000).
46. N. Eradat, J. D. Huang, Z. V. Vardeny, A. A. Zakhidov, and R. H. Baughman, in *Photonic Crystals and Light Localization*, edited by C. M. Soukoulis (Kluwer, Dordrecht, 2001).
47. Y. A. Vlasov, N. Yao, and D. J. Norris, Adv. Mater. **11**, 165 (1999).
48. H. Míguez, A. Blanco, F. Meseguer, C. López, H. M. Yates, M. E. Pemble, V. Fornés, and A. Mifsud, Phys. Rev. B **59**, 1563 (1999).
49. P. V. Braun and P. Wiltzius, Nature **402**, 603 (1999).
50. A. Blanco, E. Chomski, S. Grabtchak, M. Ibisate, S. John, S. W. Leonard, C. López, F. Meseguer, H. Míguez, J. P. Mondia, G. A. Ozin, O. Toader, and H. M. v. Driel, Nature **405**, 437 (2000).
51. H. Míguez, F. Meseguer, C. López, M. Holgado, G. Andreasen, A. Mifsud, and V. Fornés, Langmuir **16**, 4405 (2000).
52. S. G. Romanov, T. Maka, C. M. Sotomayor Torres, M. Muller, and R. Zentel, Appl. Phys. Lett. **79**, 731 (2001).
53. Y. A. Vlasov, X. Z. Bo, J. C. Sturm, and D. J. Norris, Nature **414**, 289 (2001).
54. H. S. Sözüer, J. W. Haus, and R. Inguva, Phys. Rev. B **45**, 13962 (1992).
55. K. Busch and S. John, Phys. Rev. E **58**, 3896 (1998).
56. M. Megens, J. E. G. J. Wijnhoven, A. Lagendijk, and W. L. Vos, J. Opt. Soc. Am. B **16**, 1403 (1999).
57. A. F. Koenderink, L. Bechger, H. P. Schriemer, A. Lagendijk, and W. L. Vos, Phys. Rev. Lett. **88**, 143903/1 (2002).
58. M. J. A. de Dood, Ph.D. thesis, Utrecht University, 2002.

59. A. P. Alivisatos, Science **271**, 933 (1996).
60. M. Nirmal and L. E. Brus, Acc. Chem. Res. **32**, 407 (1999).
61. A. Eychmüller, J. Phys. Chem. B **104**, 6514 (2000).
62. C. B. Murray, C. R. Kagan, and M. G. Bawendi, Ann. Rev. Mater. Sci. **30**, 545 (2000).
63. V. N. Astratov, Y. A. Vlasov, O. Z. Karimov, A. A. Kaplyanskii, Y. G. Musikhin, N. A. Bert, V. N. Bogomolov, and A. V. Prokofiev, Phys. Lett. A **222**, 349 (1996).
64. S. G. Romanov, N. P. Johnson, A. V. Fokin, V. Y. Butko, H. M. Yates, M. E. Pemble, and C. M. Sotomayor Torres, Appl. Phys. Lett. **70**, 2091 (1997).
65. H. M. Yates, M. E. Pemble, H. Míguez, A. Blanco, C. López, F. Meseguer, and L. Vázquez, J. Cryst. Growth **193**, 9 (1998).
66. C. B. Murray, D. J. Norris, and M. G. Bawendi, J. Am. Chem. Soc. **115**, 8706 (1993).
67. J. E. Bowen Katari, V. L. Colvin, and A. P. Alivisatos, J. Phys. Chem. **98**, 4109 (1994).
68. O. I. Micic, J. R. Sprague, C. J. Curtis, K. M. Jones, J. L. Machol, A. J. Nozik, H. Giessen, B. Fluegel, G. Mohs, and N. Peyhambarian, J. Phys. Chem. **99**, 7754 (1995).
69. A. A. Guzelian, U. Banin, A. V. Kadavanich, X. Peng, and A. P. Alivisatos, Appl. Phys. Lett. **69**, 1432 (1996).
70. M. A. Hines and P. Guyot-Sionnest, J. Phys. Chem. B **102**, 3655 (1998).
71. D. J. Norris, N. Yao, F. T. Charnock, and T. A. Kennedy, Nano Lett. **1**, 3 (2001).
72. Z. A. Peng and X. Peng, J. Am. Chem. Soc. **123**, 168 (2001).
73. D. J. Norris, A. Sacra, C. B. Murray, and M. G. Bawendi, Phys. Rev. Lett. **72**, 2612 (1994).
74. M. Nirmal, D. J. Norris, M. Kuno, M. G. Bawendi, A. L. Efros, and M. Rosen, Phys. Rev. Lett. **75**, 3728 (1995).
75. D. J. Norris and M. G. Bawendi, Phys. Rev. B **53**, 16338 (1996).
76. D. J. Norris, A. L. Efros, M. Rosen, and M. G. Bawendi, Phys. Rev. B **53**, 16347 (1996).
77. M. A. Hines and P. Guyot-Sionnest, J. Phys. Chem. **100**, 468 (1996).
78. X. Peng, M. C. Schlamp, A. V. Kadavanich, and A. P. Alivisatos, J. Am. Chem. Soc. **119**, 7019 (1997).
79. B. O. Dabbousi, J. Rodriguez-Viejo, F. V. Mikulec, J. R. Heine, H. Mattoussi, R. Ober, K. F. Jensen, and M. G. Bawendi, J. Phys. Chem. B **101**, 9463 (1997).
80. V. L. Colvin, M. C. Schlamp, and A. P. Alivisatos, Nature **370**, 354 (1994).
81. B. O. Dabbousi, M. G. Bawendi, O. Onitsuka, and M. F. Rubner, Appl. Phys. Lett. **66**, 1316 (1995).
82. N. Tessler, V. Medvedev, M. Kazes, S. Kan, and U. Banin, Science **295**, 1506 (2002).
83. H.-J. Eisler, V. C. Sundar, M. G. Bawendi, M. Walsh, H. I. Smith, and V. Klimov, Appl. Phys. Lett. **80**, 4614 (2002).
84. C. B. Murray, C. R. Kagan, and M. G. Bawendi, Science **270**, 1335 (1995).
85. L. Brus, J. Phys. Chem. **98**, 3575 (1994).
86. M. Brust, M. Walker, D. Bethell, D. J. Schiffrin, and R. Whyman, J. Chem. Soc. Commun., 801 (1994).
87. C. P. Collier, R. J. Saykally, J. J. Shiang, S. E. Henrichs, and J. R. Heath, Science **277**, 1978 (1997).
88. S. Sun, C. B. Murray, D. Weller, L. Folks, and A. Moser, Science **287**, 1989 (2000).
89. Y. A. Vlasov, V. N. Astratov, A. V. Baryshev, A. A. Kaplyanskii, O. Z. Karimov, and M. F. Limonov, Phys. Rev. E **61**, 5784 (2000).
90. Z.-Y. Li and Z.-Q. Zhang, Phys. Rev. B **62**, 1516 (2000).
91. N. D. Denkov, O. D. Velev, P. A. Kralchevsky, I. B. Ivanov, H. Yoshimura, and K. Nagayama, Nature **361**, 26 (1993).
92. A. S. Dimitrov, C. D. Dushkin, H. Yoshimura, and K. Nagayama, Langmuir **10**, 432 (1994).
93. C. D. Dushkin, G. S. Lazarov, S. N. Kotsev, H. Yoshimura, and K. Nagayama, Colloid Polym. Sci. **277**, 914 (1999).
94. A. S. Dimitrov and K. Nagayama, Langmuir **12**, 1303 (1996).
95. P. Jiang, J. F. Bertone, K. S. Hwang, and V. L. Colvin, Chem. Mater. **11**, 2132 (1999).
96. E. Yablonovitch and T. J. Gmitter, Phys. Rev. Lett. **63**, 1950 (1989).
97. V. Yannopapas, N. Stefanou, and A. Modinos, J. Phys.: Condens. Matter **9**, 10261 (1997).
98. D. J. Norris and Y. A. Vlasov, in *Photonic Crystals and Light Localization*, edited by C. M. Soukoulis (Kluwer, Dordrecht, 2001), p. 229.
99. K. W.-K. Shung and Y. C. Tsai, Phys. Rev. B **48**, 11265 (1993).
100. Y. A. Vlasov, M. Deutch, and D. J. Norris, Appl. Phys. Lett. **76**, 1627 (2000).
101. C. B. Murray, S. Sun, W. Gaschler, H. Doyle, T. A. Betley, and C. R. Kagan, IBM J. Res. & Dev. **45**, 47 (2001).
102. F. Chen, K. L. Stokes, W. Zhou, J. Fang, and C. B. Murray, Mat. Res. Soc. Symp. Proc. **691**, G10.2.1 (2002).

Index

MIX
Papier aus verantwortungsvollen Quellen
Paper from responsible sources
FSC® C105338

If you have any concerns about our products,
you can contact us on
ProductSafety@springernature.com

In case Publisher is established outside the EU,
the EU authorized representative is:
Springer Nature Customer Service Center GmbH
Europaplatz 3, 69115 Heidelberg, Germany

Printed by Libri Plureos GmbH
in Hamburg, Germany